2015
EPD Congress

TMS2015

144th Annual Meeting & Exhibition

March 15-19, 2015 • Walt Disney World • Orlando, Florida, USA

EPD Congress 2015

Proceedings of symposia sponsored by
the Extraction & Processing Division (EPD) of
The Minerals, Metals & Materials Society (TMS)
held during

TMS2015
144th Annual Meeting & Exhibition
March 15-19, 2015
Walt Disney World • Orlando, Florida, USA

Edited by:

James Yurko • Antoine Allanore
Laura Bartlett • Jonghyun Lee • Lifeng Zhang
Gabriella Tranell • Yulia Meteleva-Fischer
Shadia Ikhmayies • Arief Suriadi Budiman
Prabhat Tripathy • Guy Fredrickson

Editors
James Yurko
Antoine Allanore
Laura Bartlett
Jonghyun Lee
Lifeng Zhang
Gabriella Tranell
Yulia Meteleva-Fischer
Shadia Ikhmayies
Arief Suriadi Budiman
Prabhat Tripathy
Guy Fredrickson

ISBN 978-3-319-48609-3 ISBN 978-3-319-48214-9 (eBook)
DOI 10.1007/978-3-319-48214-9

Chemistry and Materials Science: Professional

Published by Springer International Publishers, Switzerland, 2016
Reprint of the original edition published by John Wiley & Sons, Inc., 2015, 978-1-119-08245-3

Printed on acid-free paper

This Springer imprint is published by Springer Nature
The registered company is Springer International Publishing AG
The registered company address is: Gewerbestrasse 11, 6330 Cham, Switzerland

TABLE OF CONTENTS
EPD Congress 2015

Materials Processing Fundamentals

Extractive Materials Processing

Casting, Solidification, and Steel Processing

Materials Processing

Poster Session

Solar Cell Silicon

Silicon Production and Refining

Crystallisation and Mechanical Properties

High-Temperature Electrochemistry II

Molten Salt Technology

Nuclear and Rare Earth Technology

Sensors and Advanced Materials

Energy Storage Devices, Corrosion and Molten Salt Science

EDITORS

Materials Processing Fundamentals

James A. Yurko

James A. Yurko is Vice President, Technology and R&D with Materion Beryllium and Composites (Elmore, Ohio), the global leader in beryllium and non-beryllium based metal matrix composites. Dr. Yurko leads an engineering group focused on beryllium processing and materials processing and product development in areas including bulk metallic glass alloys, investment casting, and aluminum metal matrix composites. He is also a co-founder of Boston Electrometallurgical Corporation and Electrolytic Research Corporation, start-up companies dedicated to commercializing molten oxide electrolysis (MOE) technology. Before working with ERC, Dr. Yurko was the R&D team leader and staff metallurgist of BuhlerPrince, Inc. where he was responsible for commercializing the Semi-Solid Rheocasting (SSR) process and various die casting development projects of aluminum, magnesium, and bulk-metallic glass alloys.

Dr. Yurko received a Ph.D. in metallurgy from the Massachusetts Institute of Technology and a B.S.E. in materials science and engineering from the University of Michigan. He is currently a member of TMS and ASM International and serves on the University of Michigan Materials Science and Engineering External Advisory Board. In 2010, Dr. Yurko was selected as the TMS EPD Young Leader Professional Development Award winner and in 2012 was awarded by the TMS Foundation the Vittorio de Nora Prize for Environmental Improvements in Metallurgical Industries. Jim is the current TMS Process Technology and Modeling chairperson and has been an organizer and editor of a number of TMS symposia and proceedings.

Antoine Allanore

Antoine Allanore received his higher education in Nancy (France) where he earned a chemical process engineer diploma from Ecole Nationale Supérieure des Industries Chimiques and a M.Sc and Ph.D. from Lorraine University. Dr. Allanore joined the Massachusetts Institute of Technology (USA) in 2010 as a postdoctoral researcher after several years of service as a research engineer for ArcelorMittal. In 2012, he was appointed the T.B. King Assistant Professor of Metallurgy in the Department of Materials Science & Engineering at MIT, where his research group aims at developing sustainable materials extraction and manufacturing processes. In particular, his laboratory investigates the electrochemical properties of metals and molten compounds, e.g. oxides and sulfides. He teaches thermodynamics and sustainable chemical metallurgy at both the undergraduate and graduate level. He received the Young Leader Professional Development Award from the TMS Extraction and Processing Division in 2011 and the Vittorio de Nora Award from TMS in 2012.

Laura Bartlett

Laura Bartlett received her B.S. in Metallurgical Engineering from Missouri University of Science and Technology in 2008 and a Ph.D. in Metallurgical Engineering from Missouri University of Science and Technology in 2013. Since January 2013, she has served as Assistant Professor of Engineering Technology at Texas State University in San Marcos, Texas, and is the Foundry Education Foundation Key Professor of Metalcasting Technology. Dr. Bartlett is the current director of the Advanced High Strength Materials Lab at Texas State, a university/industry collaboration which involves the development of novel high strength and lightweight metallic materials and other industry relevant research. Dr. Bartlett has published numerous articles on the subject of advanced high strength steel research and has won many technical awards including the Metalcasting Congress Steel Division Best Paper Award in 2011 and 2012. Dr. Bartlett is a member of AIST and is the 2014-2015 recipient of the AIST Kent D. Peaslee Junior Faculty Award. She is also a long time member of AFS and currently serves as Secretary of the AFS Steel Casting Division 9 Technical Committee. Some of her research interests include melt processing and solidification phenomena in ferrous and non-ferrous alloys, the experimental and theoretical aspects of phase transformation and mechanical property relationships in high strength alloys, development of next generation advanced high strength steels, alloy development for extreme environments, metal matrix nanocomposites, and coating technologies.

Jonghyun Lee

Jonghyun Lee is a research assistant professor at the University of Massachusetts Amherst. He received a B.S. in mechanical engineering from Inha University in South Korea and M.S./Ph.D. degrees in the same discipline from the University of Massachusetts Amherst. Dr. Lee has been working on the characterization of thermophysical properties of high-temperature materials, magnetohydrodynamic flow in an electromagnetically levitated molten metal droplet, and cold spray of titanium alloys.

Lifeng Zhang

Lifeng Zhang currently is a professor and the dean of the School of Metallurgical and Ecological Engineering at University of Science and Technology Beijing. He received his Ph.D. from University of Science and Technology Beijing in 1998 and has 17 years teaching and research work at different universities including University of Science and Technology Beijing, Missouri University of Science and Technology, Norwegian University of Science and Technology, University of Illinois at Urbana-Champaign, Technical University of Clausthal, and Tohoku University. Dr. Zhang has compound backgrounds in primary production, refining, casting, and recycling of metals, recycling of electronic wastes and solar grade silicon, and process modeling for metallurgical processes. He has published over 300 papers and has given more than 180 presentations at meetings and conferences. He is Key Reader of *Metallurgical and Materials Transactions A* and *Metallurgical and Materials Transactions B*, Advisory Board member of ISIJ International, Member of Editorial Board of *Revue de Métallurgie*, *Journal of Metals and Metallurgy*, and *Advances in Manufacturing*, and Member of Board of Review of Iron and Steel Technology (*AIST Transactions*). Dr. Zhang is a member of TMS, AIST, and ISIJ. He has received several best paper awards from TMS and AIST, and has been a reviewer for over 50 journals.

Solar Cell Silicon

Gabriella Tranell

Gabriella Tranell received her Ph.D. in Materials Science and Engineering from the University of New South Wales, Australia in 1999. After 10 years as senior scientist and research manager with SINTEF in Trondheim, Norway, she is currently Associate Professor and Deputy Head with the Department of Materials Science and Engineering, NTNU, Norway. She is also acting director of the Norwegian Centre for Renewable Energy, a collaboration between NTNU, SINTEF, the University in Oslo and the Institute for Energy Technology. Her research interests include among others: production and refining of silicon, GHG and fume emissions from metallurgical industry, and recovery of REE from waste materials.

Yulia Meteleva-Fischer

Yulia Meteleva-Fischer earned her Ph.D. in Physical Chemistry from the Institute of Problems of Chemical Physics at the Russian Academy of Sciences (IPCP RAS) in 2002. Later work in cooperation with German and Dutch universities allowed for development as a materials scientist with expertise in synthesis and properties of thin films for photovoltaic and memory system applications, nanocomposites (zeolite and polymer matrices) and silicon production technology (thermodynamics, hydro- and pyrometallurgy). Dr. Meteleva-Fischer was honored by a number of national and international awards and research grants (CORDIS, DAAD and RBFR). Materials science of silicon, development of metallurgical routes to solar grade silicon and the thermodynamics of silicon contamination processes form the scope of her current research activities. She is currently an R&D engineer with Silicon Products GmbH, Germany.

Since 2012, she has been involved in the organization of the Solar Cell Silicon symposium, and since 2013 she has been a member of the Recycling and Environmental Issues Committee within the Extraction and Processing Division.

Shadia Ikhmayies

Shadia Ikhmayies received a B.Sc. from the physics department in the University of Jordan in 1983, an M.Sc. in molecular physics from the same university in 1987, and a Ph.D. in producing CdS/CdTe thin film solar cells from the same university in 2002. Now she works in Al Isra University in Jordan as an associate professor. Her research is focused on producing and characterizing semiconductor thin films and thin film CdS/CdTe solar cells. She also works in characterizing quartz in Jordan for the extraction of silicon for solar cells and characterizing different materials by computation. She has published 33 research papers in international scientific journals, three chapters in books, and 49 research papers in conference proceedings. Dr. Ikhmayies is a member of the Jordanian Renewable Energy Society (JRES), The Minerals, Metals & Materials Society (TMS) and the World Renewable Energy Network (WREN), where she became a member of the steering committee in 2014. She is a member of the international organizing committee and the international scientific committee in the Third European Conference on Renewable Energy Systems (ECRES2015). She was an associate editor in the journal of *Physics Express* for Simplex Academic Publishers and is an associate editor in the journal *Peak Journal of Physical and Environmental Science Research* (*PJPESR*) for Peak Journals. She is a member of the editorial board of the *International Journal of Materials and Chemistry* for Scientific & Academic Publishing, and the editor-in-chief of a book for Research Signpost. She was the technical advisor/subject editor for *JOM* as a representative of the Materials Characterization Committee for the year 2014. She is a guest editor for a special issue in the *Journal of Electronic Materials,* "S.I.: ECRES 2015." Dr. Ikhmayies is a reviewer in 21 international journals and she is a member of the Materials Characterization Committee of TMS.

Arief Suriadi Budiman

Arief Suriadi Budiman received his B.S. in mechanical engineering from Institute of Technology,Bandung (ITB), Indonesia, his M.Eng.Sc. in materials engineering from Monash Univ., Australia, and his Ph.D. in Materials Science and Engineering from Stanford University, California in 2008. During his doctoral candidacy at Stanford's Department of Materials Science & Engineering under the supervision of Professor William D. Nix (MRS Von Hippel Award 2007), Dr. Budiman received several research awards (MRS Graduate Silver Award 2006, MRS Best Paper 2006) and contributed to several high-impact journal publications, including *Acta Materialia*, *Applied Physics Letters*, and *Journal of Electronic Materials*. He gave two symposium invited talks as well in the MRS spring and fall meetings in 2006. More recently Dr. Budiman has been awarded the prestigious Los Alamos National Laboratory (LANL) Director's Research Fellowship to conduct top strategic research for the energy and national security missions of Los Alamos National Laboratory.

At the Center for Integrated Nanotechnologies (CINT) at Los Alamos, Dr. Budiman's research program involves nanomaterials for extreme environments with potential applications in advanced energy systems including for next generation nuclear power reactors. Currently, at Singapore University of Technology & Design (SUTD), Dr. Budiman is leading a group researching nanomaterials and nanomechanics and their implications for extending the extreme limits of materials as well as their applications in the next generation energy technologies (solar PV, extreme environments, energy storage, etc.). His work has also recently been listed in the Lawrence Berkeley National Laboratory's *Scientific Highlights* twice in May 2010 and June 2013, with the latter focusing on the novel and innovative characterization technique that enables thin silicon solar PV technology). Dr. Budiman has contributed a book chapter titled "Electromigration in Thin Films and Electronic Devices: Materials and Reliability," for a volume published by Woodhead Publishing, Cambridge, 2011. He holds two U.S. Patents and one patent pending.

High-Temperature Electrochemistry II

Prabhat Tripathy

Prabhat Tripathy is a research scientist with more than 20 years of research experience in the fields of materials chemistry, process metallurgy of reactive and refractory metals, high-temperature electrochemistry, synthesis and characterization of advanced materials, energy-efficient manufacturing process, chemical processing of waste (including nuclear) materials, and lean ore bodies. He earned his M.Sc. (Chemistry), M. Tech. (Materials Science & Engineering) and Ph. D (Chemistry) degrees in Chemistry from Utkal University (Bhubaneswar, India), Indian Institute of Technology (Kharagpur, India) and University of Mumbai (Mumbai, India) respectively. He started his professional career at Bhabha Atomic Research Centre (Mumbai, India) and subsequently spent stints at University of Cambridge (UK) and Massachusetts Institute of Technology (USA) before joining Idaho National Laboratory. Currently, he also holds an adjunct professorship in the metallurgical engineering department at the University of Utah, Salt Lake City (USA). The scientific underpinning of his research activities is to understand metals and materials through the application of novel techniques. He has about 70 publications to his credit.

Guy Lawrence Fredrickson

Guy Fredrickson is a research scientist with over 20 years of experience in the field of extractive metallurgy with extensive experience in pyro-, hydro-, and electro-metallurgical operations at research, pilot, and industrial scales. He earned his B.Sc., M.Sc., and Ph.D. in Metallurgical and Materials Engineering from the Colorado School of Mines. He started his professional career at the Reynolds Metals Company were his work was related to the continued development and improvement of Hall-Heroult cell technology. Dr. Fredrickson is now the manager of the Separations Department, Materials and Fuels Complex (MFC) at Idaho National Laboratory (INL). This department researches pyrometallurgical and electrometallurgical technologies for the purpose of reprocessing spent nuclear fuel for the Department of Energy's advanced fuel cycle mission. His focus area is actinide/lanthanide electrochemistry.

EPD COUNCIL 2014-2015

Corby G. Anderson
Membership & Student Development Committee Representative
Harrison-Western Professor of Metallurgical & Materials Engineering
Colorado School of Mines
Golden, CO, USA

Michael L. Free
Program Committee Representative
Professor
University of Utah
Salt Lake City, UT, USA

Shijie Wang
Program Committee Representative
Principal Advisor
Rio Tinto Kennecott Utah Copper Corporation
South Jordan, UT, USA

Christina E. Meskers
Public & Government Affairs Committee Representative
Business Development
Umicore Precious Metals Refining
Antwerp, Belgium

Donna Guillen
Energy Committee Chair
Distinguished Research Engineer
Idaho National Laboratory
Idaho Falls, ID, USA

Neale R. Neelameggham
Hydrometallurgy & Electrometallurgy Committee Chair
IND LLC
South Jordan, UT, USA

John S. Carpenter
Materials Characterization Committee Chair
Technical Staff Member
Los Alamos National Laboratory
Los Alamos, NM, USA

James A. Yurko
Process Technology & Modeling Committee Chair
Materion
Elmore, OH, USA

Gerardo Alvear
Pyrometallurgy Committee Chair
Principal Metallurgist
Xstrata Technology
Vancouver, BC, Canada

Randolph Kirchain
Recycling & Environmental Technologies Committee Chair
Principal Research Scientist
Massachusetts Institute of Technology
Cambridge, MA, USA

EPD HONORS AND AWARDS 2015

Distinguished Lecturer Award
Uday Pal
Boston University

Distinguished Service Award
Adrian Deneys
Praxair Inc

Science Award
"Progress in Slag Foaming in Metallurgical Processes",
Metallurgical and Materials Transactions B, August 2012, Vol 43B

Tai Xi Zhu
McMaster University

Kenneth S. Coley
McMaster University

Gordon A. Irons
McMaster University

Matthew Peter King
ArcelorMittal Dofasco

Technology Award
"Mutual Separation of Rare Earths Using Chemically Modified Chitosan Immobilized with Functional Groups of Chelating Agents,"
Rare Metal Technology 2014

Katsutoshi Inoue
Saga University

Shafiq Alam
University of Saskatchewan

2015 EPD Congress

MATERIALS PROCESSING FUNDAMENTALS

2015 EPD Congress

SYMPOSIUM: MATERIALS PROCESSING FUNDAMENTALS

Extractive Materials Processing

Antoine Allanore,
Massachusetts Institute of Technology

EPD Congress 2015
Edited by: James Yurko, Antoine Allanore, Laura Bartlett, Jonghyun Lee, Lifeng Zhang, Gabriella Tranell, Yulia Meteleva-Fischer, Shadia Ikhmayies, Arief Suriadi Budiman, Prabhat Tripathy, and Guy Fredrickson
TMS (The Minerals, Metals & Materials Society), 2015

ARSENIC AND ANTIMONY REMOVAL FROM COPPER CONCENTRATES BY DIGESTION WITH NaHS-NaOH

Maria C. Ruiz[1], Felipe Daroch[2] and Rafael Padilla[1]

[1]Department of Metallurgical Engineering, University of Concepción,
[2]Department of Chemical Engineering, University of Concepcion
Edmundo Larenas 285, Concepcion, 4070371, Chile

Keywords: Arsenic removal, NaHS-NaOH, Alkaline digestion

Abstract

Copper concentrates with high arsenic contents must be pretreated before conventional smelting to prevent environmental pollution with arsenic compounds. In this work, experimental results concerning the selective removal of arsenic and antimony from copper concentrates are presented. The process consists of an alkaline digestion using concentrated NaHS-NaOH solutions to transform the arsenic and antimony sulfides into soluble compounds. A water leaching follows the digestion to dissolve the arsenic and antimony, leaving clean copper sulfide solid residues. The laboratory scale tests were carried out using a copper–arsenic concentrate with 15.05% As and 1.42% Sb. The results showed that the most important digestion variables were temperature and concentrations of NaHS and NaOH. Over 97% of arsenic and 92% of antimony could be removed in 10 min of digestion using 8.9 M NaOH and 100% excess of NaHS at 80 °C. The subsequent water leaching was performed at 80°C for 20 min.

Introduction

The presence of arsenic minerals in copper concentrates complicates their treatment by the conventional smelting-converting process. The sulfides and oxides of arsenic are highly volatile. Thus, in the pyrometallurgical treatment of copper concentrates with high arsenic content, there is a risk of producing arsenic emissions to the atmosphere. Therefore, the smelters impose heavy penalties on copper concentrates that contain more than 0.5% arsenic [1]. The arsenic in copper concentrates is usually present as enargite (Cu_3AsS_4) and also tennantite ($Cu_{12}As_4S_{13}$). Antimony is also a common impurity that can affect the quality of the final copper product. In copper concentrates the antimony is usually associated with the arsenic in tennantite where antimony can replace the arsenic up to the composition of tetrahedrite ($Cu_{12}Sb_4S_{13}$). Due to the presence of large amount of enargite and tennantite in some copper ores, the concentration plants cannot produce clean copper concentrates without losing a significant amount of copper. Thus, the concentrates must be pretreated to lower their arsenic content before smelting. The conventional method to reduce the content of arsenic and other deleterious impurities in copper concentrates is roasting, where the arsenic is removed by volatilization as either sulfide or oxide. Although roasting is very effective for arsenic elimination, the removal of other impurities which are also present in the concentrates such as antimony and bismuth is less effective. In addition, roasting plants may also have problems to comply with the increasingly stringent norms that restrict the arsenic emissions to the atmosphere. Therefore, there is a growing interest in the copper industry for alternative processes to remove arsenic and antimony from copper concentrates.

In this work, a process for selective dissolution of arsenic and antimony from a copper concentrate containing enargite and tennantite is discussed. The process includes an alkaline sulfide digestion with a concentrated solution of NaHS-NaOH followed by leaching of the digested pulp with water.

Unlike conventional leaching, the digestion is carried out using a pulp with a high solid concentration; therefore, the soluble arsenic and antimony compounds formed by reaction with NaHS and NaOH will precipitate at least partially. In the subsequent water leaching the precipitated arsenic and antimony will be dissolved and separated from the solids by filtration. The ongoing authors' research on the digestion of copper concentrates containing enargite with Na_2S-NaOH has established that the alkaline digestion, followed by water leaching, is an efficient method for fast removal of arsenic, leaving essentially all the copper in the solid residues as copper sulfide [2].

In the present paper the behavior of a complex copper concentrate containing arsenic and antimony using NaHS-NaOH is discussed.

Chemistry of the alkaline sulfide leaching

The direct leaching of enargite with Na_2S-NaOH solutions have been studied by several researchers and the following reaction has been proposed [3, 4, 5]:

$$2Cu_3AsS_4 + 3Na_2S \rightarrow 3Cu_2S(s) + 2Na_3AsS_4 \quad (1)$$

For tennantite and tetrahedrite the proposed reactions are [6, 7]:

$$Cu_{12}As_4S_{13}\,(s) + 6Na_2S \rightarrow 5Cu_2S_{(s)} + 2CuS_{(s)} + 4Na_3AsS_3 \quad (2)$$

$$Cu_{12}Sb_4S_{13}(s) + 6Na_2S \rightarrow 5Cu_2S_{(s)} + 2CuS_{(s)} + 4Na_3SbS_3 \quad (3)$$

When NaHS is used as the source of sulfide, the following neutralization reaction occurs in the solution:

$$NaHS + NaOH = Na_2S + H_2O \quad (4)$$

On adding reaction (4) to reactions (1) through (3) the following overall reactions can be written when enargite [8], tennantite and tetrahedrite are reacted with NaHS-NaOH solutions:

$$2Cu_3AsS_{4\,(s)} + 3NaHS + 3NaOH \rightarrow 3Cu_2S_{(s)} + 2Na_3AsS_4 + 3H_2O \quad (5)$$

$$Cu_{12}As_4S_{13\,(s)} + 6NaHS + 6NaOH \rightarrow 5Cu_2S_{(s)} + 2CuS_{(s)} + 4Na_3AsS_3 + 6H_2O \quad (6)$$

$$Cu_{12}Sb_4S_{13\,(s)} + 6NaHS + 6NaOH \rightarrow 5Cu_2S_{(s)} + 2CuS_{(s)} + 4Na_3SbS_3 + 6H_2O \quad (7)$$

As seen in these overall reactions, the stoichiometric amounts of NaHS and NaOH required to dissolve all of the arsenic or antimony from these minerals are equal to 1.5 moles per mole of As or Sb.

These overall reactions (5) through (7) will be assumed to occur in the digestion step of the process studied here.

Experimental Work

Materials

The concentrate used in this research was prepared by manual selection from a primary sample of natural crystals of arsenic rich minerals. The selected crystals were crushed, ground, and classified into narrow size fractions using a U.S. sieve series. The size fraction -150+106 μm was used in the experimental work. The chemical analysis of this size fraction indicated 37.4% Cu, 15.05% As, 1.42% Sb, 9.69% Fe and 35.0% S.

The X-ray diffraction analysis of the concentrate is presented in Figure 1, which showed that the main minerals present were enargite, tennantite and pyrite.

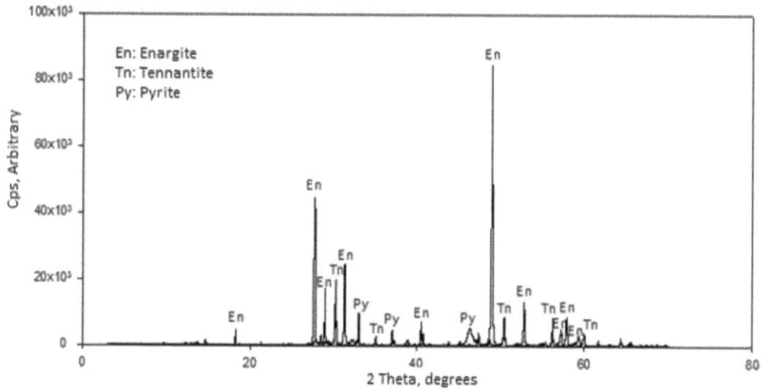

Figure 1. X-ray diffraction spectrum of the concentrate size fraction -150 + 106 μm.

A few individual particles of the concentrate were also analyzed by energy dispersive spectrometry (EDS). All the particles analyzed showed the presence of antimony ranging from 1 to 3.2%. In some of the particles, presumably tennantite, the presence of iron was also detected in concentrations ranging from 5 to 7.5%.

The chemical reagents used to prepare the digestion solution were $NaHS{\cdot}xH_2O$ from Sigma Aldrich and NaOH from Merck.

Experimental Procedure

The alkaline digestion experiments were carried out with 4 g of the concentrate sample -150 + 106 μm prepared as described earlier. A small volume of freshly prepared solution (usually 4 or 5 ml) containing the desired amounts of NaHS and NaOH was preheated over a hot plate in a 100 ml steel crucible until it reached the temperature selected for the test. The alkaline digestion was then started by adding the solid preheated sample. During digestion the pulp was vigorously mixed using a paddled stir rod for the predetermined time. Once the digestion period ended, the pulp was immediately poured into a glass reactor containing preheated water, to carry out the water leaching. In all the tests the leaching was performed using 500 ml of water at the constant temperature of 80 °C and 385 rpm. The leaching time was usually 20 min. The reason for using a large volume of water (500 ml) in the leaching step was to prevent further reaction between the remaining enargite-tennantite minerals and the unreacted NaHS. The leaching reactor was a 2-liter glass round bottom flask with mechanical agitation system. After the water leaching, the solids were filtered, washed and finally dried at 80 °C for chemical analysis.

Results

The experimental work included the study of the main digestion variables: temperature, NaHS and NaOH concentrations and time. Regarding the water leaching, time was the only variable studied.

Effect of the Digestion Temperature

Experiments were carried out at digestion temperatures in the range of 60 to 90 °C, using 4 ml of a solution with NaHS and NaOH concentrations of 3.3 M and 5.8 M, respectively. The digestion time was 6 min. The water leaching time was 20 min for all the experiments. The removal of arsenic and antimony obtained at the various temperatures are shown in Figure 2.

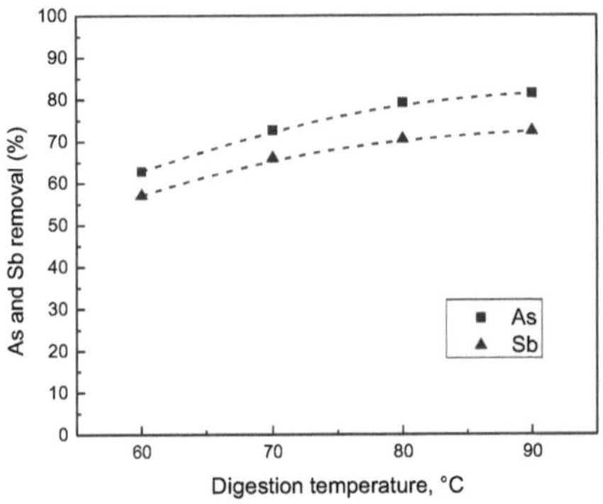

Figure 2. Effect of digestion temperature on arsenic and antimony removal. Digestion conditions: 4 g of solid, 4 ml of solution NaHS 4.8 M and NaOH 5.7 M, 6 min.

As seen in this figure, the digestion temperature has a significant effect on the extent of arsenic and antimony removal from the concentrate. We can also see that the removal of arsenic is higher than the removal of antimony at all the temperatures. An X-ray diffraction analysis was also carried out of a sample digested for 6 min at 80 °C using a solution with a NaHS concentration of 3.3 M and NaOH 5.8 M. The X- ray spectrum obtained is presented in Figure 3.

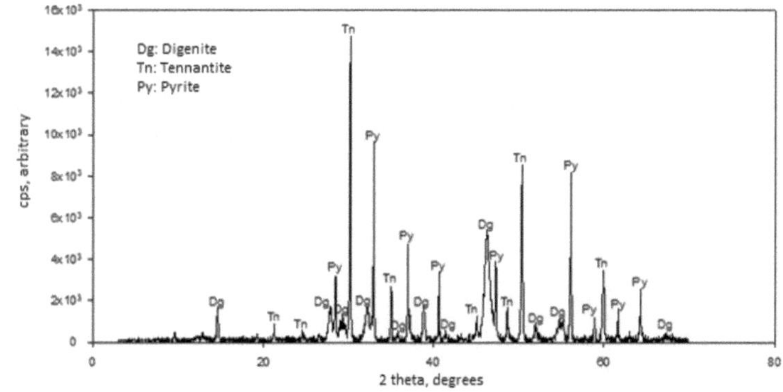

Figure 3. X-ray diffraction spectrum of the residues obtained in digestion for 6 min at 80 °C, using 4 ml of a solution 3.3 M in NaHS and 5.8 M in NaOH, followed by water leaching.

As seen in this figure the main diffraction lines observed in this partially digested sample correspond to tennantite, pyrite and digenite ($Cu_{1.8}S$). The enargite lines are absent indicating that this mineral reacted faster than tennantite. This result also suggests that the fraction of the antimony present in tennantite is larger than the antimony in enargite, which would explain the lower antimony removal observed in all the experimental conditions tested. The formation of digenite as the copper sulfide product is in agreement with the results obtained in an earlier study on the alkaline digestion and water leaching of enargite [9].

Effect of the concentration of the reagents

Previous investigations on the alkaline sulfide leaching of enargite and also tetrahedrite have shown that the concentration of both, the sulfide salt and the hydroxide, have a significant effect on the extent of arsenic and antimony removal [5, 8, 10, 11]. The results of experiments performed with concentrations of NaHS in the range of 3.3 to 8.0 M and NaOH concentrations in the range of 5.7 to 10.5 M are summarized in Table I. The third and fourth columns in this table indicate the excess NaHS and excess NaOH used in the digestion. These values were calculated considering a stoichiometric requirement of NaHS and NaOH of 1.5 moles per mole of arsenic plus antimony in the concentrate, according to reactions (5) through (7). As seen in the table, a larger excess of NaOH than NaHS was used in all the experiments, in order to allow for the neutralization of the HS^- to S^{2-} according to reaction (4) plus an extra amount to maintain a highly alkaline digestion solution. All the digestion experiments were carried out at 80 °C, using 4 ml of solution and digestion times of 6 min.

Table I. Effect of reagents concentration in the digestion on arsenic and antimony removal.

Concentrations		Reagent excess		Removal after 6 min	
NaHS, M	NaOH, M	NaHS, %	NaOH, %	As, %	Sb, %
3.3	5.8	0	79	72.28	59.48
4.8	5.7	50	79	79.30	70.57
6.4	8.9	100	179	95.82	92.17
8.0	10.5	150	229	95.47	92.09

As seen in this table, the use of a high concentration of NaHS and NaOH in the digestion step is necessary for an efficient removal of arsenic and antimony from the concentrate.

Effect of digestion time

The effect of digestion time was studied at the temperature of 80 °C using 4 ml of solution and for two levels of reagents concentration. The results are shown in Figure 4.

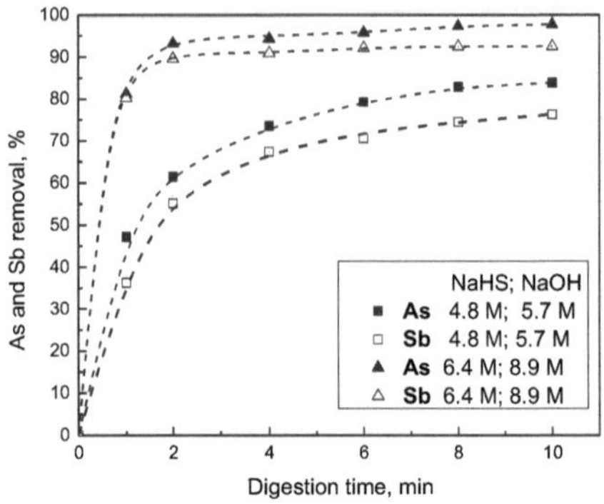

Figure 4. Effect of digestion time on arsenic and antimony removal.
Digestion conditions: 4 ml of solution and 80 °C.

It can be observed in this figure that for NaHS concentration of 6.4 M, and NaOH concentration of 8.9 M the rate of reaction was so fast that in two minutes of digestion the removal of arsenic and antimony from the concentrate was about 93% and 90%, respectively. As seen in Table I, the NaHS excess was 50% when the concentration was 4.8 M and 100% excess when the concentration was 6.4 M. The corresponding NaOH excesses were 79% for 5.7 M and 179 % for 8.9 M. From these data the advantage of using a small volume of solution in the digestion is apparent. This means that in a small volume we can have high concentrations of NaHS and NaOH, with the resulting high levels of arsenic and antimony removal without using large amounts of reagents.

Effect of the leaching time

The purpose of the water leaching step of the process studied here was to dissolve the digestion products: thioarsenate, thioarsenite or thioantimonite that precipitated during the digestion. Thus the water leaching separates the dissolved arsenic and antimony from the solid copper sulfides that constitute the main part of the solid residue. For the determination of the arsenic and antimony dissolved in the digestion, 500 ml of water was used in the leaching to prevent further reaction of the minerals through a large dilution of the pulp. However, in a practical application of the process the leaching could be carried out with a small liquid/solid ratio. In addition, since the arsenic and antimony that precipitated in the digestion are very soluble compounds, the dissolution in water should be a very rapid process and thus the leaching time could be shorter than 20 min. To verify this possibility, experiments were carried out under identical digestion conditions and leaching times in the range of 10 min to 40 min. The arsenic and antimony removal obtained are presented in Figure 5.

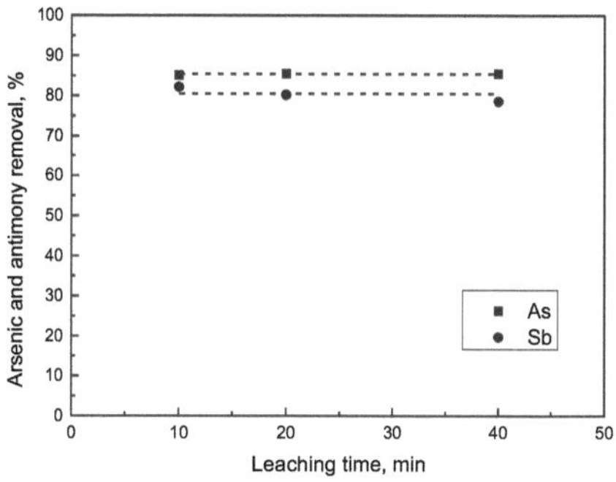

Figure 5. Effect of leaching time on arsenic and antimony removal from the concentrate. Digestion conditions: 5 ml of solution 5.4 M in NaHS and 7.4 M in NaOH, 8 min.

As seen in Figure 5, the extents of arsenic and antimony removal from the concentrate were not affected by the leaching time in the range of 10 to 40 min. Therefore, 10 min of leaching are sufficient to efficiently solubilize all the arsenic and antimony that precipitated during the digestion step.

Conclusions

- The digestion with NaHS-NaOH followed by water leaching is an efficient method for rapid removal of arsenic and antimony from an enargite-tetrahedrite concentrate, containing 15.05% As and 1.42% Sb.
- The copper in the concentrate remains in the solid phase as digenite.
- The variables that affect the arsenic and antimony removal are the digestion temperature, the concentrations of NaHS and NaOH, and the digestion time.
- At a digestion temperature of 80 °C, over 97% of the arsenic and 92% of the antimony could be removed in 10 min of digestion using a digestion solution with a NaOH concentration of 8.9 M and NaHS 6.5M (corresponding to 100% excess of NaHS over the theoretical stoichiometric requirement).
- A leaching with water at 80 °C for 10 min is sufficient to ensure complete dissolution of all the arsenic and antimony compounds that precipitated during the digestion.

Acknowledgments

Support from the National Fund for Scientific and Technological Development (FONDECYT) of Chile under Project no. 1120585 is gratefully acknowledged.

References

1. K. Baxter, H. Scriba and I. Vega, "Treatment of high-arsenic copper-gold concentrates – An options review," *Proceedings of Copper 2010*, ed. GDMB (Clausthal-Zellerfeld, Germany: Society for Mining, Metallurgy, Resources and Environmental Technology, 2010), 1783-1802.

2. M.C. Ruiz, R. Bello and R. Padilla, "Removal of arsenic from enargite rich copper concentrates," *Materials Processing Fundamentals*, ed. L. Zhan A. Allanore, J.A. Yurko and J. Crapps (Hovoken, NJ: John Wiley & Sons, 2013), 217-223.
3. R.M. Nadkarni and C.L. Kusic, "Hydrometallurgical removal of arsenic from copper concentrates," *Arsenic Metallurgy, Fundamentals and Applications*, ed. R.G. Reddy, J.L. Hendrix and P.B. Queneau (Warrendale, PA: The Minerals Metals and Materials Society, 1988), 263-286.
4. C.G Anderson and L.G. Twidwell, "The alkaline sulfide hydrometallurgical separation, recovery and fixation of tin, arsenic, antimony, mercury and gold," *Lead and Zinc 2008* (Johannesburg, South Africa: The Southern African Institute of Mining and Metallurgy, 2008), 121-132.
5. L. Currelli, C. Garbarino, M.Ghiani and G. Orrù, "Arsenic leaching from a gold bearing enargite flotation concentrate," *Hydrometallurgy*, 96 (2009), 258-263.
6. D. Filippou, P. St-Germain, T. Grammatikopoulos, "Recovery of metal values from copper arsenic minerals and other related resources," *Mineral Processing and Extractive Metallurgy Review*, 28 (2007), 247-298.
7. S.A. Awe, C. Samuelson and A. Sandström, "Dissolution kinetics of tetrahedrite mineral in alkaline sulphide media," *Hydrometallurgy*, 103 (2010), 167-172.
8. W. Tongamp, Y. Takasaki and A. Shibayama, "Selective leaching of arsenic from enargite in NaHS–NaOH media," *Hydrometallurgy*, 101 (2010), 64-68.
9. M.C. Ruiz, L. Grandon and R. Padilla, "Selective arsenic removal from enargite by alkaline digestion and water leaching," *Hydrometallurgy*, 150 (2014), 20-26.
10. W. Tongamp, Y. Takasaki and A. Shibayama, "Arsenic removal from copper ores and concentrates through alkaline leaching in NaHS media," *Hydrometallurgy*, 98 (2009), 213-218.
11. F. Parada, M.I. Jeffreys and E. Asselin, "Leaching kinetics of enargite in alkaline sodium sulphide solutions," *Hydrometallurgy*, 146 (2014), 48-58.

EPD Congress 2015
Edited by: James Yurko, Antoine Allanore, Laura Bartlett, Jonghyun Lee, Lifeng Zhang, Gabriella Tranell, Yulia Meteleva-Fischer, Shadia Ikhmayies, Arief Suriadi Budiman, Prabhat Tripathy, and Guy Fredrickson
TMS (The Minerals, Metals & Materials Society), 2015

RATE OF METAL DEPOSITION FROM AQUEOUS SOLUTIONS

Anne-Marie Suriano[1], Stanley Howard[1], Sadegh Safarzadeh[1]

[1]South Dakota School of Mines and Technology; 501 E Saint Joseph St.; Rapid City, SD 57701, USA

Keywords: Kinetics, Deposition, Electrodeposition, Extraction

Abstract

The electrodeposition of metal ions from aqueous electrolyte solutions has been a well-known process for the last half century. With numerous industrial applications such as electroplating, electrowinning and electrorefining, the kinetics of electrochemical reactions involved in the process has been extensively researched. Arrhenius-type rate equations, the Butler-Volmer equation, and the empirical Tafel equation are among the most important models to explain the rate of electrochemical reactions. The study of the kinetics of metal deposition is critical to understanding the underlying mechanisms through which a certain morphology of metal deposit is formed on the cathode. The purpose of this paper is to elucidate the influence of experimental parameters such as pH and additives on the rate and the reaction mechanism determining the rate limiting step of the electrodeposition reactions.

Introduction

Electrodeposition as a method by which aquated metal is deposited onto a metal surface can be described by the charge transfer equation

$$M_{aq}^{+} + ze^{-} \rightarrow M_s \quad (1)$$

where M_{aq}^{+} is the metal ion in solution and M_s is the deposited solid metal. The electrodeposition of metal ions from an aqueous solution onto an electrode is a multi-step process consisting of

1) Mass transport of metal ions from aqueous electrolyte solution bulk to the metal-solution double layer, analogous to a boundary layer
2) Diffusion through the boundary layer
3) Integration onto the interface
4) Bulk metallic deposit growth

Commonly employed electrodeposition kinetics relations used to describe the driving forces behind metal deposition rate include the Butler-Volmer, generalized Frumkin-Butler-Volmer (gFBV), and Tafel equations. While the Tafel equation has been developed empirically, the former two equations rely upon ion concentration models of the reaction surface's diffusion layer. This layer is referred to as the Double Layer (DL) and shown to determine the rate for the many electrocrystallization cases which are controlled by a slow electron transfer step [1].

<u>Electrodeposition Experimental Set-up</u>

Figure 1 shows a typical electrolytic cell with two anodes and a central cathode. The cathode is rotating such that convection currents in the electrolyte create steady state conditions. A rotation

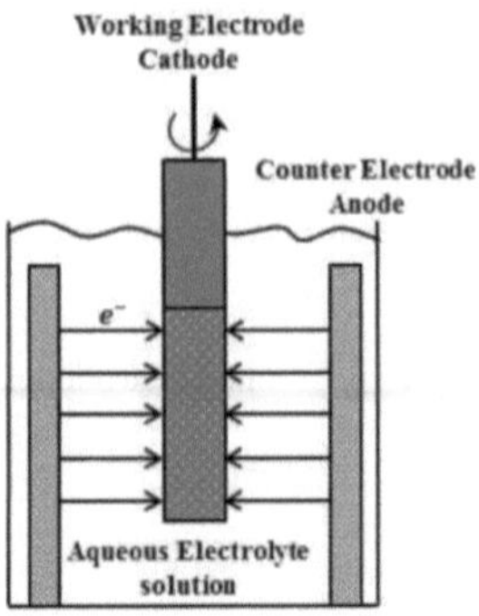

Figure 1: A Schematic of the rotating cylindrical electrolytic cell.

rate of approximately 1000 rpm is needed to achieve steady state [2, 3, 4, 5]. Shown is a Rotating Cylindrical Electrode (RDE). A Rotating Disk Electrode (RDE) which uses a uniform flat electrode may also be used. So long as the same solution flow is induced, the method of inducing it does not factor into subsequent equations and calculations.

If the anode is assumed to have already dissolved sufficiently into the solution so as to reach a steady rate of deposition onto the working electrode, the cell is potentiostatically controlled and reversible potentials used for deposition may be calculated using the Nernst equation,

$$E = E^0 + \left(\frac{RT}{nF}\right) \ln\{\frac{a_{ox}^p}{a_{red}^q}\} \tag{2}$$

where F is Faraday's constant, R is the gas constant, T is temperature, E is the electrode potential, E^o is the standard electrode potential, and a_{ox} and a_{red} are the activities of the oxidized and reduced species with stoichiometric coefficients p and q [1]. For a constant bulk concentration of ions in solution the plating potential is also constant, as shown in the Nernst equation.

Determination of Rate Determining Step

There are many variables in an electrolytic bath and set up which would affect the kinetic rates and the subsequent slowest step. Convection, pH, concentration, current density, and additives have all been shown to effect rates [1-13].

Mass Transfer Rate Equations

The Nernst-Plank relation for ion transport in electrolyte,

$$-\nabla \cdot J_i = \nabla \cdot [D_i(c_i + c_i z_i f \nabla V)] \tag{3}$$

where J_i is the flux, D_i diffusivity, z_i ion valence, c_i ion concentration and V voltage potential, is has been used to describe mass transport for a dilute electrolyte [6]. It however does not allow for advection or chemical reactions in solution. These effects may be addressed by the Poisson equation, which relates electrical potential with charge density [6],

$$\nabla \cdot (\varepsilon \nabla V) = -F \sum_i z_i c_i. \quad (4)$$

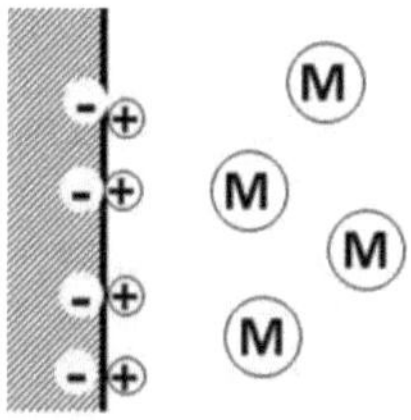

Figure 2: Classical Butler-Volmer double lay model for + solvent surface coverage and M ions in solution.

The large boundary layer potential as described in the next section must also be considered when considering the effect of the electrical potential in Equ. (4). For a steady state system the Poisson-Nernst-Plank equation describes mass transfer in the bulk solution by combining Eqs. (3) and (4) with boundary layer potential effects giving the ion concentration in solution at a position and time,

$$c(x, \tau) = 1 + 2i_{app}\{0.5 - x - \sum_{n=1}^{n=\infty} f_n \cos(\tfrac{2Nx}{L})\} \quad (5)$$

where f_n and τ are functions of time, I_{app} is the applied current, and L and x are spatial positions [6].

Since ion flux and thus solution convection currents determine mass transfer rate, the speed of the rotating electrode is important. The slowest rate step, whether diffusion or chemical reaction, may be determined by monitoring the effect of cathode rotation speed on the rate of metal deposition [7]. The mass transfer coefficient increases with flow rate [5] while the current density decreases indicating a shift from electron-transfer control to diffusion control [3].

Charge Transfer Rate Equations

The Butler-Volmer description of the classic Double Layer (DL) model is shown in Fig. 3. Here the metal ions in solution are considered to have finite size and the reaction surface is coated in a layer of solvent molecules. Since the ions are restricted in their surface approach by their size both a charge free and a charge covered surface layer are created. These act as a capacitor where the ion layer is one plate and referred to as the outer Helmholtz plane and the surface of the metal-aqueous solution is the other plate called the reaction plane. Ions in solution undergo at least one electron transfer reaction at the DL, which is described by a Faradaic charge transfer relation. In order to balance the charge and maintain the operating potential, ions transfer charge and replace the solvent atoms shown in Fig. 2. This charge-transfer adsorption is called specific adsorption [1] and generally occurs before bulk metallic growth. Since the behavior of an electrochemical reaction is characterized by a polarization resistance at the interface, the capacitance of the DL determines the rate [8]. The potential drop between the working electrode and the bulk electrolyte solution, which is analogues to the overpotential, drives the reaction [6] and leads to the general Butler-Volmer equation,

$$i = -nFAk_{Red}C_{Red}exp(\alpha nfE) \quad (6)$$

for

$$f = F/RT \quad (7)$$

where A is the electrode surface area, k_{Red} the rate constant and α the transfer coefficient [1]. It should be noted that the Butler-Volmer equation shown is in terms of current, through measurement of which rate is determined [5]. Commonly deposited are higher oxidized metal ions, such as Cu^{2+}, which must go through multiple consecutive electron reactions at the DL. In such cases the slower reaction is generally found to be the first electron reaction [4, 8].

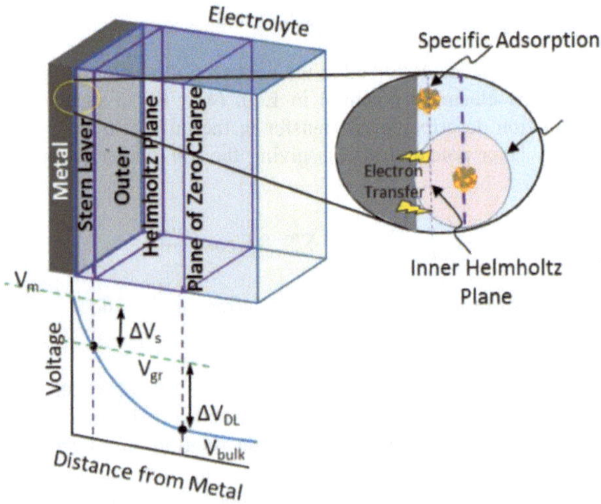

Figure 3: Fumkin-Butler-Volmer double layer model. Courtesy of John Suriano.

The generalized Frumkin-Butler-Volmer (gFBV) equation is extension of the Butler-Volmer model, which goes farther to take into effect upon the rate the DL composition [6]. In this model, which is shown in Fig. 3, a diffuse layer with non-zero charge is included between the reaction plane and the zero-charge zone causing a change in potential drop across the DL, which is assumed to be linear in the Butler-Volmer model. This contributes to the total cell potential although the ion concentration of the bulk remains the same as considered in the simpler model. Therefore this impacts the charge transfer rate and the electrical potential. The driving force is the ion concentration at the reaction plane and the electric field strength describing the potential drop across the charge-free layer,

$$J_F = K_R c_{O,bulk} \exp\{-(\alpha *)nf(V_m - V_{bulk})\} - K_O c_{R,bulk} \exp\{(1 - \alpha *)nf(V_m - V_{bulk})\} \quad (8)$$

where K_i are rate constants, V_i are voltages, c_i are concentrations, and the effective transfer coefficient, $\alpha *$, is the transfer coefficient adjusted for dependence upon the diffusion layer to zero-charge layer capacitance ratio [6]. From this it can be seen that for zero-charge layer thickness << diffusion layer thickness, the latter will dominate the charge transfer rate necessitating the use of the gFBV equation. For zero-charge layer thickness >> diffusion layer thickness, the former dominates and $\alpha *$ will approach the value of the classical transfer coefficient such that the Butler-Volmer equation may be used [6].

Charge Transfer Adsorption

Before metal ions grow in the bulk metal solid phase, they are generally adsorbed on the electrode surface. This is described by the Frumkin isotherm for charge-transfer adsorption,

$$\left[\frac{\theta}{1-\theta}\right] expf\theta = kCexp(\frac{\gamma nFE}{RT}) \tag{9}$$

where (1-θ) is the fraction of active adsorption sites, n is number of electrons, k & f are constants, C is adsorbate concentration, and □ is the electrosorption valency which accounts for DL charge changes during adsorption [1]. The rate and manner of which the layers of metal ions are adsorbed and consequently deposited may vary for the first monolayer deposited upon the working electrode surface and the subsequent multilayers deposited onto the bulk metal, since the composition and crystal structure of the two substrates are generally different [1].

Adsorption is very important in the case of organic surfactants being added to the electrolyte for nucleation enhancement. Studies found that the addition of organics which do not affect the metal ion's deposition reaction pathway inhibit rate of the deposition due to adsorption [4]. This is shown by the decreasing limiting current density with increase in organic concentration. Organic molecules in the electrolyte solution follow the Langmuir isotherm,

$$\frac{\theta}{1-\theta} = KC \tag{10}$$

for equilibrium constant K [7]. Therefore, the metal ions and the organic molecules will vie for active adsorption sites and inhibit the adsorption and deposition rate [4, 7].

Nucleation

Alternately, the interface conditions of the electrocrystallization process in an electrolytic cell may result from nucleation rather than adsorption. The rate of nucleation in such processes is generally found empirically by evaluating the current-time transients over potential steps [1].

Experimental Methods

The slowest mechanism is determined experimentally. Voltammetry analysis can be used to show if a deposition process with given parameters is diffusion controlled. A voltammetry sweep is conducted over a determined range, giving potential regions at different current densities. A kinetic peak indicates the mass transfer process and lack of any diffusion peak indicates charge-transfer rate control [9]. Chronoamperometric analysis is also used. Current transients I-$t_{1/2}$ are plotted and linear behavior indicate mass-transfer rate control. These experiments may be conducted with respect to varying parameters such that the effect of the parameter upon rate determining mechanism may be analyzed. As is modeled by the Butler-Volmer equation where the current density drives the equation, much investigation into the RDS

is done through observing current density behavior relative to a given electrolytic system. The Tafel Equation,

$$E = a - b\log|i| \tag{11}$$

where a and b are the Tafel parameters determined experimentally [1], relates the current to the potential and are used to determine the effect of current efficiency [10, 11].

Conclusions

The electrolytic cell controlling electrodeposition of metal ions from aqueous solutions to bulk metal solids is complex and has many variables that may affect the rate of deposition. Considerable experimental and theoretical work has been done to determine the effect of varying parameters on the rate determining mechanism. Current density and overpotential examinations give data sufficient to determine the slowest step for most electrodeposition systems.

References

1. I.M. Ritchie, "The metal-solution interface," *Adv. In Colloid and Interf. Sci.*, Amsterdam: Elsevier, 80 (1999), 183-231.

2. E.J. Podlaha, and D. Landolt, "An Experimental Investigation of Ni-Mo Alloys," *J. Electrochem. Soc.*, 143 (1996), 885-892.

3. E.J. Podlaha, and D. Landolt, "A Mathematical Model Describing the Electrodeposition of Ni-Mo Alloys," *J. Electrochem. Soc.*, 143 (1996), 893-899.

4. S. Varvara, "Kinetics of copper electrodeposition in the presence of triethyl-benzyl ammonium chloride," *J. of Appl. Electrochemistry*, Netherlands: Kluwer Academic Publishers, 33 (2003), 685-692.

5. T. Subbaiah, S.C. Das, and R.P.Das, "Mass transfer rates in an electrochemical cell," *Hydrometallurgy*, Amsterdam: Elsevier, 33 (1993), 153-163.

6. M.V. Soestbergen, "Frumkin-Butler-Volmer Theory and Mass Transfer in Electrochemical Cells," *Russian J. of Electrochem.*, Pleiades Publishing Inc., 48 (2012), 570-579.

7. A.M.M. Ahmed, "Effect of Organic Solvents on the Electrodeposition of Copper from Acidified CuSO4," *J. of Dispersion Sci. and Tech.*, Taylor & Francis, 33 (2012), 898-912.

8. V.S. Protsenko, "Applying a Theory of Generalized Variables to Electrochemical Kinetics," *Protection of Metals*, Pleiades Publishing Inc., 43 (2007), 398-406.

9. M. Torabi, "A Kinetic study on the electrodeposition of nickel nanostructure and its electrocatalytic activity for hydrogen evolution reaction," *J. Appl. Electrochem.*, Springer, 40 (2010), 1941-1947.

10. O.E. Kongstein, F.M. Haarberg, and J. Thonstad, "Current efficiency and kindetics of cobalt electrodeposition in acid chloride solutions," *J. Appl. Electrochem.*, Springer, 37 (2007), 669-674.

11. O.E. KongsteinF.M. Haarberg, and J. Thonstad, "Pt II. The influence of chloride and sulphate concentrations," *J. Appl. Electrochem.*, Springer, 37 (2007), 675-680.

12. I. Giannopoulou, D. Panias, I. Paspaliaris, "Electrochemical modeling and study of copper deposition from concentrated ammoniacal sulfate solutions," *Hydrometallurgy*, Amsterdam: Elsevier, 99 (2009), 58-66.

EPD Congress 2015
Edited by: James Yurko, Antoine Allanore, Laura Bartlett, Jonghyun Lee, Lifeng Zhang, Gabriella Tranell, Yulia Meteleva-Fischer, Shadia Ikhmayies, Arief Suriadi Budiman, Prabhat Tripathy, and Guy Fredrickson
TMS (The Minerals, Metals & Materials Society), 2015

ELECTROCHEMICAL REDUCTION OF COBALT TETRAFLUOROBORATE IN 1-BUTYL-3-METHYLIMIDAZOLIUM TETRAFLUOROBORATE IONIC LIQUID

Min Li,[1,2] Zhongning Shi,[1] Zhaowen Wang,[1] and Ramana G. Reddy[2*]

[1]School of Materials and Metallurgy, Northeastern University, Shenyang 110819, China
[2]Department of Metallurgical and Materials Engineering, The University of Alabama, Tuscaloosa, Alabama, 35487, USA

Abstract

Electrodeposition of cobalt from cobalt tetrafluoroborate ($Co(BF_4)_2$) was investigated using 1-butyl-3-methylimidazolium tetrafluoroborate ($BMIMBF_4$) ionic liquid. The experiments were conducted at 353 K. Chronoamperometry experiments confirm that the electrodeposition of cobalt on tungsten electrode proceeds via three-dimensional instantaneous nucleation with diffusion-controlled growth process. The average diffusion coefficient of Co(II) was found to be 7.5×10^{-8} cm^2s^{-1} at 353 K, which is in good agreement with the estimated value from cyclic voltammetry. The electrochemical deposit was characterized using SEM-EDS and XRD methods. The SEM image shows formation of a dense and compact deposit at - 0.75 V. The EDS and XRD analysis confirm that the obtained deposit is pure cobalt metal.

Keywords: Cobalt tetrafluoroborate, Electroreduction, $BMIMBF_4$ ionic liquid

Introduction

Cobalt and its alloys are very important functional materials due to their magnetic properties, and resistance to corrosion. Electrodeposition of Co is relatively difficult from the aqueous solutions due to hydrogen evolution. In recent years, ionic liquids (ILs) have received greater attention as a novel media for the electrodeposition of metals and alloys. Ionic liquids have several advantages over conventional aqueous solutions as electrolytes. They are non-flammable, display high thermal stability with negligible vapor pressure and are good solvents for numerous salts and polymers. Their wider electrochemical window and relatively high conductivities allowed the electrodeposition of less noble metals, semiconductors and alloys, which otherwise are very difficult to deposit in aqueous electrolytes that possess limited electrochemical windows and poor thermal stabilities [1-7].

Among the various air-stable room temperature ionic liquids (RTILs), 1-butyl-3-methylimidazolium tetrafluoroborate ($BMIMBF_4$) ionic liquid have been extensively studied and used for electrochemical applications like electrodeposition and batteries. At ambient temperatures, dissolution of commonly occurring cobalt salt ($CoCl_2$) in $BMIMBF_4$ is very difficult because of the poor coordination capability of BF_4^- anions. Therefore, specific cobalt salts need to be synthesized for desired electrochemical applications. Sun *et al.* [8-12] found that

* Corresponding Author, Tel: (205) 348-4246, E-mail: rreddy@eng.ua.edu

when excess of EMICl is present in the IL melt, free chloride ions are produced and forms a complex with the metal ions or metal chlorides in order to facilitate the dissolution of salt species in the melt. Intermetallic alloys like on Pd-Ag, Pd-In, Pd-Cu, Sb-In and Gd-Te were studied in in BMIM-Cl-BF_4 systems [8-12].

Electrodeposition of cobalt and cobalt alloys have been primarily focused on using chloroaluminate-based ILs because of their adjustable Lewis acidity or basicity of the ILs. The chronoamperometric transient behavior showed that pure cobalt deposition proceeded via 3D progressive nucleation with diffusion controlled growth [13-15]. An *et al.* [16] studied the electrodeposition behavior of cobalt in $ZnCl_2$-EMIC-$CoCl_2$ system and their results showed that the electrodeposition of cobalt metal proceeds through diffusion-controlled growth process. Hsieh *et al.* [17] investigated the speciation and coordination of cobalt-chloride-based ionic liquid, containing different mole percentages of $CoCl_2$ in EMIC melts. The coordination number and the mean Co-Cl bond length decreases with the increase in concentration of $CoCl_2$ salt in the IL. The results showed that various Co(II) chloride compounds such as $CoCl_4^{2-}$, $Co_2Cl_5^-$, and $Co_3Cl_7^-$ are formed for different molar ratio of $CoCl_2$ and EMIC in the melt. Tulodziecki *et al.* [18] studied the importance of double layer structure formation on electrochemical deposition of Co from ILs containing soluble Co(II)-based precursors. The authors provided an insight on the mechanism of Co^{2+} reduction that taking place on the electrode surface at elevated temperature. In another study, Tulodziecki *et al.* [19] revealed the electrodeposition behavior of zinc-cobalt (Zn-Co) alloy in deep eutectic solvent system i.e., choline chloride/urea (1:2 molar ratio) containing 0.11 M $ZnCl_2$ and 0.01 M $CoCl_2$. The CV results showed preferential reduction of Co and no anomalous codeposition of Zn-Co alloy occurred using this solvent. Chronoamperometric (CA) investigations combined with field emission scanning electron microscopy (FE-SEM) indicated that the electrodeposition of Zn-Co alloys followed the mechanism of instantaneous nucleation. Recently, our research group have studied the electrodeposition of Co from eutectic mixture of urea and choline chloride [20].

In the present study, chronoamperometry technique was used to study the nucleation behavior of cobalt(II) ions in $BMIMBF_4$ ionic liquid. The electrodeposits were characterized using scanning electron microscope (SEM), energy dispersive spectroscopy (EDS) and X-ray diffraction (XRD) methods.

Experimental Method

The 1-butyl-3-methylimidazolium tetrafluoroborate ($BMIMBF_4$) ionic liquid of ≥98% (Sigma Aldrich) was used as received without any further purification or drying. Prior to the starting of the experiments, anhydrous $Co(BF_4)_2$ was obtained from the dehydration of $Co(BF_4)_2 \cdot 6H_2O$ (96%, Alfa Aesar) under vacuum at 393 K for more than 24 h. The electrolyte solution was obtained by dissolving the stoichiometric amount of $Co(BF_4)_2$ in $BMIMBF_4$ IL under dry Ar atmosphere and with a constant stirring until the dissolution of $Co(BF_4)_2$ is complete at 353 K.

The tungsten wire (0.11 cm^2, 99.95%, Alfa Aesar) was used as the working electrode, while platinum wire (>99.997%, Alfa Aesar) and silver wire (99.9%, Alfa Aesar) were used as counter and quasi-reference electrodes respectively. The tungsten wire was polished successively with increasingly finer grades of emery paper and then rinsed well for several times with ethanol and distilled water and then dried in air. All the electrochemical experiments were performed using EG&G PARC model 273A Potentiostat/Galvanostat under an Ar atmosphere. Electrochemical experiments were performed using a three-electrode cell system [20].

Electrodeposition experiments were carried out on copper foil (0.25 cm^2, Alfa Aesar) as cathode electrode. The as-deposited sample was washed using anhydrous ethanol to remove any salt contaminants that adhere to the cathode surface. A high resolution scanning electron microscope (SEM, JEOL 7000, Japan) was used to investigate the surface morphology and energy dispersive spectroscope (EDS), attached to the SEM, for determining the elemental composition of the electrodeposits. The X-ray diffraction (XRD, Philips APD 3720, Netherlands) was used to record the phase and structure of the electrodeposits.

Results and Discussion

Cyclic Voltammetry

Figure 1 shows the overlay of the CVs of tungsten electrode in $BMIMBF_4$ containing 0.20 M of $Co(BF_4)_2$ at different temperatures (333 to 363 K). As shown in Figure 1, the cathodic peak current increases significantly with the increase in temperature. Also, the cathodic peak potential and the reduction onset potential of Co(II) shifts positively with the increase in temperature. It may be due to higher mobility of ions at higher temperature.

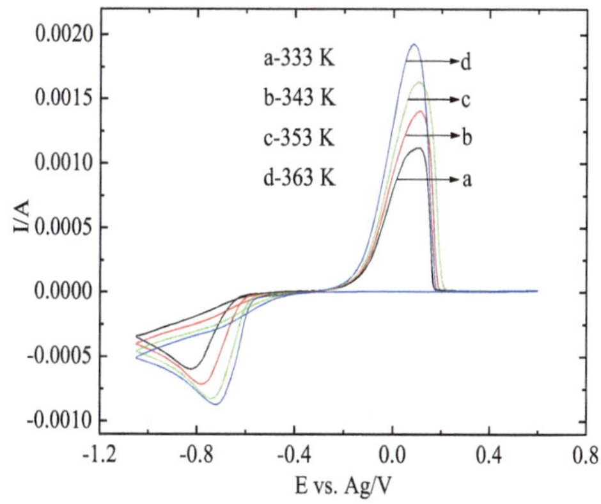

Figure 1. Cyclic voltammograms of tungsten electrode in $BMIMBF_4$ containing 0.20 M of $Co(BF_4)_2$ at different temperatures of 333, 343, 353 and 363 K respectively.

Chronoamperometry

In order to investigate the nucleation and growth mechanism, chronoamperometric measurements were carried out by stepping the potential of the working electrode from a value where no reduction of cobalt occurs to those sufficiently negative potentials that induces the reduction process. This technique have been extensively investigated for understanding the electrodeposition mechanism of metals by comparing the obtained experimental data with theoretical models [21, 22]. Typical current-time transient curves for the experiments at different voltages (-0.60 to -0.72 V) are shown in Figure 2. The curves are characterized by (a) initial sharp current peaks at the onset potentials due to the double layer charging, (b) a subsequent rise in the current due to the formation and growth of cobalt nuclei on the electrode surface and after passing through the maximum (t_{max}, I_{max}) point, (c) the current decreases as a function of time ($t^{-1/2}$) to a constant value, thus indicating a diffusion controlled process given by Cottrell equation [23]:

$$I = nFAD^{1/2}C(\pi t)^{-1/2} \quad [1]$$

where I is the current passing through electrolyte in A, n is the number of transferred electrons, F is the Faraday constant, 96485 Cmol^{-1}, A is the electrode area in cm^2, C is the bulk concentration in molcm^{-3}, t is the time in s and D is the diffusion coefficient of electroactive species in cm^{2}s^{-1}. As shown in Figure 3, the plot of I vs. $t^{-1/2}$ gives a straight line. From Eq. [1] and using the slopes of the straight lines of I vs. $t^{1/2}$, the diffusion coefficient of Co(II) ions in $BMIMBF_4$ lies between 11.0×10^{-8} cm^{2}s^{-1} and 4.5×10^{-8} cm^{2}s^{-1} and the values are listed in Table I. The average value of diffusion coefficient of Co(II) ions is calculated to be 7.5×10^{-8} cm^{2}s^{-1},which is in good agreement with the estimated value (7.6×10^{-8} cm^{2}s^{-1}) obtained from the cyclic voltammetry.

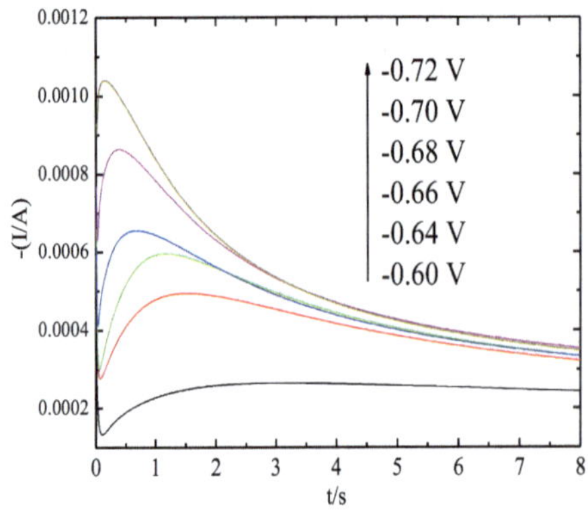

Figure 2. Current-time transients of chronoamperometric experiments for $BMIMBF_4$ containing 0.20 M of $Co(BF_4)_2$ on tungsten electrode at 353 K.

Table I. Diffusion coefficient of Co(II) at different potentials.

Potential (V)	**D (cm^{2}s^{-1})**
-0.64	11.0×10^{-8}
-0.66	9.4×10^{-8}
-0.68	6.1×10^{-8}
-0.70	6.7×10^{-8}
-0.72	4.5×10^{-8}
Average	7.5×10^{-8}

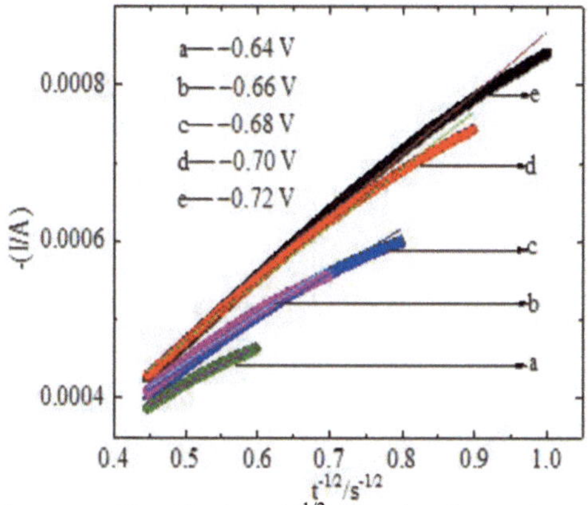

Figure 3. Variation of current (I) with time ($t^{-1/2}$) for the decreasing portion of current-time transients in Figure 2 during the electrodeposition of cobalt on tungsten electrode at 353 K.

The equation, based on 3D nucleation and growth process is controlled by diffusion of the electroactive species (Co (II) ions) in the electrolyte has two limiting cases i.e., instantaneous and progressive nucleation. To identify which of these two mechanisms govern the electrodeposition of cobalt, it is important to compare the dimensionless experimental current–time transients with the dimensionless transients of 3D instantaneous (Eq. [2]) and progressive (Eq. [3]) nucleation processes [21, 22]:

$$\left(\frac{i}{i_m}\right)^2 = 1.9542\left(\frac{t_m}{t}\right)\left(1-\exp\left[-1.2564\left(\frac{t}{t_m}\right)\right]\right)^2 \qquad [2]$$

$$\left(\frac{i}{i_m}\right)^2 = 1.2254\left(\frac{t_m}{t}\right)\left(1-\exp\left[-2.3367\left(\frac{t}{t_m}\right)^2\right]\right)^2 \qquad [3]$$

The experimental and calculated plots are shown in Figure 4. Compared to the calculated data, the experimental data plots can be explained better by an instantaneous nucleation rather than progressive nucleation process. This suggests that the initial stage of the cobalt deposition on a tungsten electrode follows instantaneous nucleation under diffusion controlled growth of cobalt nuclei.

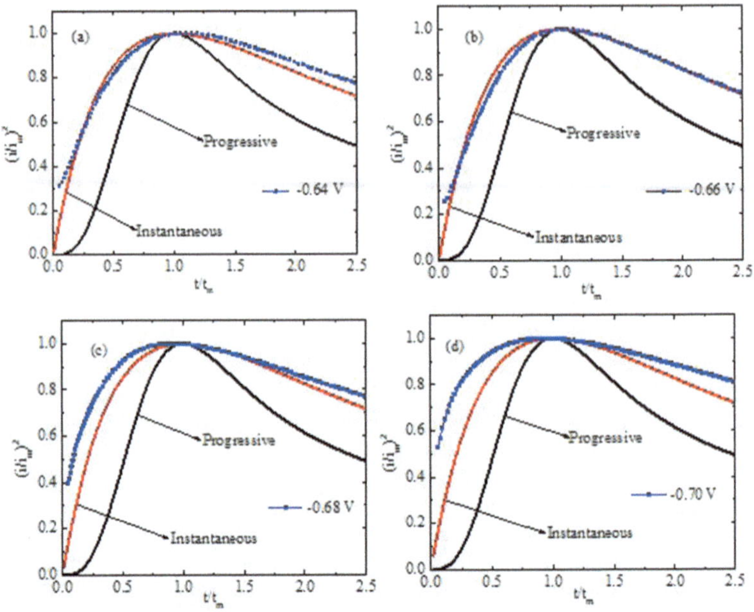

Figure 4. Comparison of the experimental dimensionless current-time transients with the calculated 3D nucleation process for the chronoamperometric electrodeposition of cobalt on tungsten electrode at (a) -0.64 V, (b) -0.66 V, (c) -0.68 V and (d) -0.70 V. All the experiments are conducted using $BMIMBF_4$ containing 0.20 M of $Co(BF_4)_2$ at 353 K.

Electrodeposition and characterization of cobalt

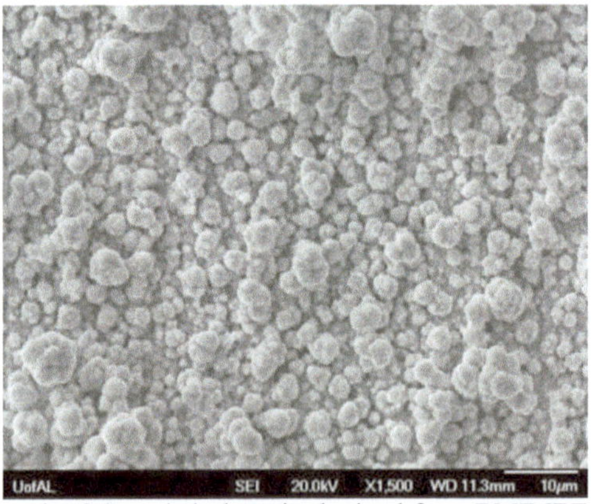

Figure 5. SEM micrograph of the cobalt deposit obtained from $BMIMBF_4$ containing 0.20 M of $Co(BF_4)_2$ on copper cathode electrode substrate at -0.75 V and at 353 K.

Electrodeposition of cobalt on a copper (0.5 cm^2) cathode electrode was performed in $BMIMBF_4$ containing 0.20 M of $Co(BF_4)_2$ at 353 K and by applying a constant potential of -0.75 V between the electrodes. After electrolysis, the deposited Co sample was rinsed thoroughly with ethanol, completely dried and then examined using SEM, EDS and XRD techniques.

The SEM micrograph of the electrodeposited cobalt obtained from $BMIMBF_4$ ionic liquid containing 0.20 M of $Co(BF_4)_2$ is shown in Figure 5. The cobalt deposit obtained by constant potential electrolysis consists of flat and compact surface with the small cauliflower like structures. EDS spectrum of the Co deposit is shown in Figure 6. The electrodeposit contain pure cobalt with no trace of other impurities such as trapped B and F atoms from the ionic liquid. Figure 7 shows that the XRD pattern of as-deposited Co metal. Except the peak due to the Cu substrate, all other peaks in the XRD spectrum are related to Co metal. No presence of cobalt oxide peak in the XRD spectrum indicates that the electrodeposit contain pure cobalt metal.

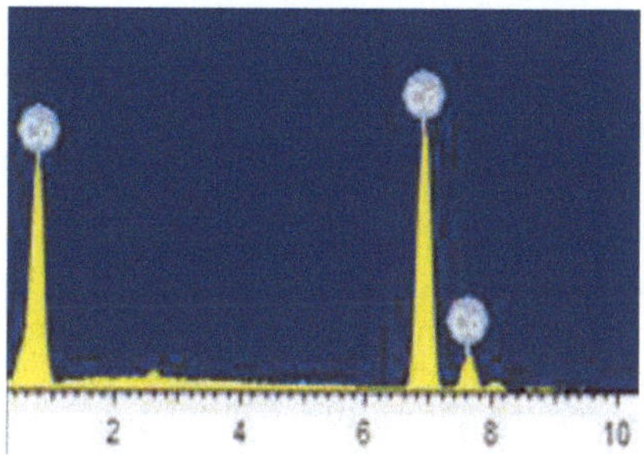

Figure 6. EDS spectrum of the cobalt electrodeposit obtained from $BMIMBF_4$ containing 0.20 M of $Co(BF_4)_2$ on copper at -0.75 V and at 353 K.

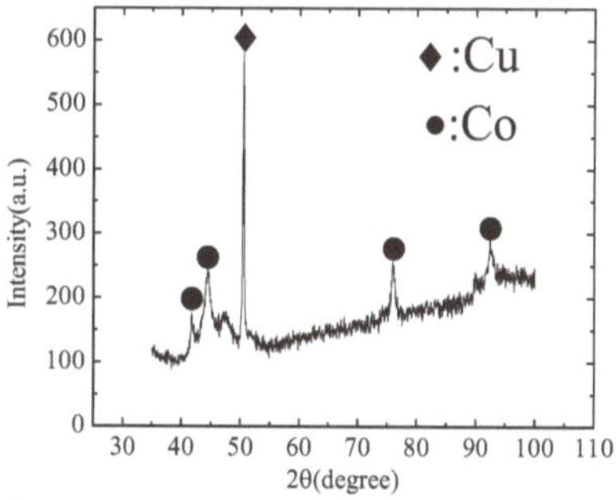

Figure 7. XRD spectrum of the cobalt electrodeposit obtained from $BMIMBF_4$ containing 0.20 M of $Co(BF_4)_2$ on copper at -0.75 V and at 353 K

Conclusions

The cyclic voltammetry and chronoamperometry experiments were conducted to reduce $Co(BF_4)_2$ in $BMIMBF_4$ ionic liquid containing 0.20 M of $Co(BF_4)_2$ at 353 K. Chronoamperometric experiments shows that the electrodeposition of cobalt is a 3D progressive nucleation process under diffusion control. From Cottrell equation, the average diffusion coefficient of Co(II) ions in the melt was calculated to be 7.5×10^{-8} cm^2s^{-1} at 353 K, which is in good agreement with the estimated value (7.6×10^{-8} cm^2s^{-1}) from cyclic voltammetry. A dense and compact electrodeposit of cobalt was obtained under the experimental conditions. EDS and XRD results show that the electrodeposit is mainly composed of cobalt metal.

Acknowledgements

The author (Min Li) is thankful for the award of scholarship by China Scholarship Council (CSC) to pursue study as a visiting student at the Department of Metallurgical and Materials Engineering, The University of Alabama, Tuscaloosa, Alabama, United States of America. Authors are pleased to acknowledge the financial support provided by ACIPCO and NSF Grant No. DMR-1310072 for this research project. We also thank The University of Alabama for providing the experimental and analytical facilities needed for this research work.

References

1. A. Aal et al., "Interfacial electrochemistry and electrodeposition from some ionic liquids: In situ scanning tunneling microscopy, plasma electrochemistry, selenium and macroporous materials," *Electrochim. Acta*, 56 (2011), 10295-10305.

2. F. Endres et al., "Do salvation layers of ionic liquids influence electrochemical reactions?," *Phys. Chem. Chem. Phys.*, 12 (2010), 1724-1732.

3. F. Endres and S.Z.E. Abedin, "Air and water stable ionic liquids in physical chemistry," *Phys. Chem. Chem. Phys.*, 8 (2006), 2101-2116.

4. W. Freyland et al., "Nanoscale electrodeposition of metals and semiconductors from ionic liquids," *Electrochim. Acta*, 48 (2003), 3053-3061.

5. P. Hapiot and C. Lagrost, "Electrochemical Reactivity in Room-Temperature Ionic Liquids," *Chem. Rev.*, 108 (2008), 2238-2264.

6. W. Simka, D. Puszczyk, and G. Nawrat, "Electrodeposition of metals and from non-aqueous solutions," *Electrochim. Acta*, 54 (2009), 5307-5319.

7. Y.-Z. Su et al., "The Electrode/Ionic Liquid Interface: Electric Double Layer and Metal Electrodeposition," *Chem Phys Chem*, 11 (2010), 2764-2778.

8. C.-C Tai, F.-Y Su, and I.-W. Sun, "Electrodeposition of palladium-silver in a Lewis basic 1-ethyl-3-methylimidazolium chloride-tetrafluoroborate ionic liquid," *Electrochim. Acta*, 50

(2005), 5504-5509.

9. S.-I Hsiu, C.-C Tai, and I.-W. Sun, “Electrodeposition of palladium-indium from 1-ethyl-3-methylimidazolium chloride tetrafluoroborate ionic liquid,” *Electrochim. Acta*, 51 (2006), 2607-2613.

10. L. S. Jou et al., “Electrodeposition of Palladium-Copper Films from 1-Ethyl-3-methylimidazolium Chloride-Tetrafluoroborate Ionic Liquid on Indium Tin Oxide Electrodes,” *J. Electrochem. Soc.*, 156(6) (2009), D193-D197.

11. M. H Yang, M.-C Yang, and I.-W. Sun, “Electrodeposition of Indium Antimonide from the Water-Stable 1-Ethyl-3-methylimidazolium Chloride/Tetrafluoroborate Ionic Liquid,” *J. Electrochem. Soc.*, 150(8) (2003), C544-C548.

12. S. I. Hsiu and I.-W. Sun, “Electrodeposition behavior of cadmium telluride from 1-ethyl-3-methylimidazolium chloride tetrafluoroborate ionic liquid,” *J. Appl. Electrochem.*, 34 (2004), 1057-1063.

13. R. T. Carlin, P.C. Trulove, and H.C.D. Long, “Electrodeposition of Cobalt-Aluminum Alloys from Room Temperature Chloroaluminate Molten Salt,” *J. Electrochem. Soc.*, 143 (9) (1996), 2747-2758.

14. R. T. Carlin et al., “Microelectrode Evaluation of Transition Metal-Aluminum Alloy Electrodepositions in Chloroaluminate Ionic Liquids,” *J. Electrochem. Soc.*, 145 (5) (1998), 1598-1607.

15. J. A. Mitchell et al., “Electrodepositions of Cobalt and Cobalt-Aluminum Alloys from a Room Temperature Chloroaluminate Molten Salt,” *J. Electrochem. Soc.*, 143 (11) (1996), 3448-3455.

16. M. Z. An et al., “Electrodepositions of Cobalt in an Ionic Liquid Electrolyte at Ambient Temperature,” *Chin. J. Chem.* 26 (7) (2008) 1219-1224.

17. Y. T Hsieh et al., “Speciation of cobalt-chloride-based ionic liquids and electrodeposition of Co wires,” *Electrochim. Acta*, 117 (2014), 217-223.

18. M. Tulodziecki et al., “Importance of double layer structure in the electrochemical deposition of Co from soluble Co^{2+}-based precursors Ionic Liquid media,” *Electrochim. Acta*, 134 (2014), 55-66.

19. Q. W. Chu, J. Liang, and J. C. Hao, “Electrodeposition of zinc-cobalt alloys from choline chloride-urea ionic liquid,” *Electrochim. Acta*, 115 (2014), 499-503.

20. M. Li, Z. W. Wang, and R. G. Reddy, “Cobalt electrodeposition using urea and choline chloride,” *Electrochim. Acta*, 123 (2014), 325-331.

21. G. Gunawardena et al., “Electrochemical nucleation part I. general considerations,” *Electrochim. Acta*, 138 (1982), 225-239.

22. B. Scharifker and G. Hills, “Theoretical and experimental studies of multiple nucleation,” *Electrochim. Acta*, 28(7) (1983), 879-889.

23. J. Bard and R.L. Faulkner, *Electrochemical methods fundamentals and applications* (Bei Jing, BJ: Chemical Industry Press, 2006), 163.

EPD Congress 2015
Edited by: James Yurko, Antoine Allanore, Laura Bartlett, Jonghyun Lee, Lifeng Zhang, Gabriella Tranell, Yulia Meteleva-Fischer, Shadia Ikhmayies, Arief Suriadi Budiman, Prabhat Tripathy, and Guy Fredrickson
TMS (The Minerals, Metals & Materials Society), 2015

EXTRACTING ALUMINA FROM COAL FLY ASH WITH CONCENTRATED SULFURIC ACID SINTERING AND ULTRASOUND AIDED LEACHING

Wenbo LUO[2], Jilai Xue[1, 2], Jun Zhu[1, 2], Kang Liu[2], Chunlei Yang[2], Fusheng Mao[2].

[1] State Key Laboratory of Advanced Metallurgy, University of Science and Technology Beijing, Xueyuan Road 30, Beijing 100083, China
[2] School of Metallurgical and Ecological Engineering
University of Science and Technology Beijing
Xueyuan Road 30, Beijing 100083, China

Keywords: Coal Fly Ash, Alumina, Ultrasound, Leaching.

Abstract

Extracting alumina from coal fly ash (CFA) is of great interest for both environmental protection and recycling valuable alumina content. In this study, ultrasonic aided leaching process as an effective method was investigated to enhance extracting Al_2O_3 from CFA via concentrated sulfuric acid sintering. The influence of ultrasonic power and processing time were tested under the conditions as the same as those without ultrasound. It is found that the alumina extraction rate can reach 90.5 % with ultrasound assistance, while the ultrasonic treatment can reduce almost 15 ~ 30 min in leaching time and 5 ~ 10 °C in leaching temperature than those without ultrasonic operation at the same Al_2O_3 extraction rate. In addition, the ultrasonic can reduce calcium content by 6 ~ 8 % dissolved into the leaching liquid. SEM analysis also shows that the residue particle size with ultrasound is smaller than that without ultrasonic treatment.

Introduction

Extracting alumina from coal fly ash (CFA) is of great interest for both environmental protection and recycling valuable alumina content. There are two major methods which are currently applied for extracting alumina from CFA, alkali-sintering and acid leaching [1]. In the alkali-sintering process, the alumina extraction rate is high (>90 %), but the operating temperature must be ensured to be more than 1000 °C) with large volume of solid residues [2]. By contrast, the acid-leaching process needs lower energy consumption and produces less residues, but its Al_2O_3 extraction rate is lower (82 ~ 85 %) [3].

There are a number of publications available on improving Al_2O_3 extraction rate in the acid-leaching process for treating coal fly ash. Nayak [4] used different concentrations of dilute sulfuric acid to leach coal fly ash, where Al_2O_3 extraction rate reached 85 %. MU [1] mixed coal fly ash together with concentrated sulfuric acid (98 %) before sintering, and finally could obtain 87.64 % Al_2O_3. Most of the leaching temperatures were kept in the range of 80 ~ 90 °C [3, 5, 6]. As known from the open literature, the raising leaching temperature and the increasing concentration of sulfuric acid were the major common methods to increase Al_2O_3 extraction rate from CFA, but those could face technical constrains in possible increased chemical corrosion in equipment lining materials. Therefore, further research is needed to look for other technical means and better understanding of processing parameters.

Ultrasonic can be a powerful tool for enhancing the leaching process [7, 8]. Nakui [4] used ultrasonic treatment to accelerate hydrazine degradation which increased from 1 μmol $L^{-1}\cdot min^{-1}$ to 3.5 μmol $L^{-1}\cdot min^{-1}$ under ultrasonic – an increase by 250 %. Bese [9] observed the effect of ultrasonic treatment on the dissolution of copper from copper converter slag by acid leaching, and the ultrasound enhanced the dissolution of Fe and other metallic. Belviso [10] also found that the ultrasonic treatment facilitated the formation of zeolites at a lower-temperature (25 °C) than did no sonication (40 or 60 °C).

In this work, the concentrated sulfuric acid sintering combined with ultrasonic aided leaching process was applied to enhance extracting Al_2O_3 from CFA. The leaching behaviors of Al^{3+} and other metallic (Fe^{3+} / Fe^{2+} and Ca^{2+}) were also investigated. The technical data obtained will facilitate further improvements in the recycling efficiency and product purity control in the acid-sintering process of extracting alumina from CFA.

Experimental

Materials and Chemicals

Table I is the chemical compositions of coal fly ash collected from a thermal power plant in Inner Mongolia, China. The specimens used for chemical analysis of the composition were taken from a mixture of CFA in order to get a good reprehensive result for the raw materials. The composition dada presented here was applied to the process design. Major parts of CFA were SiO_2, Al_2O_3, Fe_2O_3 and CaO, which were taken into account for the leaching tests, while the other small amounts of components were not considered. The particle sizes of about 60 % CFA raw material were -150 meshes.

Table I. Chemical Compositions of Coal Fly Ash.

Composition	Content/wt %	Composition	Content/wt %
SiO_2	54.55	TiO_2	1.44
Al_2O_3	33.54	K_2O	0.98
Fe_2O_3	3.59	MgO	0.82
CaO	3.80	SO_3	0.28
P_2O_5	0.28	Na_2O	0.31

Figure 1 is XRD pattern of coal fly ash as raw material for performing alumina extraction. The main crystalline phase in coal fly ash is mullite and quartz. In the leaching experiments, H_2SO_4 (GR) was used as leaching agent and reactant. The pure water used was produced in our own laboratory.

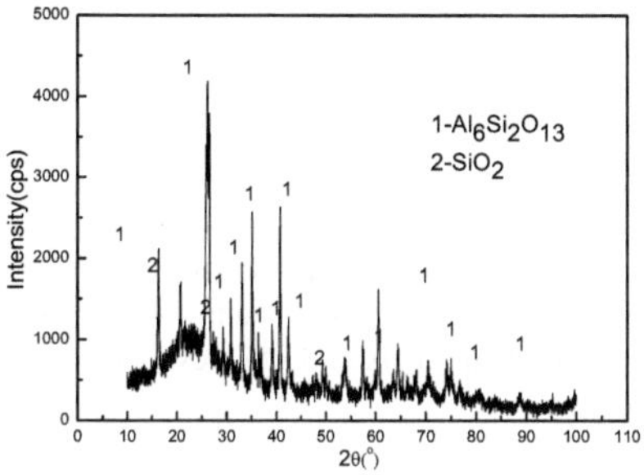

Figure 1. XRD spectrum of coal fly ash as raw material

Leaching Process and Analysis

The coal fly ash which was first ground for 3 h, then sintered with H_2SO_4 (80 %) at 1: 1.5 mass ratio for 1 h at 290 °C. The sintered mass was mixed with pure water to make aqueous solution in a container placed in a heating water tank, and the liquid-solid ratio was kept 12: 1, while the liquid was treated with ultrasonic unit (Ningbo Scientz biotechnology CO.). Ultrasonic operation was conducted with an alternative interval between 2 s-on and 3 s-off for varying period of time and temperature. After that, all leaching tests took place in the heating water tank for a fixed period of 30 min and at the end the liquid was filtrated to separate solid residues.

Al^{3+}, Ca^{2+}, Fe^{3+} and Fe^{2+} in the liquid were determined by titration analysis method. The chemical compositions of leaching residues were measured by X-ray fluorescence analyzer (Shimadzu Corp.). The crystalline phases were measured by Rigaku X-ray diffraction analyzer (RigakuD: MAX-RB12KW, scanning from 10° to 100°, the rate was 0.02° per second, Cu (40kV, 40mA)). And the leaching residue was examined using analytical scanning electron microscope (JSM-6510A, JEOL CO., Ltd.).

Results and discussion

Leaching Behavior of Al^{3+}

Temperature, ultrasonic treatment time and power were studied as variables in the leaching process, in where the reaction in leaching process may be described as follows:

$$Al_2(SO_4)_3 - 2Al^{3+} + 3SO_4^{2-} \tag{1}$$

Al extraction rate, r, can be calculated by equation (2):

$$r = \frac{C \times 50 \times \frac{V}{2}}{W \times Al_2O_3 wt\% \times \frac{54}{102}} \times 100\% \tag{2}$$

where C, V and W are Al concentrations in leaching solution measured by ICP, the volume of filtrate in a leaching test and the mass of raw CFA used, respectively.

Figure 2 shows Al extraction rate against varying ultrasonic power applied in the pre-treatment before the leaching process. It is obvious that the ultrasonic pre-treatment to the aqueous solution can enhance the reaction (1) to release Al^{3+} into the solution, while 30 ~ 50 % of power input to the ultrasonic unit is proved sufficient to produce such an effect under this testing condition. Therefore, more experiments only using a fixed 35 % power input (333 W) were performed in the later part of work.

For comparison, the effects of varying ultrasonic pre-treatment time and temperature on Al extraction rate are presented in Figure 3. In general, the prolonged time in ultrasonic pre-treatment can improve the Al extraction rate more or less. While the higher operating temperatures may add even more assistance to this action, for instances, Al extraction rate can reach 88 ~ 91 % in a temperature range of 80 °C to 90 °C. At the same time, as the temperature rises to higher than 80 °C, Al extraction rate is already up to about 85 % within a period of just about 10 min, which means a time-saving or production rate increase for the process.

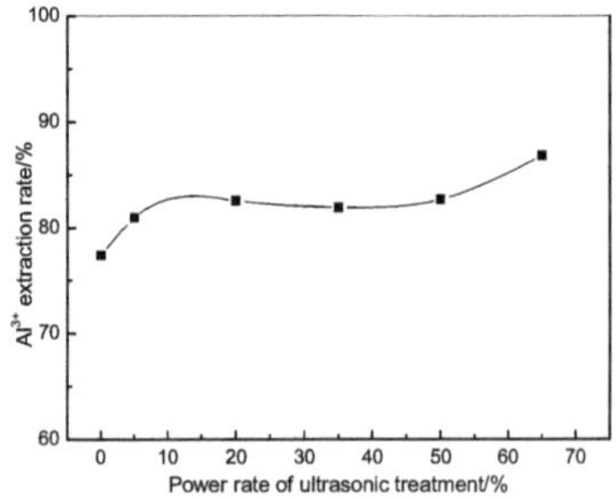

Figure 2. Al extraction rate vs. ultrasonic power at leaching temperature of 85 °C

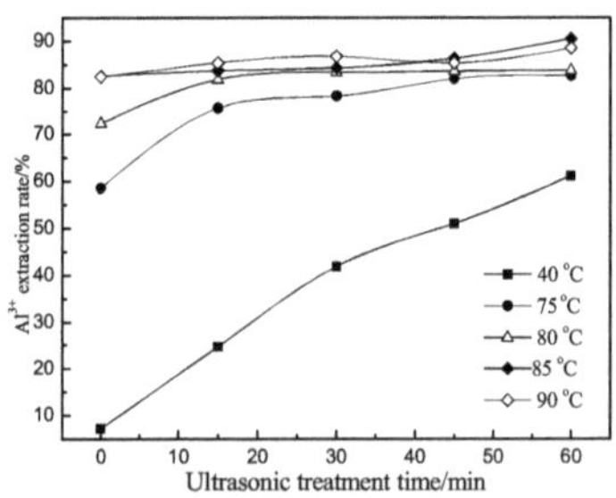

Figure 3. Al extraction rate vs. the time of ultrasonic pre-treatment with various temperatures at 35 % power input

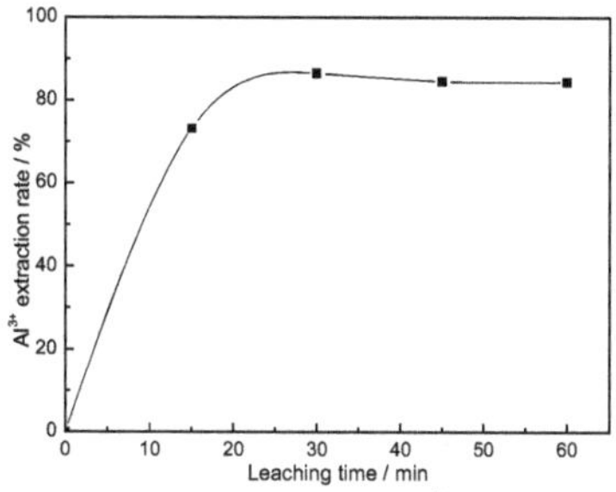

Figure 4. Al extraction rates vs. leaching time at 85 °C leaching temperature after 15 min ultrasonic treatment with 35 % (333W) power input of ultrasonic unit

<u>Leaching Behaviors of $Fe^{3+}$and Fe^{2+}</u>

Understanding of leaching behaviors of Fe^{3+} and Fe^{2+} will be useful for better control of these impurities. The possible reactions can be as follows:

$$FeSO_4 – Fe^{2+} + SO_4^{2-} \tag{3}$$

$$Fe_2(SO_4)_3 – 2Fe^{3+} + 3SO_4^{2-} \tag{4}$$

Figure 5 shows that higher temperatures make both Fe^{3+} and Fe^{2+} being dissolved more into the liquid in the process of leaching, while the longer time of ultrasonic treatment makes more Fe^{2+} entered into liquid at various temperatures. More Fe^{3+} and Fe^{2+} are leached from sintered coal fly ash with ultrasonic treatment than those without it.

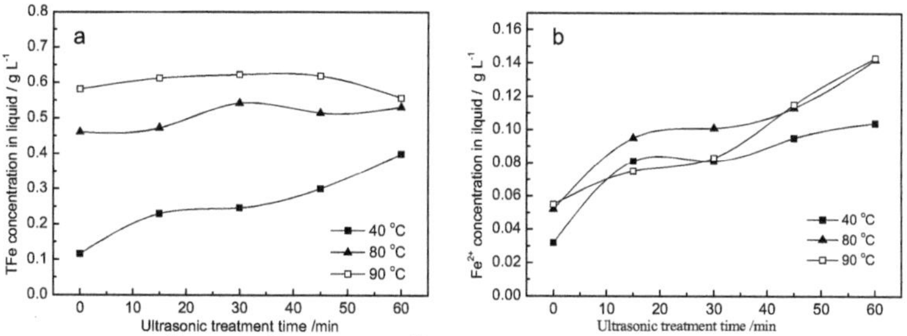

Figure 5. (a) TFe (Fe^{3+} and Fe^{2+}) and (b) Fe^{2+} contents vs. ultrasonic time at different leaching temperatures with 35 % (333W) power input of ultrasonic unit

It is known that ·OH radical that is strong oxidable will be produced under ultrasonic treatment in aqueous solutions [7, 8, 11]. This means that during leaching process in aqueous solution, the reactions (5) to (7) may take place:

$$\cdot OH + \cdot OH - H_2O_2 \tag{5}$$

$$\cdot OH + Fe^{2+} - OH^- + Fe^{3+} \tag{6}$$

$$H_2O_2 + Fe^{2+} - OH^- + \cdot OH + Fe^{3+} \tag{7}$$

As Anbar reported [10] where H_2O_2 yield was 4.2×10^{-6} mol L^{-1} min^{-1} in water, the concentration of H_2O_2 could be estimated as 2.52×10^{-4} mol L^{-1} after 60 min ultrasonic treatment in aqueous solution. This implies that H_2O_2 concentration from the reaction (5) could also exist in our leaching solution to make the reaction (6) and reaction (7) occurred, and thus may result in an increase of Fe^{3+} content. For example, as Fe^{2+} increased from about 0.05 g L^{-1} to 0.10 g L^{-1} after 30 min of ultrasonic treatment at 80 °C, TFe (Fe^{3+} and Fe^{2+}) did from 0.48 g L^{-1} to 0.53 g L^{-1} at the same time. However, such an effect could be limited as·OH radical and H_2O_2 amount may be far less than the ferrous ions in the leaching solution.

Leaching Behavior of Ca^{2+}

In Figure 6, ultrasonic treatment hinders Ca^{2+} into the liquid. In the leaching solution, the reaction (8) and equation (9) may be present as follows:

$$CaSO_4 - Ca^{2+} + SO_4^{2-} \quad K_{sp}^{\theta}(298K) = 4.93\times10^{-5} \tag{8}$$

$$J = \{C(Ca^{2+})\}\cdot\{C(SO_4^{2-})\} = 2.8\times10^{-5} < K_{sp}^{\theta} \tag{9}$$

where J is for the product of $C(Ca^{2+})$ and $C(SO_4^{2-})$ in the dissolution of calcium sulfate, and K_{sp}^{θ} is solubility product of reaction (8) that is relative stable with changing temperature. The ultrasonic treatment may lead to an increase in J value, which may result in some precipitation of $CaSO_4$ and hence reduce Ca^{2+} content in the leaching solution. However, this interesting point needs further investigation to explore its details and reaction mechanism.

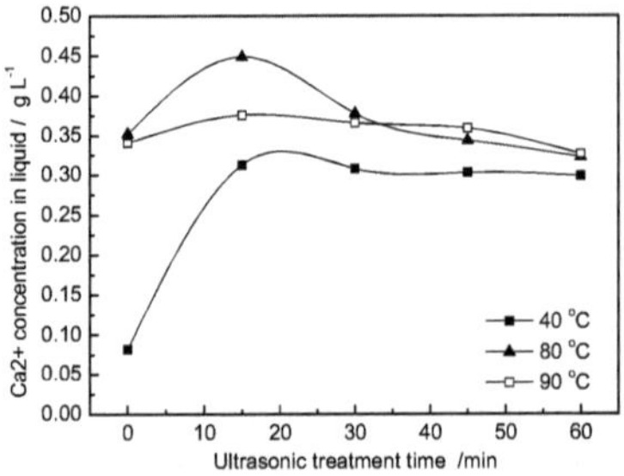

Figure 6. Ca^{2+} content vs. ultrasonic time at various leaching temperatures with 35 % (333W) power input of ultrasonic unit

It should be pointed out that Fe^{3+}, Fe^{2+} and Ca^{2+} are impurities that have to be removed from the liquid solution. Appropriate control of the balance between Al and these impurities would be very important for process economy and product quality. Through adjusting the time of ultrasonic treatments in the pre-leaching or the leaching process, for instances, a large amount of Al^{3+} could be obtained with a small or almost no increase of Fe^{3+}, Fe^{2+} contents and reduced Ca^{2+}content in the leaching solution.

Leaching Residues

Figure 7 presents SEM micrographs of the leaching residues, where the uniform, fine particles are found with ultrasonic treatment. The smaller particles may be due to a faster precipitation process occurring with ultrasonic treatment that can enhance the kinetic process of mass transportation in liquid solution.

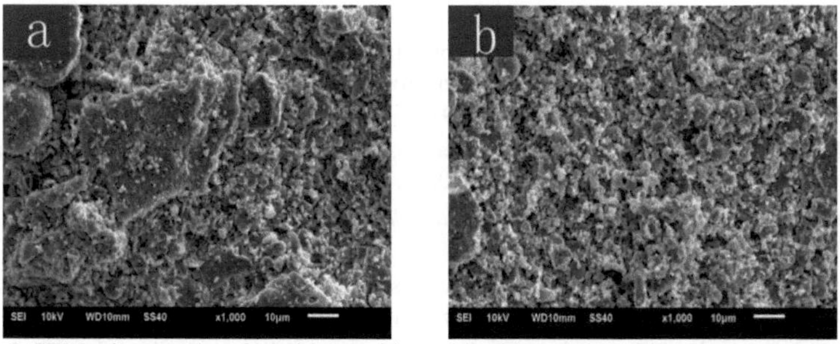

Figure 7. SEM micrographs of leaching residues: (a) without ultrasonic treatment and (b) with ultrasonic treatment

Conclusion

1. Ultrasonic aided leaching process as an effective method can enhance extracting Al_2O_3 from CFA via concentrated sulfuric acid sintering. With ultrasound assistance, the alumina extraction rate can reach 90.5 %, while both leaching time and leaching temperature can reduce 15 ~ 30 min and 5 ~ 10 °C, respectively, for the same rate of Al_2O_3 extraction.

2. Fe^{3+} and Fe^{2+} contents in leaching solutions increase with ultrasonic treatment, while Ca^{2+} content can decrease at the same time.

3. The particle size of leaching residue with ultrasonic treatment is smaller than that without ultrasonic treatment.

Acknowledgements

Part of support from aluminum and carbon industries in China is gratefully acknowledged. Most of analysis work was carried out in the Central Laboratory of Metallurgical School at University Science and Technology Beijing.

Reference

1. W.N. Mu, Y.C. Zhai, "Preparation of α-alumina from coal fly ash by sintering with sulfuric acid," *Advanced Materials research*, 113-114(2010), 2039-2044.
2. P.P. Ge et al., "Progress on extraction of alumina from fly ash," *Inorganic chemicals industry*, 42(7) (2010), 1-4.
3. R.C. Wang, Y.C. Zhai, Z.Q. Ning, "Thermodynamics and kinetics of alumina extraction from fly ash using an ammonium hydrogen sulfate roasting method," *International journal of minerals, metallurgy and materials*, 21(2) (2014), 144-149.
4. H. Nakui et al., "Effect of coal ash on hydrazine degradation under stirring and ultrasonic irradiation conditions," *Ultrasonics Sonochemistry*, 15 (2008), 472-477.
5. N. Nayak, C.R. Panda, "Aluminium extraction and leaching characteristics of Talcher Thermal Power Station," *Fuel*, 89 (2010), 53-58.
6. K. Liu, J.L. Xue, W.B. Luo, "Effects of ultrasound on Al_2O_3 Extraction rate during acid leaching process of coal fly ash", (Paper presented at 144_{th} Annual meeting & Exhibition, San Diego, California, 20 February 2014), 251-258.
7. J.L. Luque-Garcia, M. D. Luque de Castro, "Ultrasound: a powerful tool for leaching," *Trends in analytical chemistry*, 22(1) (2003), 41-47.
8. Pankaj, *Theoretical and Experimental Sonochemistry Involving Inorganic Systems* (Germany: Springer, 2010), 213-263.
9. A.V. Bese, "Effect of ultrasound on the dissolution of copper from copper converter slag by acid leaching," *Ultrasonics Sonochemistry*, 14 (2007), 790-796.
10. C. Belviso et al., "Effects of ultrasonic treatment on zeolite synthesized from coal fly ash," *Ultrasonics Sonochemistry*, 18 (2011), 661-668.
11. M. Anbar, I. Pecht, "On the Sonochemical Formation of Hydrogen Peroxide in Water," *The Journal of Physical Chemistry*, 62(2) (1964), 352-355.
12. G.J. Price et al., "Acoustic emission from cavitating solutions: implications for the mechanisms of sonnochemical reactions," *The journal of physical chemistry B*, 109(38) (2005), 17799-17801.

2015 EPD Congress

SYMPOSIUM: MATERIALS PROCESSING FUNDAMENTALS

Casting, Solidification, and Steel Processing

Laura Bartlett,
Texas State University

EPD Congress 2015
Edited by: James Yurko, Antoine Allanore, Laura Bartlett, Jonghyun Lee, Lifeng Zhang, Gabriella Tranell, Yulia Meteleva-Fischer, Shadia Ikhmayies, Arief Suriadi Budiman, Prabhat Tripathy, and Guy Fredrickson
TMS (The Minerals, Metals & Materials Society), 2015

A CELLULAR AUTOMATON MODEL COUPLED WITH FINITE ELEMENT METHOD FOR SOLIDIFICATION PROCESS OF BEAM BLANK CONTINUOUS CASTING

Yan Zhao[1], Dengfu Chen[2], Mingtu Ma[1], Xinming Wan[1], Yang Li[1], Rongshan Qin[3]

[1]Automotive lightweight Engineering Research Center, Chinese Automotive Engineering Research Institute Co. Ltd;
9 Jinyu Road, Chongqing 401122, P.R. China.
[2] College of Materials Science and Engineering, Chongqing University;
174 Shazheng Road, Chongqing 400030, P.R. China.
[3] Department of Materials, Imperial College London;
Exhibition Road, London SW7 2AZ, United Kingdom.

Keywords: Cellular automaton, Finite element, beam blank continuous casting, microstructure evolution

Abstract

A two dimensional cellular automaton (CA) model coupled with finite element (FE) method has been developed for simulating the formation of solidified microstructure during beam blank continuous casting, which represents columnar and equiaxed dendrites growth in the case of detailed secondary cooling boundary conditions. In this model, a new adaptive mesh division is proposed to couple CA cubic lattice and irregular FE mesh of beam blank. A double-accuracy liner interpolation method is employed to obtain micro CA temperature field via macro FE calculation. The nucleation process and the velocity of dendrite tip are calculated by continuous nucleation model and KGT model, respectively. Moreover, the preferred orientation of dendrite growth is determined using a probabilistic model. The effects of super heat and casting speed on the microstructure characteristics are analyzed. This model can be applied to optimize processing parameters for beam blank continuous casting basing on microstructure evaluation.

Introduction

H section steel has excellent performance and is widely used in different industrial fields as important materials. Continuously casting beam blank as well as its rolling process can obviously reduce production cost as an advanced technology. However, owing to the irregular cross-section shape, homogeneous cooling is difficult to fulfill during beam blank continuous casting, as a result, the solidification shrinkage is complex. Therefore, the surface and internal cracks are prominent in final product. The solidification microstructure has significant influence on the quality of casting beam blank. Most of studies on continuous casting beam blank mainly focus on heat transfer and mechanical behavior rather than its microstructure [1-5]. There are few literature reported on the microstructure evolution of beam blank. However, the relationship between microstructure and process parameter is important for improving final quality of product. In this paper, 2D cellular automaton (CA) model coupled with finite element (FE) method is developed to simulate the microstructure evolution during beam blank continuous casting. The temperature of beam blank is calculated using FE method, and the detailed secondary cooling condition is taken into account on both longitudinal and cross directions of beam blank. On the basis of the macro temperature field, an adaptive mesh division as well as a double-accuracy liner interpolation method is introduced to calculate the temperature of micro CA lattice. The nucleation and the velocity of solid/liquid (S/L) interface are calculated via continuous nucleation model and KGT model, respectively. The present model is applied to simulate microstructure formation at various casting temperatures and speeds during beam blank continuous casting process.

Model description

The multi-scale temperature model

A 1/4 cross-section of beam blank is selected as the computational region. Two meshes are employed to calculate the temperature fields of macro FE and micro CA models. The computational region is composed of 42×33 nodes and 1312 elements with irregular shapes. Generally, CA lattice is square, thus cannot be equally divided in these elements. Moreover, the evolution rule of CA model is difficult to define due to the complicated profile of beam blank if the section consists of square FE elements. Thus, an adaptive mesh division is introduced to solve this problem. The square CA lattices are directly divided in rectangle area that beam blank locates rather than along with profile of beam blank. The entire region consists of 1250×812 CA lattices and the size of each lattice is 0.2mm. Once a coordinate of CA lattice center is confirmed, the lattices out of computational region are assigned a higher temperature than liquidus and no crystal grows in them. The method of mesh division is illustrated in Fig.1.
In order to describe the actual beam blank continuous casting, accurate boundary conditions of heat transfer in the secondary cooling zone are considered, including spray water cooling, water evaporation cooling, radiation cooling and roll contact cooling on the casting direction. The non-uniform distribution of spray water flow and various heat transfer patterns on the cross-section is also taken into account. The model's details have been reported in Ref. [6, 7]. The time step is set as 0.01s in macro heat transfer model and 0.0001s in micro CA model, respectively. It is assumed that the heat flux of each node on the boundary maintained constant during every 100 steps in CA model and is obtained from macro FE model.

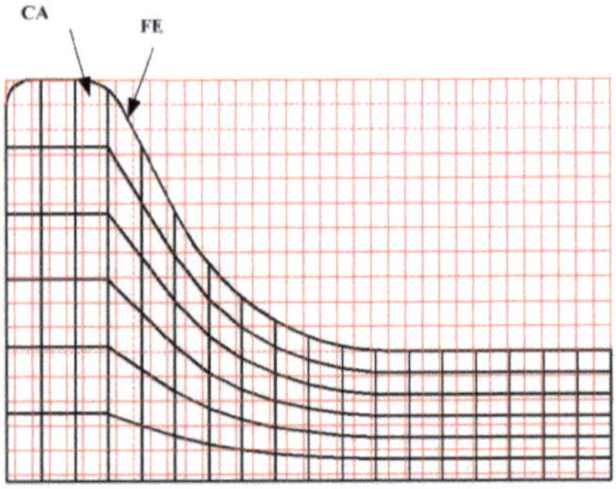

Figure1. Mesh generation in FE-CA model

First, it is essential to confirm the corresponding FE mesh of each CA lattice. The FE mesh is also called the parent mesh of CA lattice, and obtained via their coordinates. The center of each CA lattice is given by:

$$T(r) = b_1 + b_2\xi(r) + b_3\eta(r) + b_4\xi(r)\eta(r) \tag{1}$$

where b_1, b_2, b_3 and b_4 are interpolation parameters, ζ (r) and η (r) are local x-coordinate and y-coordinate of the CA lattice in its parent FE mesh.
Once any CA lattice solidifies, the latent heat is on average released onto the four nodes of its parent FE mesh.

Nucleation model

The nucleation density during a certain undercooling is regarded as a continuous process, which is obtained via [8, 9]:

$$n(\Delta T) = \int_0^{\Delta T} \frac{dn}{d(\Delta T')} d(\Delta T') \tag{2}$$

It is supposed that the undercooling increases δT at each time step, thus the nucleation density increment of each volume unit is evaluated by:

$$\delta n = n(\Delta T + \delta T) - n(\Delta T) \tag{3}$$

The nucleation probability in each volume unit with a certain undercooling is represented by:

$$P_n = \delta n V_c \tag{4}$$

in which, V_c is volume unit. In this CA model, when the temperature of the CA lattice is lower than the liquidus and this lattice is still in molten state, a random parameter q between 0 to 1 is generated and used to compare with P_n. The CA lattice will solidify if $P_n > q$, otherwise it remains liquid.

Dendrite growth scheme of CA model

The four-neighbor Von Neumann rule is used in the CA model to represent the migration of S/L interface. During solidification process, each CA lattice has three possible states: liquid, solid and interface. At the beginning of solidification, all of the CA lattices are molten, once a lattice solidified due to nucleation or other case, its four neighbors will transfer into interface states. Dendrite of cubic metal often prefers to grow along with [1 0] orientation. The present model brings a capturing probability method to describe this phenomena, i.e. in each time step, a random number between 0 to 1 is assigned to the interface lattice. If this number is smaller than the capturing probability, the neighbor lattice will be captured by solidified lattice and the S/L interface starts to move in it. The probability expression is as follows:

$$P_c = \frac{1}{\sqrt{\tan^2\theta + 1}} \tag{5}$$

where θ is the angle between [1 0] orientation and X axis.
The velocity of S/L interface in the interface lattice is calculated using the KGT model [10].

Yamazaki [11] has given a numerical approximation written as:

$$V(\Delta T) = 2.0\times10^{-4}\Delta T^4 + 4.6\times10^{-3}\Delta T^3 + 3.54\times10^{-3}\Delta T^2 + 7.56\times10^{-2}\Delta T \quad (6)$$

The solid fraction in the interface lattice is expressed via:

$$\Delta f_s = \frac{\Delta t}{a}(V_x + V_y - V_x V_y \frac{\Delta t}{a}) \quad (7)$$

where V_x and V_y are the velocities of S/L interface on the direction of X axis and Y axis, respectively. a is the size of CA lattice, Δ t is time step.
When the solid fraction of interface lattice reaches 1 or more, it will solidify and continue to capture its liquid or interface neighbors.

Result and Discussion

During continuous casting process, final quality of product is mainly affected by solidified microstructure, for which superheat degree and casting speed are key [12,13]. The microstructure evolution of Q235 steel during beam blank continuous casting is simulated using the developed FE-CA model, in which the processing parameters are from Ref. [6].

The effect of casting speed on solidified micro structure

Fig.2 shows the solidified morphology at the end of Segment 3 with various super heat degrees. As shown in Fig.2, the whole beam blank has already completely solidified. From surface to center, fine crystal zone, fine-columnar Transition (FCT) crystal zone, columnar crystal zone and equiaxed zone can be clearly found at web and fillet. With the rising superheat, the thickness of fine crystal zone varies little, while that of FCT zone increases. Some equiaxed dendrites precipitate at solidified end, whose amount reduces as superheat rises. As seen in Fig. 2(a)-(c), the columnar zone in beam blank accounts for the largest proportion, and few equiaxed dendrites appear at the internal center of web owing to intensive cooling and small thickness. At the flange tip region, coarse inter-dendrites grow along with different directions and promote the formation of gap, which will result in the gather of bubbles or inclusion. Such observations have been reported in practical production.

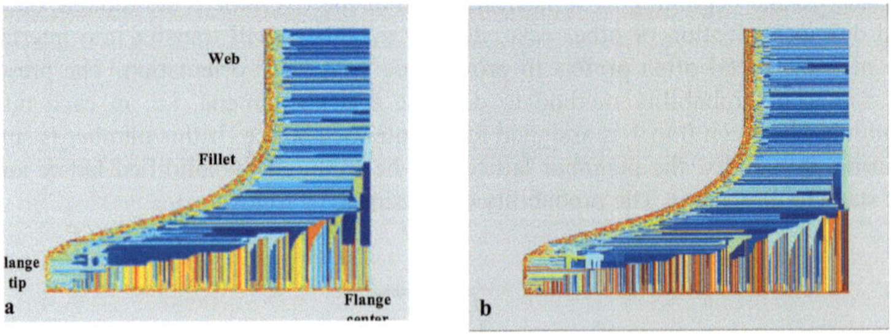

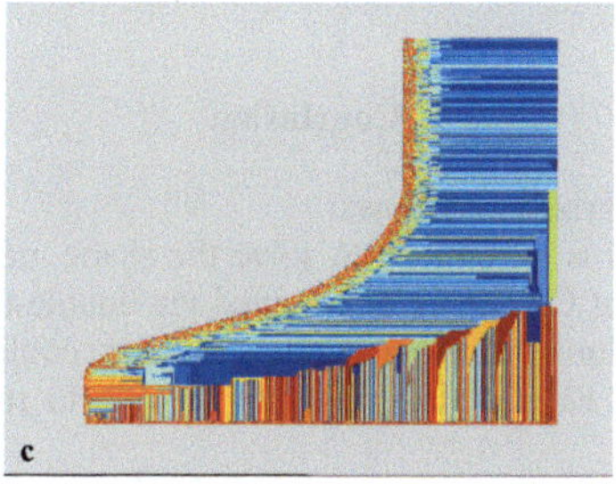

Figure2. Solidification structures at the end of Segment 3 with different superheats: (a) 10℃; (b) 20℃; (c) 30℃.

The effect of casting speed on solidified micro structure

Fig.3 illustrates the microstructure of beam blank at the end of Segment 3 with various casting speeds. Fig. 3 shows that, beam blank has completely solidified at various casting speeds, and equiaxed dendrites appear at the solidified end. The columnar dendrites at web restrict the formation of equiaxed dendrites, the amount of the latter being apparently less than in other locations. The FCT zone has a larger thickness at the surface of web when casting speed is 1.0 m/min. The percent of equiaxed dendrites is a little higher at 1.1 m/min than that at other casting speeds. The columnar dendrite tends to be finer as casting speed increases.

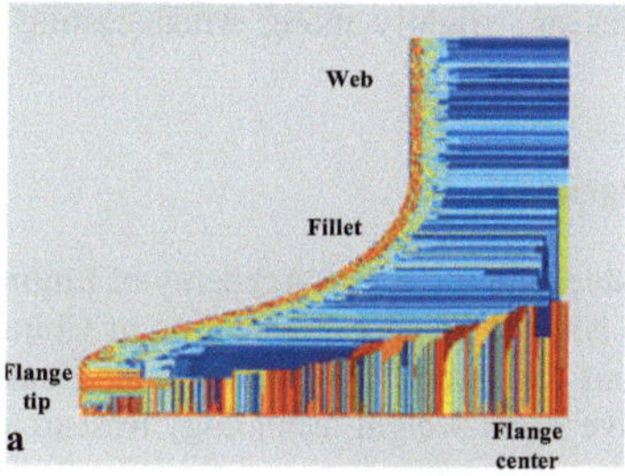

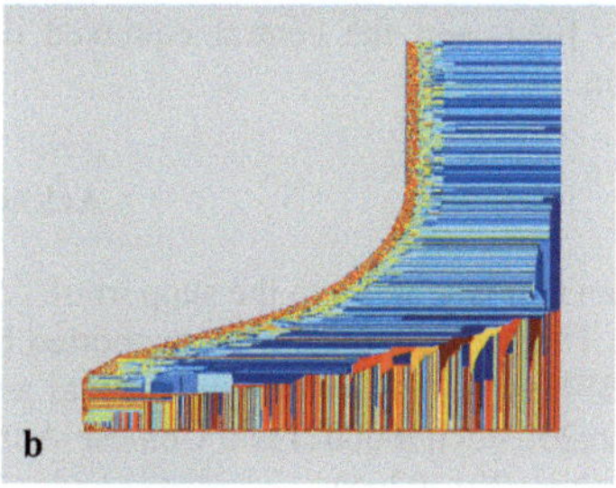

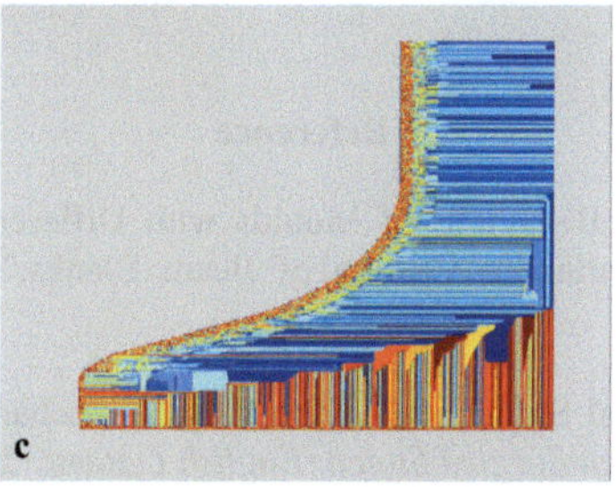

Figure3. Solidification structures at the end of Segment 3 with different casting speeds: (a) 1.0m/min, (b) 1.1m/min, (c) 1.2m/min.

Conclusion

The present research is summarized as follows:
1. An adaptive mesh division is introduced to solve the shape matching between FE mesh and CA lattice. The temperature of CA lattice is obtained via double-accuracy interpolation method and macro FE temperature. Moreover, the accurate secondary cooling boundary condition is also taken into account in macro FE simulation, which is able to reveal the actual solidification process of casting beam blank.
2. The nucleation is evaluated using continuous nucleation model, while the velocity of S/L interface is calculated via KGT model. On the basis of above models, the CA rule representing migration of S/L interface is defined via Von Neumann neighbor rule, and the preferred growth direction of dendrite is represented using a probability method.
3. The solidified microstructure of Q235 steel is simulated using the developed FE-CA model, and the effects of casting speed and super heat degree on the solidification structure are studied. The results show that: the columnar dendrite in beam blank account for much larger percent than conventional billet or slab, while equiaxed dendrite only appears at solidified end, whose proportion is very small. Surface fine crystal and FCT zones mainly form at web and fillet rather than other locations. At flange tip, inter-dendrites grow along with different directions and trend to be coarse. With the increase of super heat degree, the thickness of FCT zone rises, the columnar crystals tend to be coarse, and the proportion of equiaxed dendrites decreases. As casting speed increases, the thickness of FCT zone gradually decreases, the columnar crystals becomes finer, and the central equiaxed dendrites are slightly more, when casting speed is 1.1m/min.

Acknowledgement

Rongshan Qin is grateful to the support of TATA steel and the Royal Academy of Engineering at the UK. This work is financially supported by the Natural Science Foundation of China (NSFC, NO.:51374260), Chongqing Fundamental and Front Research Plan- Outstanding Youth Project (No.: cstc2013jcyjjq60001) and Chinese Automotive Light-weight technology Research Institute Co. Ltd (No.: 20130205).

Reference

1. H.L. Xu et al., "Thermal Behaviour of Moulds with Different Water Channels and Their Influence on Quality in Continuous Casting of Beam Blanks," *Ironmaking & Steelmaking*, 37(5)(2010), 380-386.

2. W. Chen et al., "Numerical Simulation of the Thermo-mechanical Process for Beam Blank Continuous Casting," *Acta Metallurgica Sinica (English Letters),* 20(4)(2007), 241-250.

3. W. Chen et al., "Thermomechanical Analysis and Optimisation for Beam Blank Continuous Casting," *Ironmaking & Steelmaking*, 35(2)(2008), 129-136.

4. W. Luo et al., "Improvement to Secondary Cooling Scheme for Beam Blank Continuous Casting," *Ironmaking & Steelmaking*, 39(2)(2012), 125-132.

5. M. Onishi et al., "Continuous Casting of Beam Blanks," (Report NO.3, Kawasaki Steel, 1981).

6. Y. Zhao et al., "Two-dimensional Heat Transfer Model for Secondary Cooling of Continuously Casting Beam Blanks," *Ironmaking & Steelmaking*, 41(5)(2014), 377-386.

7. Y. Zhao, "The Investigation of Cooling Pattern of Solidification during Beam Blank Continuous Casting and Dendritic Growth Simulation" (Ph.D. thesis, Chongqing University, 2013), 41-47.

8. P. Thevoz, J.L. Desbiolles, and M. Rappaz, "Modeling of Equiaxed Microstructure Formation in Casting," *Metallurgical Transactions A-Physical Metallurgy and Materials Science*, 20(2)(1989), 311-322.

9. M. Rappaz and C.A. Gandin, "Probabilistic Modeling of Microstructure Formation in Solidification Processes," *Acta Metallurgica et Materialia*, 41(2)(1993), 345-360.

10. W. Kurz, B. Giovanola and R. Trivedi, "Theory of Microstructural Development during Rapid Solidification," *Acta Metallurgica*, 34(5)(1986), 823-830.

11. M. Yamazaki et al., "Numerical Simulation of Solidification Structure Formation during Continuous Casting in Fe-0.7mass%C Alloy Using Cellular Automaton Method," *ISIJ International*, 46(6)(2006), 903-908.

12. Y. Zhao, R.S. Qin, and D.F. Chen, "A Three-dimensional Cellular Automata Model Coupled with Finite Element Method and Thermodynamic Database for Alloy Solidification," *Journal of Crystal Growth*, 377(2013), 72-77.

13. Y. Zhao et al., "A Three Dimensional Cellular Automata Model for Dendrite Growth with Various Crystallographic Orientations during Solidification," *Metallurgical and Materials Transactions B-Process Metallurgy and Materials Processing Science*, 45 (2014), 719-725.

EPD Congress 2015
Edited by: James Yurko, Antoine Allanore, Laura Bartlett, Jonghyun Lee, Lifeng Zhang, Gabriella Tranell, Yulia Meteleva-Fischer, Shadia Ikhmayies, Arief Suriadi Budiman, Prabhat Tripathy, and Guy Fredrickson
TMS (The Minerals, Metals & Materials Society), 2015

Numerical Simulation of the Coupled Turbulent Flow, Heat and Solute Transport in the Turbulent Flow Region of Slab Continuous Casting

Huabiao Chen, Dengfu Chen, Lintao Gui, Mujun Long, Yunwei Huang and Youguang Ma

College of Materials Science and Engineering, Chongqing University, Chongqing 400044, China.
Contact e-mail: chendfu@cqu.edu.cn.

Keywords: Continuous casting, Transport phenomenon, Macrosegregation, Multicomponent system, Numerical simulation

Abstract

In the current work, a three-dimensional model coupling turbulent flow, heat and solute transport was developed to investigate the solute transport and redistribution in the mold of slab continuous casting where the Reynolds number is very high and the molten steel flows strongly. The conservation equations of momentum, energy and species for a multicomponent system which includes solute element C, Si, Mn, P, S were solved with the commercial software ANSYS Fluent. The parameters used in the model are based on the actual continuous casting. The fluid flow, temperature distribution and solute elements distribution were analyzed. Results showed that negative segregation occurred near the strand surface, and the species concentration reached a peak value at the solidification front during solidification. The segregation degree of solute element S is higher than the other solute elements because of its lower equilibrium partition coefficient. The species concentration in the liquid pool is homogeneous because of the turbulent flow effect.

Introduction

Nowadays, continuous casting (CC) has become the most common process for prodution of steel. As the demand for steel products increasing, defect-free continuous casting is the main research direction in the future. Macrosegregation is one of the most severe internal defects encountered in slab continuous casting, which represents the non-uniformity of chemical composition in the cast section and will lead to poor properties of steel products. Usually, as for a common alloy, the solubility in the solid phase is smaller than in the liquid phase. So the solute will be rejected to the liquid phase during alloys solidification. This is the cause of segregation formation [1]. Macrosegregation cannot be eliminated by the subsequent heat treatment. Researchers proposes many techniques to decrease macrosegregation, such as low temperature casting [2], intensive cooling [3], uniformity second cooling [4], electromagnetic stirring (EMS) [5, 6], soft reduction [7], *etc.*

Continuous casting is a very complex process. It is almost impossible to measure the macrosegregation inside the mold directly during the continuous casting because of the high temperature molten steel. In addition, the measurements cannot meet the demand of steel production because of its low efficiency. Therefore, numerical simulation has become an important method to predict and investigate the macrosegregation. As early as 1967, Flemings and co-workers [8, 9] proposed a model which based on the mass equilibrium principle and ignored the effect of melt flow in the mushy zone to describe the segregation. In order to describe the effect of melt flow in the mushy zone on the solutes re-distributed, Mehrabian *et al.* [10] treated the mushy zone as a porous media and calculated the fluid flow by Darcy's law. The fluid flow status in the mushy zone can be described with permeability. Permeability is directly related to the dendritic structure and usually can be calculated by Carmon-Koseny formula. In order to precisely describe permeability, Felicelli *et al.* [11] pointed out that the permeability is anisotropic in the columnar-dendritic structure region. The permeability parallel and vertical to the primary dendrite arms are described with different components. Moreover, in order to couple the microscopic and macroscopic phenomena and achieve the macrosegregation status during alloy solidification, in 1987, Bennon & Incropera [12, 13] first employed the classic mixture theory to establish the continuous model and employed the model to describe the flow, heat and solute transport phenomena in the binary solidification system.

In the present study, based on the models above, a three-dimensional model which coupled turbulent flow, heat and solutes transport is established and performed using the commercial software ANSYS Fluent. Based on the simulation results, the velocity, temperature and solutes distributions are investigated. The formation process of slab centerline segregation during continuous casting is also discussed.

Model Description

(a) Assumptions of Model

The following assumptions were made in this model.

(1) The continuous casting process was assumed a steady state.

(2) The liquid steel in mold was assumed to be incompressible Newtonian fluid.

(3) The strand curvature, bulging, oscillation, mold taper, mold powder and air gap were ignored.

(4) The radiative heat transfer and the heat transfer on the mold surface were neglected.

(5) The latent heat of solid phase transformation was considered to be negligible.

(6) Local thermodynamic equilibrium was assumed to prevail in the model.

(b) Fluid and Turbulence Model

The molten steel flow pattern in the liquid pool was achieved through solving the continuity and momentum equations. The popular *K*-ε turbulent model [15] was chosen. The equations are:

$$\rho\frac{\partial(\rho u_j)}{\partial x_i}=0 \tag{1}$$

$$\rho\frac{\partial u_i u_j}{\partial x_j}=\frac{\partial}{\partial x_j}\left[\mu_{eff}\left(\frac{\partial u_i}{\partial u_x}+\frac{\partial u_j}{\partial x_i}\right)\right]-\frac{\partial P}{\partial x_i}+S_B+S_P \tag{2}$$

Where S_B is thermosolutal buoyancy, the S_P is the sink term of velocity, and the equations can be described respectively as:

$$S_B=\rho g_i\beta_T\left(T-T_L\right)+\sum_m \rho g_i\beta_{C,m}\left(C_{L,m}-C_{L,m,0}\right) \tag{3}$$

$$S_P=\frac{\left(1-f_l\right)^2}{\left(f_l^3+\varepsilon\right)}A_{mushy}\left(v-v_p\right) \tag{4}$$

Where $C_{L,m,0}$ is the solute concentration at the liquid temperature, A_{mush} is the constant of mushy zone [14], v_p is the casting speed, ε is a small number (0.001) to prevent division by zero.

(c) Solidification and Heat-Transfer Model

The governing equation for heat transfer and solidification is:

$$\rho u_i \frac{\partial H}{\partial x_i} = \frac{\partial}{\partial x_i}\left(\lambda_{eff} \frac{\partial T}{\partial x_i}\right) \tag{5}$$

Where H is the enthalpy of steel, which is computed as the sum of the sensible enthalpy and the latent heat.

(d) SoluteTransfer Model

The conservation equation for the multicomponent can be expressed as:

$$\frac{\partial(\rho u_i C_m)}{\partial x_i} = \frac{\partial}{\partial x_i}\left(\rho D_L^m \frac{\partial C_m}{\partial x_i}\right) + S_{dif} + S_{C,con} \tag{6}$$

Where S_{dif}, $S_{C,con}$ is diffusion source item and convection source item, respectively.

$$S_{dif} = \frac{\partial}{\partial x_i}\left[\rho f_s D_S^m \frac{\partial(C_{S,m} - C_m)}{\partial x_i}\right] + \frac{\partial}{\partial x_i}\left[\rho f_L D_L^m \frac{\partial(C_{L,m} - C_m)}{\partial x_i}\right] \tag{7}$$

$$S_{C,con} = \frac{\partial}{\partial x_i}\left[\rho(u_i - u_{s,i})(C_{L,m} - C_m)\right] \tag{8}$$

Based on the classic mixture theory:

$$C_m = f_L C_{L,m} + f_S C_{S,m} \tag{9}$$

Then the $C_{L,m}$, $C_{S,m}$ can be written as follows:

$$C_{L,m} = \frac{C_m}{1 + f_S(k_m - 1)} \tag{10}$$

$$C_{S,m} = \frac{k_m C_m}{1 + f_S(k_m - 1)} \tag{11}$$

(e) Division of Computational Domain

Based on the symmetry, only a half of mold region was simulated. The operating parameters presented in the Table I. Figure 2 gives a part of the computational meshes for the mold and the coordinate system. The mold has about 190,000 grid cells in a structured mesh.

Table I. The Operating parameters of mold

Operating Parameters	Values
Mold section	1530mm×190mm
Mold Length	800mm
Computer Length	4035mm
Inside Size of SEN, width×thickness	86mm×45mm
Outside Size of SEN, width×thickness	141mm×100mm
Port Size of SEN， width× height	45mm×73mm
Port angle	-15 deg
Casting speed	1.2 m/min
Casting Temperature	1811 K

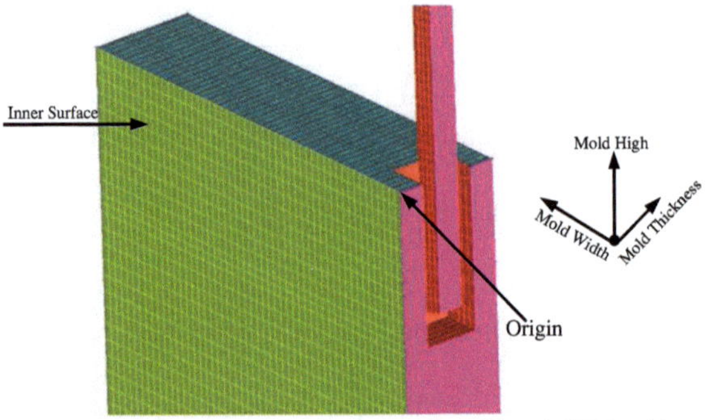

Figure 1. The schematic diagram of mold and SEN grids.

(f) Boundary Condition

The inlet velocity of the SEN was calculated through the mass conservation based on the casting speed. The value of turbulent kinetic energy and the rate of turbulent energy dissipation at the inlet are 0.00305 and 0.00668 which were estimated using the semi-empirical relations. The casting temperature is 1811 K, the heat transfer condition of the narrow face, inner surface and outer surface are all employed the heat transfer coefficient average value 1100 $W \cdot (m^2 \cdot ℃)^{-1}$. The initial concentration of the solute elements in the multicomponent is presented in the Table II.

Table II. The initial concentration of solute elements

Element	C	Si	Mn	P	S
wt pct	0.15	0.27	1.38	0.02	0.009

(g) Thermophysical Properties

The steel properties employed on the simulation are presented in Table III. The equilibrium partition coefficients k_m, the slopes of liquidus line m_m, and the solute expansion coefficient βc are presented in Table Ⅳ. Due to the lack of parameter, and the molecular weight of elements Si, P, S are similar, the solutal expansion coefficient of elements P and S were employ the same value as the element Si in the calculation. The solutal expansion coefficient of elements C, Si, Mn could be found in the publication [15].

Table III. The physical parameters of steel

Physical Properties	Values
Density, kg/m^3	7330
Specific Heat, $J \cdot (kg \cdot K)^{-1}$	$319.59+0.1934 \times T(K)$
Thermal Coefficient, $W \cdot (m \cdot K)^{-1}$	$57.524-0.0164 \times T(K)$
Viscosity, $kg \cdot (m \cdot s)^{-1}$	0.0055
Latent Heat, J/kg	255500
Thermal expansion coefficient, 1/K	2.0×10^{-4}

Table Ⅳ. Equilibrium partition coefficients, liquidus slopes and expansion coefficients of solutes

Element	C	Si	Mn	P	S
k_m	0.19	0.77	0.76	0.23	0.05
m_m	78.0	7.6	4.9	34.4	38.0
βc,1/Wt pct	1.1×10^{-2}	1.19×10^{-2}	1.92×10^{-1}	—	—

Results and discussion

Figure 2 shows half of the flow field (a) and temperature field (b) of mold at the mid-plane between wide face. The jet flow leaves the nozzle and the velocity of molten steel slows down. When the jet flow impacts on the narrow face of the mold, it is divided into two parts and forms the upper and lower recirculation regions. It is obvious that the temperature in upper region is higher than the lower region. Comparing the Figure 2 (a) and (b), the molten steel flow field and temperature field has the similar pattern. The jet flow from the nozzle impacts on the narrow face and brings the fresh molten steel to the upper region. This causes the temperature in upper region higher. It is clear that a reasonable flow field in the mold is useful to form a suitable temperature field. Generally, the mold surface requires higher temperature to increase the mold powder melting rate. At the same time, the velocity near mold surface should keep a reasonable value to avoid the slag entrainment. Therefore, a reasonable flow field and temperature filed is very important to keep the continuous casting process stable.

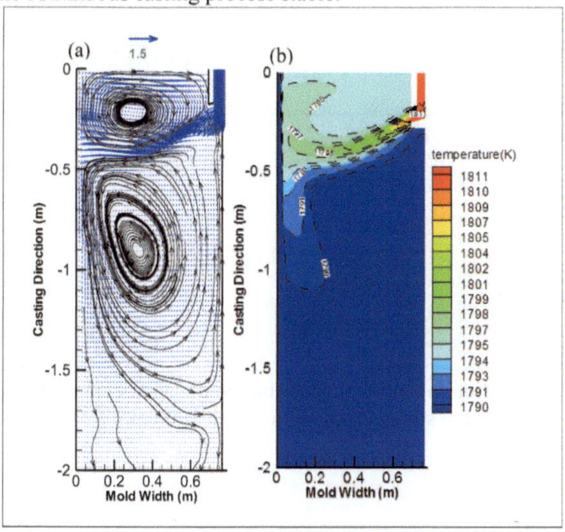

Figure 2. The flow field (a), (m/s) and the temperature field (b) at the mid-plane between WF.

Figure 3 (a) shows the distribution of liquid fraction on the strand cross section (1000 mm below meniscus). Due to the high temperature jet flow impact on the narrow face, temperature in the region near to the narrow face is higher and more difficult to solidify. Therefore, the thickness of the solidified shell in this region is thinner. Figure 3 (b) ~ (f) shows the redistribution of solute element C, Si, Mn, P, S on the cross section, respectively. These solute elements have the same redistribution law. The solute elements in the liquid pool are uniformity because of the flow effect of the molten steel in the turbulent flow region. Comparing the Figure 3 (a) and (c), it is clear that the solute element enriches at the solidification front. Because the equilibrium partition coefficient k_m is smaller than 1, solute element is rejected to the liquid during solidification. These lead to negative segregation occurs near the strand surface and the solute concentration increase in the liquid pool especially enriches at the solidification front. Figure 4 clearly shows the forming process of slab centerline segregation. The carbon segregation degree at the solidification front at the 1/4 line of 1500 mm, 2000 mm, 2500 mm, 3500 mm below meniscus is increasing. The solute elements are rejected and enrich at the solidification front during solidification. As the solidification front moving to the slab center, the solute concentration is increasing slowly. When the solidification front moves to the center, the slab solidifies completely and the solute concentration reaches to a very high value. This is the slab centerline segregation forming process.

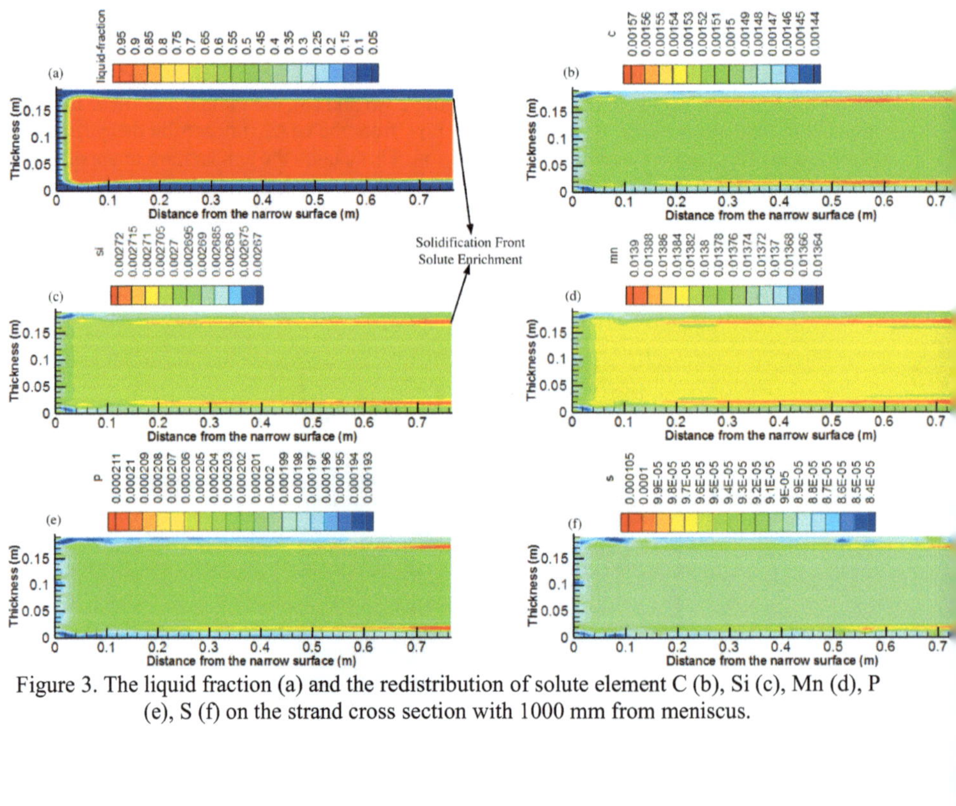

Figure 3. The liquid fraction (a) and the redistribution of solute element C (b), Si (c), Mn (d), P (e), S (f) on the strand cross section with 1000 mm from meniscus.

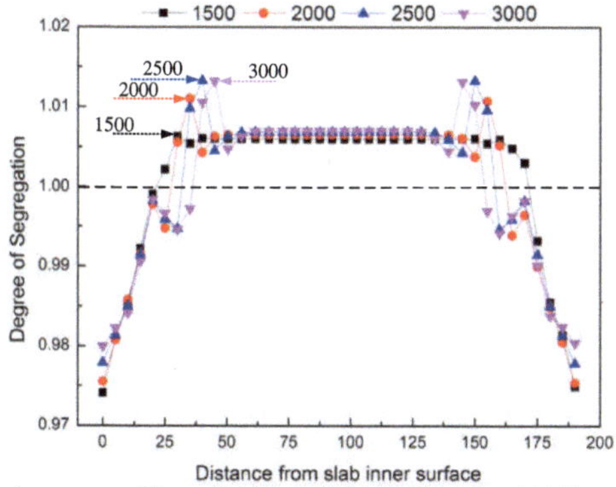

Figure 4. Segregation degree profiles of solute element C at the 1/4 line of strand wide surface with different distances (mm) from meniscus.

From Figure 5, it can be seen than the distribution of solute element C, Si, Mn, P, S has the same tendency. The solute element enriches in the mushy zone and reaches a peak value at the solidification front. The segregation degree of S is more severe than the other solute elements because of its smaller equilibrium partition coefficient. The segregation degree of solute element Si and Mn are very small due to their big equilibrium partition coefficients.

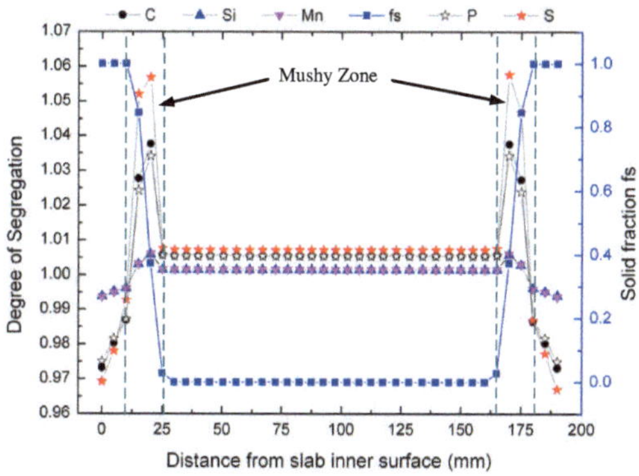

Figure 5. Comparison of the segregation degree of solute element C, Si, Mn, P, S.

Conclusions

In this paper, a three-dimensional model coupled flow, heat and solute transport has been solved and investigated the pattern of the velocity, temperature and especially the distribution of solute in the turbulent flow region. Conclusions are made as follow:

(1) The jet flow from nozzle develops into upper and lower recirculation regions. Temperature in upper recirculation region is higher because of the molten steel refreshed.

(2) The solute distribution has the tendency. The solute concentration reaches a peak at solidification front and distributes uniformly because of the turbulent flow.

(3) Negative segregation occurs near the strand surface. The solute rejected and the solidification moving lead to the formation of slab centerline segregation during solidification. The segregation

degree is related to the equilibrium partition coefficient and the segregation of solute element S is more severe than the other solute elements because of its smaller equilibrium partition coefficient.

Reference

1. R. Vertnik, B. Sarler, and B. Sencic, "Solution of Macrosegregation in Continuously Cast Billets by a Meshless Method" (Paper presented at the 3rd International Conference on Advances in Solidification Processes, Aachen, Netherlands, 7 June, 2011), 1-6.

2. M. RÉGER and S. LOUHENKILPI, "Effect of Some Technological Parameters on the Macrosegregation of CC Slabs" (Paper presented at the Int. Jubilee Conference, Budapest Polytechnic, September, 2004), 131-138.

3. A.V. Kuznetsov, "Parametric study of macrosegregation in the horizontal strip casting process for different cooling rates and different casting speeds," *Heat and Mass Transfer*, 35 (3) (1999), 197-203.

4. M.J. Long, et al., "A Mathematical Model for Mitigating Centerline Macro Segregation in Continuous Casting Slab," *Metalurgia International*, 16 (10) (2011), 19-33.

5. W.D. Du, et al., "Effect of special combined electromagnetic stirring mode on macrosegregation of high strength spring steel blooms," *Ironmaking & Steelmaking*, 35 (2) (2008), 153-156.

6. K.S. Oh and Y.W. Chang, "Macrosegregation Behavior in Continuously Cast High-Carbon Steel Blooms and Billets at the Final Stage of Solidification in Combination Stirring," *ISIJ International*, 35 (7) (1995), 866-875.

7. A. Shiraishi, et al., "Improvement of Internal Quality of Large Section Concast Bloom with Soft Reduction Technology and Induction Heater in Tundish," *Revue De Metallurgie-Cahiers D Informations Techniques*, 87 (9) (1990), 777-784.

8. M.C. Flemings and G.E. Nereo, "Macrosegregation. I.," *Transactions of the Metallurgical Society of Aime*, 239 (9) (1967), 1449-1461.

9. M.C. Flemings, "Mehrabia.R, and G.E. Nereo, "Macrosegregation. II.," *Transactions of the Metallurgical Society of Aime*, 242 (1) (1968), 41-49.

10. R. Mehrabian, M. Keane, and M. Flemings, "Interdendritic fluid flow and macrosegregation; influence of gravity," *Metallurgical and Materials Transactions B-Process Metallurgy and Materials Processing Science*, 1 (5) (1970), 1209-1220.

11. D.R. Poirier, "Permeability for flow of interdendritic liquid in columnar-dendritic alloys," *Metallurgical and Materials Transactions B-Process Metallurgy and Materials Processing Science*, 18 (1) (1987) 245-255.

12. W.D. Bennon and F.P. Incropera, "The Evolution of Macrosegregation in Statically Cast Binary Ingots," *Metallurgical and Materials Transactions B-Process Metallurgy and Materials Processing Science*, 18 (3) (1987), 611-616.

13. W.D. Bennon and F.P. Incropera, "A Continuum Model for Momentum, Heat and Species Transport in Binary Solid Liquid-Phase Change Systems. 1. Model Formulation," *International Journal of Heat and Mass Transfer*, 30 (10) (1987), 2161-2170.

14. A.D. Brent, V.R. Voller, and K.J. Reid, "Enthalpy-Porosity Technique for Modeling Convection-Diffusion Phase-Change-Application to the Melting of a Pure Metal," *Numerical Heat Transfer*, 13 (3) (1988), 297-318.

15. H.L. Yang, et al., "Mathematical simulation on coupled flow, heat, and solute transport in slab continuous casting process," *Metallurgical and Materials Transactions B-Process Metallurgy and Materials Processing Science*, 29 (6) (1998) 1345-1356.

EPD Congress 2015
Edited by: James Yurko, Antoine Allanore, Laura Bartlett, Jonghyun Lee, Lifeng Zhang, Gabriella Tranell, Yulia Meteleva-Fischer, Shadia Ikhmayies, Arief Suriadi Budiman, Prabhat Tripathy, and Guy Fredrickson
TMS (The Minerals, Metals & Materials Society), 2015

INTERPHASE EVOLUTION AND DEFECT FORMATION DURING HORIZONTALLY DIRECTIONAL SOLIDIFICATION PROCESS OF Sn-Zn ALLOYS

Alex I. Kociubczyk[1], Wilky Desrosin[1], Lucía Mabel Boycho[1], Carlos E. Schvezov[1, 2], Alicia E. Ares[1, 2]

[1] Materials Institute of Misiones, IMAM (CONICET-UNaM), University of Misiones; 1552 Azara Street, Posadas, Misiones, 3300 Argentina.
[2] Member of CIC of the National Research Council (CONICET) of Argentina.

Keywords: CET, Sn-Zn Alloys, Horizontal Solidification, Thermal Parameters, Structures.

Abstract

In the present research, commercially pure Sn and Zn, and Sn-Zn alloys (Sn-1wt.%Zn, Sn-2wt.%Zn, Sn-4wt.%Zn, Sn-8wt.%Zn and Sn-8.9wt.%Zn, weight percent) were obtained by a horizontal directional solidification process with two opposite senses and heat extraction.
The solidification process was realized using a horizontal furnace with two heat extraction systems at both ends. The temperature was measured using eight K-type thermocouples and an electronic recorder of temperature data. The resulting structures were analyzed using optical microscopy. From the solidification process the thermal and metallographic parameters were determined in all samples. The presence of defects in the solidified pieces was observed. Internal defects pretended to be dependent not only on the composition of the alloys under consideration but also on the size of the structures formed, also, on the velocities and accelerations of interphases, and on the variation of thermal gradients. A model of the phenomenon from the first principles is presented.

Introduction

Zinc and tin metals are widely used in different branches of industry as coatings in order to protect steel against corrosion [1, 2]. Under normal conditions, the steel has an anodic behavior with respect to tin, zinc while facing the steel acts as a cathode. The disadvantage that presents in the tin-coating or hot dip by electro-galvanizing is porosity, which allows the corrosion to dissipate through the pores over time so that it promotes the corrosion of steel. However, zinc acts as a sacrificial anode, which itself is consumed during the process, and the oxidation of the steel is inevitable after a certain amount of zinc has been consumed.
In order to reduce the porosity of tin coatings, alloys of both metals was used as electrodeposited alloys can have of finer grains than pure metals deposited under comparable conditions, and less porosity is expected in finer grains. Furthermore, in addition to alloying metals, properties of electrochemical coatings are modified, which provides a means of adjusting the reactivity of the cover [5-7].
In the present research, commercially pure Sn and Zn and Sn-Zn alloys (Sn-1wt.%Zn, Sn-2wt.%Zn, Sn-4wt.%Zn, Sn-8wt.%Zn and Sn-8.9wt.%Zn, weight percent) were obtained by a horizontally directional solidification process with two opposite senses and heat extraction.

The solidification process was realized using a horizontal furnace with two heat extraction systems at both ends. The temperature was measured using eight K-type thermocouples and an electronic recorder of temperature data. The resulting structures were analyzed using optical microscopy. From the solidification process the thermal and metallographic parameters were determined in all samples. The presence of defects in the solidified pieces was observed. Internal defects pretended to be dependent not only on the composition of the alloys under consideration but also on the size of the structures formed, also, on the velocities and accelerations of interphases, and on the variation of thermal gradients. A model of the phenomenon from the first principles is presented.

Experimental Procedure

Samples of Zn-Sn alloys were prepared by directional solidification. The Zn and Sn pure elements were merged into graphite molds in a muffle furnace, and then unidirectionally solidified in clay mold in the horizontal furnace with heat extraction in two opposite directions (see Figure 1). Next, the samples were grinded with SiC abrasive paper of different grain size, from # 60 to # 1500. To observe the macrostructure, the samples were subjected to a chemical attack which consisted of exposing them to a solution of 36.5% HCl at room temperature between 5 to 30 seconds. This allowed defining zones of different structures along the sample: a columnar zone of large and elongated grains and the equiaxed zone with smaller grains.

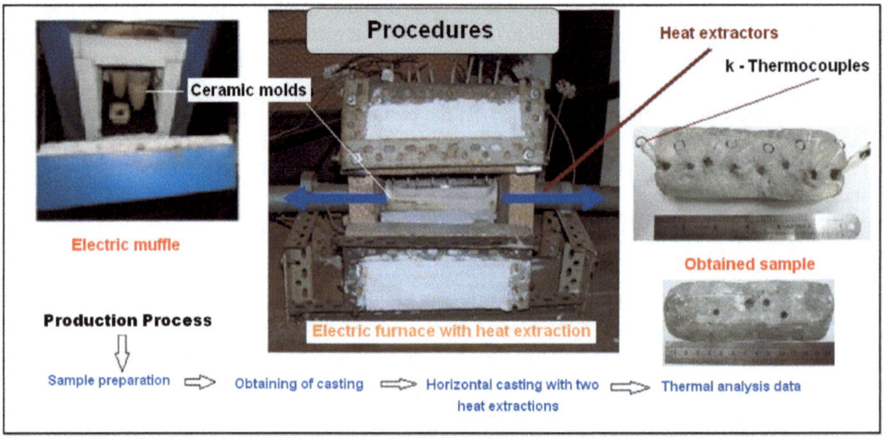

Figure 1. Schematization showing equipment, the production process, a sample and a macrostructure obtained.

Results and Discussion

From the collected temperature data, thermal parameters such as, the local solidification time in the positions of each thermocouple, temperature gradients and cooling rates, were determined.
Plotting the data of temperature versus time (Figures 2 (a) to (e) for each alloy, the presence of three zones on the graph was observed: 1st - a period of cooling of the melt (temperature overheating and the liquidus temperature). the 2nd - solidification period (between the liquidus temperature and the eutectic temperature) and 3 - a period of cooling of the solid (below the eutectic temperature).

During the solidification period, it could be noted that the average value of eutectic temperature equaled to 198.5°C. Table 1 shows the local solidification time in the positions of each thermocouple for all Sn-Zn alloys tested.

Table 1. Local solidification times. Sn-Zn alloys.

	Local solidification time, t_{SL} [s] = t_S - t_L				
	Sn-1wt.%Zn	Sn-2wt.%Zn	Sn-4wt.%Zn	Sn-8wt.%Zn	Sn-8.9wt.%Zn
T1	1650	1580	1710	1690	1440
T2	1570	1450	1620	1610	1330
T3	1570	1460	1610	1550	1270
T4	1500	1460	1570	1520	1240
T5	1580	1430	1550	1520	1280
T6	1600	1450	1610	1570	1320
T7	1570	1580	1670	1650	1400
T8	1630	1690	1700	1730	1650
$T_{Average}$	1583.8	1512.5	1630.0	1605.0	1366.3

In macrographs obtained, Figures 2 (f) - (o) the presence of fully equiaxed grains for the case of solidification of alloys Sn-1wt.%Zn, Sn-2wt.%Zn, Sn-4wt.%Zn and Sn-8.9wt.%Zn was observed, indicating a low heat removal from the ends of the samples. In contrast, in the case of Sn-8wt.%Zn, completely columnar structure was obtained (high heat removal from the ends of the sample). The average size of the equiaxed grains as measured by the ASTM E112 standard is presented in Table 2. In general, the larger grain size was obtained at the center of the samples.

Table 2. Grain sizes of equiaxed grains measures in the samples.

Alloy	Grain size, [mm] (Right end)	Grain size, [mm] (Left end)	Grain size, [mm] (Center)
Sn-1wt.%Zn	2.1	2.4	3.9
Sn-2wt.%Zn	2.5	3.0	3.2
Sn-4wt.%Zn	3.7	4.1	4,5
Sn-8,9wt.%Zn	3.9	3.3	5.2

In Figures 2 (p) - (t) we can observe the position of the liquid interphase, [L/(L + S)] and solid interphase, [(S + L)/S] in the samples vs. time. We can observe that although each pair of interphases of the same type is represented by a single line, each line corresponds to two interphases of the same type, since each interphase can not be in two points in the space at the same instant.

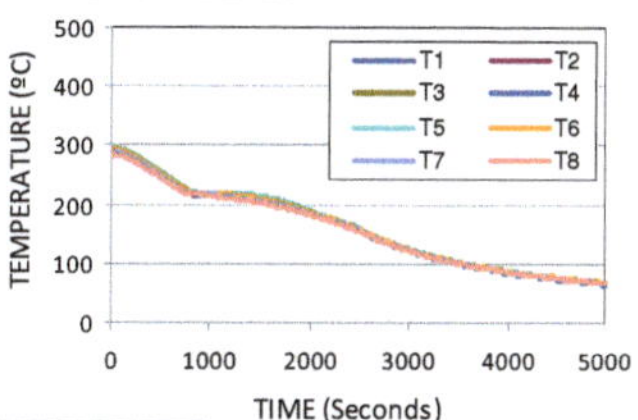

Sn-1wt.%Zn

(a) Equiaxed structure

(f)

(g)

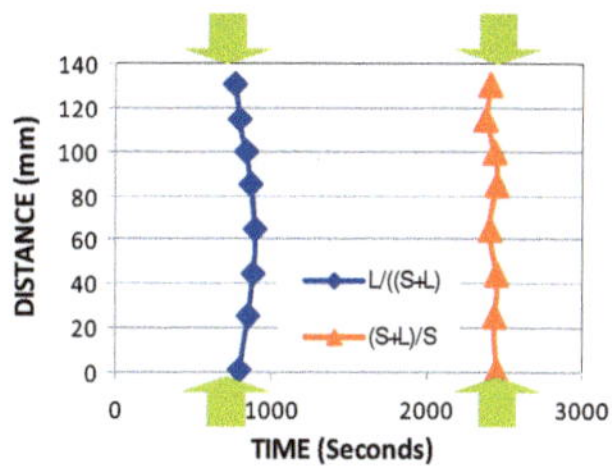

(p)

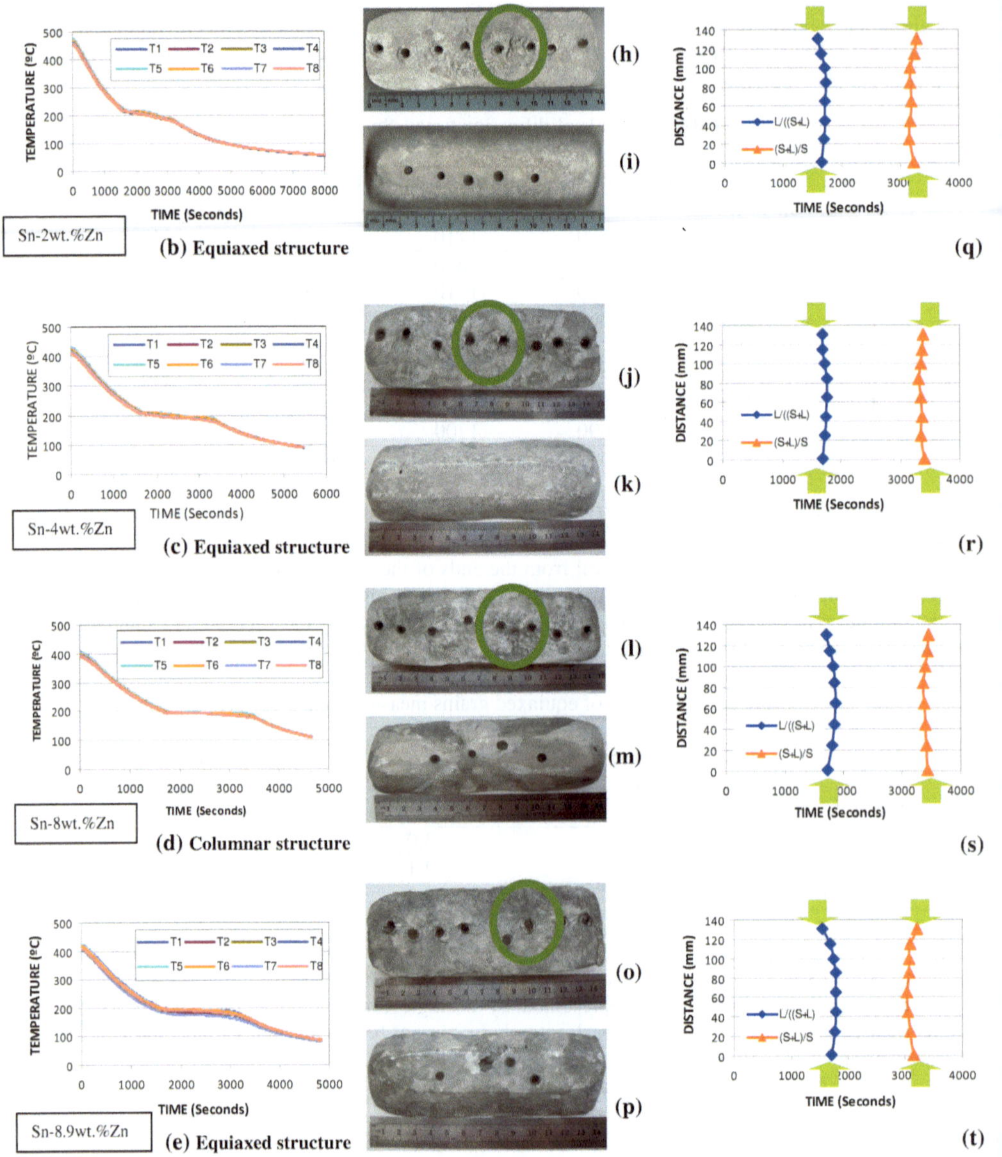

Figure 2. (a) - (e) cooling curves. (f) - (o) Macrostructures of the samples obtained. (p) - (t) Position of the interphases versus time. Sn-Zn alloys (Sn-1wt.%Zn, 2wt.%Sn-Zn, Sn-4wt.%Zn, Sn-8wt.%Zn and Sn-8.9wt.% Zn).

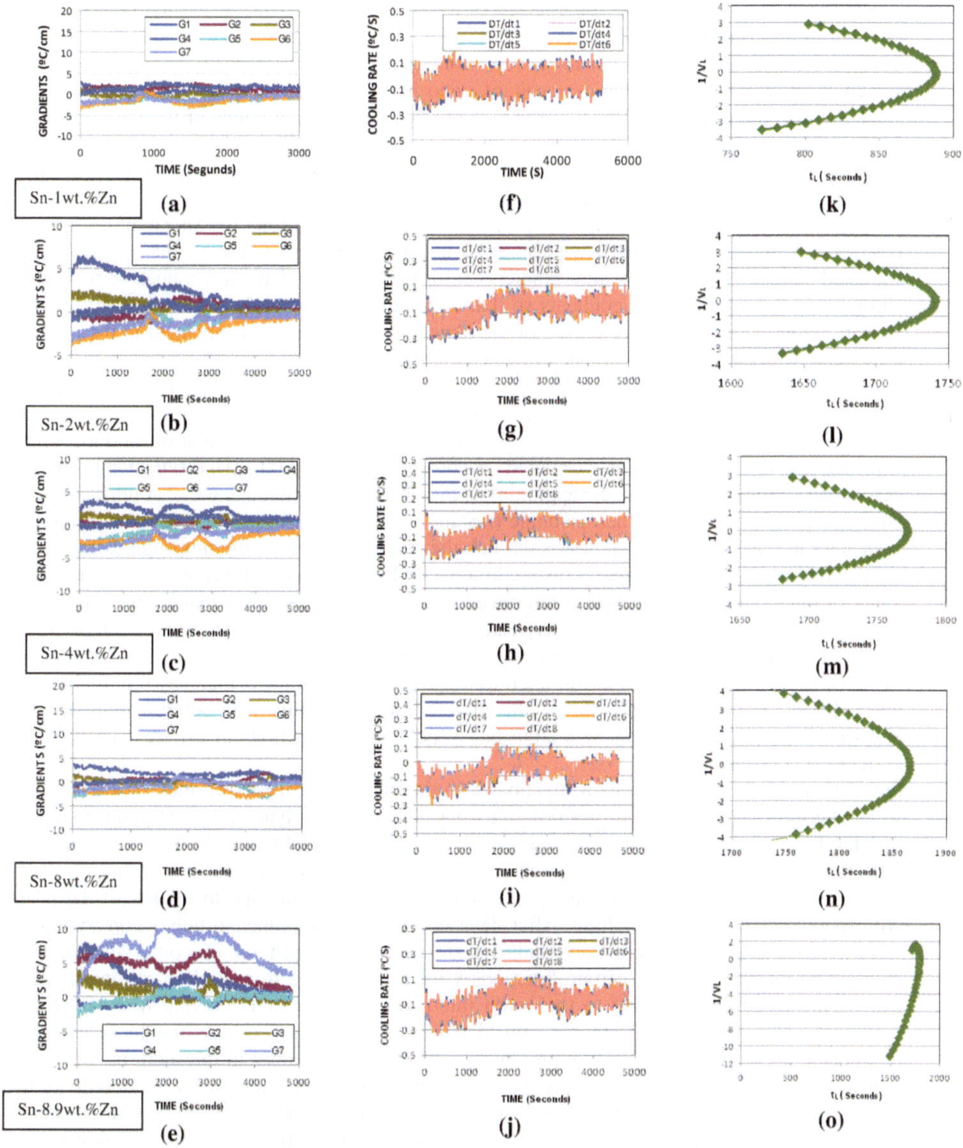

Figure 3. (a) - (e) Temperature gradients with time. (f) - (j) Cooling rates versus time from the ends of the sample. (k) - (o) Sn-Zn alloys (Sn-1wt.%Zn, Sn-2wt.%Zn, Sn-4wt.%Zn, Sn-8wt.%Zn, Sn-8.9wt.%Zn).

The moment of collision of two interphases of the same type is characterized by the maximum moment in the considered position, or the farthest corners as point "C", see Figures 3 (k) to (o). Thus, the curves in Figure 2 (p) to (t) represented four interphases but only two curves are shown. Each curve corresponds to a pair of oppositely interphases advancing and collides at some point inside the sample (green arrows). Note that the collision points of the interphases or furthermost points of the different curves "C" does not match all in the same position of the samples, which is meaning that the maximum of the curves position vs. time (x vs. t) do not coincide for different pairs of related interphases.
Figures 3 (a) - (e) shows the temperature gradients as function of time for Sn-Zn alloys, at two positions of the samples (one at each end). It is noted that in the case of the samples with equiaxed structures, the temperature gradients from the beginning have low values (< 5 °C/cm) and no minimum and critical gradients so that a transition occurs in the structure identified by growth grains. Similar behavior profile of the temperature gradients was obtained in the case of the sample with completely columnar structure (Sn-8wt.% Zn).
After obtaining the temperature versus time data for the alloy concerned, as indicated in Figures 2 (a) - (e), the cooling rates were determined in the liquid, mushy and solid alloys taking the values of the derivative of temperature with respect to time for each position of thermocouple, considered as difference quotients centered, as shown in Figures 3 (f) - (j). From changes in the derivatives of the cooling rates, the approximate moments of the onset and the of solidification in the sensing volume for each thermocouple (instant of liquid interphase, [L/(L+S)], or instant of each solid interphase, [(S+L)/S], passes through the position where the thermocouple is located]) were determined. As each sample cools predominantly from both ends, it is possible to determine a pair of interphases for each type of interphase ([L/(L+S)] or [(S+L)/S]), that is, two liquid and two solid interphases. Each pair of interphases of the same kind was moving in the longitudinal direction but having opposite directions of movement. In order to determine the moment of collision of the interphases that allow determining the existence of the completely liquid phase in the cylinder, the end of growth of the first solid that forms and solidification to the experimentally determined data was adjusted by polynomial functions of different levels with correlation coefficients close to one.
Since the position vs. time is not a function from the mathematical perspective, fit polynomials were determined for the time vs. position functions. While each polynomial corresponds to a pair of interphases, the only way to determine which points correspond to which interphase (right or left) is establishing the maximum of each polynomials, and taking all those points that are right the maximum corresponds to the interphase right and all points that are to the left of the peak corresponding to the left interphase. After determining what experimental points correspond to which interphase, independent polynomials were used to fit each set of points allocated to appropriate interphase. From certain functions x vs. t, data points from each of the interphases can be identified as belonging to the given interphase (right or left) moving in each of the directions. The end points were not considered in the settings, because the most extreme thermocouples T_1 and T_8 are only representative of one half volumes in question, but serve as points of comparison with extrapolated from the determined function values. The derivative of these functions corresponds to the inverse of the velocity of interphases, treated as a single curve. In the graphs of Figure 3 (k) - (o) the calculated values of V^{-1} from the time of polynomials constructed according to the position (expressed as a function of time) are shown. Positive values of V^{-1} represent interphases moving from left to right, and negative values of V^{-1} are interphases that move from right to left. The maximum time in each graph corresponds to the instant of collision of two oppositely advancing interphases.
This collision of interphases leaves to the formation of voids, pores and internal defects on the samples (areas marked with a green circle on macrographs of Figures 2 (f) -. (n)).

Conclusions

Experiments of horizontal solidification were performed with two directions of heat extraction coincident with the longitudinal axis of Sn-Zn alloys, determining the major parameters, namely: a) the moment of start and end of the solidification at each position considered, b) the local solidification time, c) cooling rates, d) the temperature gradient, e) the advancing average liquid velocities in opposite directions interphases f) the value of the speed when the solidification fronts collide they move in opposite directions in the samples.
The liquid interphases collide in samples in a different position than are the solid interphases.
With the temperature gradients obtained, no transition from the columnar to equiaxed grains was observed during the horizontal solidification.
The results are consistent with those obtained before in Zn-Al and Zn-Sn alloys [8, 11-13].

Acknowledgements

This work was supported by PICT-2011-1378 of the National Agency for Promotion of Science and Technology.
Thanks are due to the Argentinean Research Council (CONICET) for the financial support.

References

1. G. Reinhart, N. Mangelinck-Noël, H. Nguyen-Thi, T. Schenk, J. Gastaldi, B. Billia, P. Pino, J. Härtwig, J. Baruchel, “Investigation of Columnar–Equiaxed Transition and Equiaxed Growth of Aluminium Based Alloys by X-ray Radiography.” Mater Sci Eng A, 413–414, (2005), p. 384–388.

2. S. McFadden, D. J. Browne, C. A. Gandin, “A Comparison of Columnar-to-Equiaxed Transition Prediction Methods Using Simulation of the Growing Columnar Front”. Metall. Mater. Trans. 40A,
(2009), p. 662-672.

3. C. A. Siqueira, N. Cheung, A. Garcia, “Solidification Thermal Parameters Affecting the Columnar-to-Equiaxed Transition”. Metall Mater Trans. 33A, (2002) p. 2107-2118.

4. B. Willers, S. Eckert, U. Michel, I. Haase, G. Zouhar, “The Columnar-to-Equiaxed Transition in Pb–Sn Alloys Affected by Electromagnetically Driven Convection.” Mater. Sci. Eng. A, 402, (2005), p. 55–65.

5. R. B. Mahapatra, F. Weinberg, “The Columnar-to-Equiaxed Transition in Tin-Lead Alloys.” Metall Trans. 18B (1987), p. 425–432.

6. I. Ziv, F. Weinberg, “The Columnar-to-Equiaxed Transition in Al-3pctCu.” Metall. Trans., 20B, (1989), p.731-734.

7. C. A. Gandin, “From Constrained to Unconstrained Growth during Directional Solidification”. Acta Mater, 48, (2000), p. 2483–2501.

8. S. F. Gueijman, C. E. Schvezov, A. E. Ares, “Directional Solidification and Characterization of Zn-Al and Zn-Ag Diluted Alloys”, Materials Transactions, 51, (2010), p. 1851-1870.

9. A. E. Ares, C. E. Schvezov, Solidification Parameters during the Columnar-to-Equiaxed Transition in Lead-Tin Alloys. Metall. Mater. Trans., 31A, (2000), p. 1611-1625.

10. A. E. Ares, C. E. Schvezov, "Influence of Solidification Thermal Parameters on the Columnar-to-Equiaxed Transition of Aluminum-Zinc and Zinc-Aluminum Alloys". Metall. Mater. Trans., 38A, (2007), p.1485-1499.

11. S. F. Gueijman, C. E. Schvezov, A. E. Ares, Theory and Experiments of Interphase Formation and Structure in Horizontal Directionally Solidified Zn-Al Dendritic Alloys, In: Crystal Growth: Theory, Mechanisms and Morphology, Chapter 8, Pages: 265-293, Editors: Nicole A. Mancuso and James P. Isaac. Nova Science Publishers, Inc. ISBN: 978-1-61324-529-3 (2012).

12. M. Zurko, C. M. Rodriguez, C. E. Schvezov, C. M. Mendez, A. E. Ares, Horizontal Directional Solidification of Zn-Al Alloys, TMS Annual Meeting (2012) 153-160.

13. W. Desrosin, C. E. Schvezov, A. E. Ares, Thermal and Metallographic Parameters Evolution During Solidification of Zn-Sn Alloys, TMS Annual Meeting (2014) 501-508.

EPD Congress 2015
Edited by: James Yurko, Antoine Allanore, Laura Bartlett, Jonghyun Lee, Lifeng Zhang, Gabriella Tranell, Yulia Meteleva-Fischer, Shadia Ikhmayies, Arief Suriadi Budiman, Prabhat Tripathy, and Guy Fredrickson
TMS (The Minerals, Metals & Materials Society), 2015

EFFECT OF TECHNOLOGICAL PARAMETERS ON MOLD POWDER ENTRAINMENT BY WATER MODEL STUDY

Lizhi Zhang[1], Yugang Li[1], Qian Wang[1] and Cheng Yan[1]

[1]College of Materials Science and Engineering, Chongqing University, Chongqing 400030, P.R.

Keywords: Continuous casting, Mold powder entrainment, Water model

Abstract: In continuous casting process, mold powder has lots of metallurgical effects, such as covering the molten steel to prevent liquid steel from oxidation by air, preserving heat for top layer liquid steel, lubricating initial shell and so on. However, mold powder also can deteriorate the quality of final products obviously after the entrapment of mold powder in steel. In the present work, based on the similarity theory the effect of technological parameters on mold powder entrainment was investigated by water model. Experiment was carried out in the water mold using oil and water to simulate slag and steel, respectively. The results showed that increasing casting speed increased the water-oil interfacial fluctuation and entrainment frequency; viscosity of oil also had a great impact on the interface fluctuation and entrainment; water-oil interfacial tension effect the entrainment while not changing the interfacial fluctuation; submergence depth had little impact on entrainment and interfacial fluctuation.

Keywords: Continuous casting, Mold powder entrainment, water model

INTRODUCTION

Interface stability acts an important role during continuous casting process. Severe fluctuation of steel-slag interface will always accelerate the emulsification of slag [1, 2] and lead a vicious circle. In this case, a huge number of slag entrainment will take place, then the interfacial area will increase which returns to accelerate the emulsification. Hence, it is necessary to have an understanding of the effect of some technological parameters on slag entrainment in casting process.

As a result of interface instability, droplet may form in continuous casting process. In the past works, slag entrainment was investigated by physical and numerical simulation. As a normal form, the slag entrainment by shear stress is found relative early. Savolainen et al. [3] studied the effect of physical properties on slag entrainment due to shear force by the cold model. They found that an increasing of physical properties (viscosity of light phase, density difference, interfacial tension and thickness of light phase) resulted in a rise of critical flow velocity for slag entrainment. On the other hand, another way of slag entrainment may form due to Karman vortex which happens in the situation of high speed or asymmetrical flow. Watanabe et al. [4] investigated the effect of physical property on suction by Karman's vortex by using water model and hot model respectively. It is found that viscosity of slag and steel-slag interfacial tension had great influence on slag entrainment and the effect of viscosity is larger than that of interfacial tension. Furthermore, in order to prevent clogging of SEN port, argon is blown into the molten steel from SEN. Previous studies found that rupture of large argon bubble can float the steel-slag interface and induce entrainment when the gas flow rate reach the critical value[5-6]. Addition to that, Yoshida et al. [7] pointed out that mold powder may be sucked down along the outer surface of SEN due to the pressure difference in the longitudinal direction.

In the current work, the entrainment at the meniscus of mold is investigated. First of all, the similarity criterion for the flow field and slag was obtained. Then, experiment was carried out using water and some kinds of compound oils in the water model. At last, effect of casting speed, submergence depth, viscosity of oil and interfacial tension on entrainment and the fluctuation was discussed.

INVESTIGATION METHOD AND MODEL APPARATUS

A. Similarity Criterion

In the current work, Froude similarity criterion is used to design the inner flow filed of water model. In this case, the ratio of momentum and gravity is equal in steel caster and water model. The Froude similarity is following:

$$Fr = \frac{U^2}{gL} \tag{1}$$

Where Fr is the Froude number; U is the velocity of liquid flow; g is the gravitational acceleration rate; L is the characteristic length.

In order to simulate the mold powder, the following similarity criterion is used:

$$\frac{v_o}{v_w} = \frac{v_s}{v_m} \tag{2}$$

$$v = \frac{\mu}{\rho} \tag{3}$$

Where v_o, v_w, v_s and v_m are the kinetic viscosities of the oil, the water, the slag and the molten steel; v is the kinetic viscosity; μ is the dynamic viscosity; ρ is the density. Combining Eq. (2) and (3) gives:

$$\mu_o = \frac{\mu_s \mu_w \rho_o \rho_m}{\mu_m \rho_s \rho_w} \quad (4)$$

Where μ_o, μ_w, μ_s and μ_m are the kinetic viscosities of the oil, the water, the slag and the molten steel; ρ_o, ρ_w, ρ_s and ρ_m are the kinetic viscosities of the oil, the water, the slag and the molten steel. In the situation of 293K for water mold and 1573K for prototype, μ_w = 0.001 Pa.s, μ_s = 0.2-1.3 Pa.s, μ_m = 0.005 Pa.s, ρ_o = 876 kg/m^3, ρ_w = 997 kg/m^3, ρ_m = 7000 kg/m^3, ρ_s = 2500 kg/m^3. So for the water mold, the range of μ_o is 0.1 to 0.59 Pa.s.

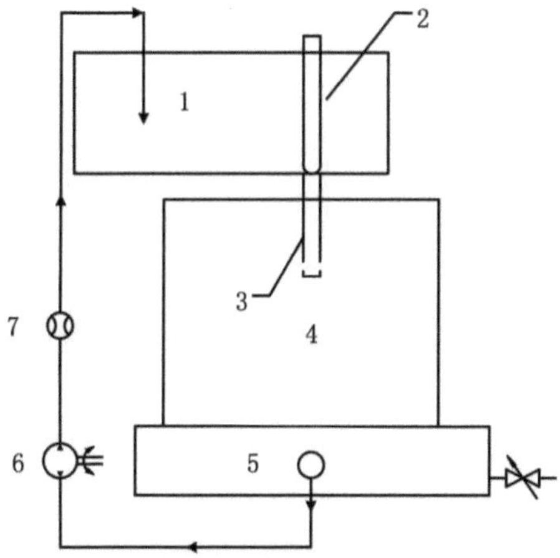

Figure 1. Schematic of the water model experiment (1: tundish, 2: stopper, 3: SEN, 4: mold, 5: water collection tank, 6: water pump, 7: water flow meter).

B. Experiment Design

Figure 1 shows the schematic diagram of the present experimental apparatus. In this experiment, the water mold made of acrylic slab was 1/2 scale of the casting mold which was 250mm in thickness, 1350mm in width. Video camera was taken to record the entrainment process during experiment.

Compound oil, which simulated the liquid mold power, was putted to cover the water in the water mold. The viscosity of the compound oil was adjusted by using a small quantity of kerosene, and the interfacial tension against the water was adjusted by the pure oleic. Viscosity (298 K) was measured by CQKJN-3 Slag Rotating Viscometer based on rotating viscometer method. Interfacial tension of water-oil interface (298 K) was measured by JJ2000B Spinning Drop Interface Tension meter based on spinning drop method.

RESULTS AND DISCUSSION

Figure 2 shows the steel-slag interfacial profiles in the situation of different casting speeds. With the casting speed increasing from 1.2 to 1.4 m/min, interface became more unstable then slag entrainment began forming. The casting speed increased from 1.4 to 1.6 m/s, entrainment frequency increased rapidly. It is mainly because that, the surface flow velocity of molten steel increases with increasing casting speed then the shear force of molten steel flow also increases. Subjecting from this enhance of shear force, the entrainment frequency got a rapid increase. The relationship between casting speed and the entrainment frequency is shown in Figure 3. The entrainment frequency at 1.6 m/s was three times as 1.4 m/s.

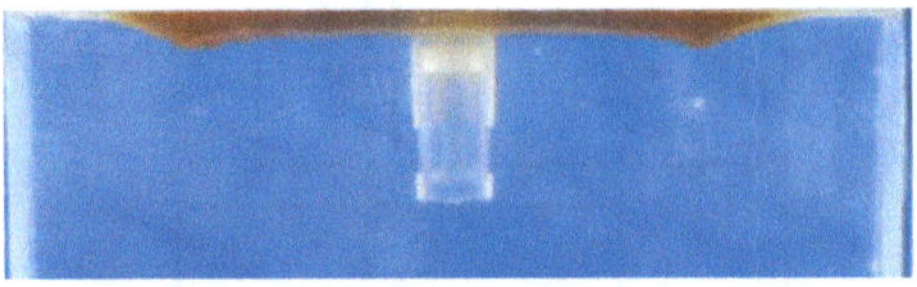

(a) 1.2 m/min

(b) 1.4 m/min

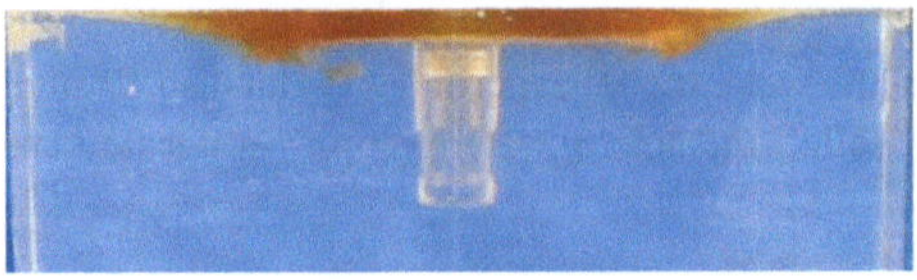

(c) 1.6 m/min

Figure 2. Effect of the casting speed on the interfacial profiles

Normally, the submergence depth of SEN has an important effect on the flow field. Lower submergence depth will decrease the space of upper flow field, and increased the jet impacting depth. However, in this study the effect of submergence depth on the slag entrainment was not obvious. Figure 4 shows the interfacial profile in the situation of different submergence depths of SEN. The submergence depth decreased from 160 mm to 100 mm, interfacial fluctuation change little and the entrainment still not formed.

Figure 5 shows the interfacial profiles in the situation of different oil-water interfacial tension. The interfacial fluctuation seemed the same in different interfacial tension, but the oil droplet was more easily to form in low interfacial tension. It is mainly because that drop down of interfacial tension makes the interfacial energy become smaller. When the momentum of oil is larger than this interfacial energy, droplet will form. Figure 6 shows the relation between the oil-water interfacial tension and the entrainment frequency. The entrainment frequency decreased for increasing interfacial tension. For the interfacial tension varying from 3.9 to 16.2 mN/m, entrainment frequency changed from 1.67 to 0.33 time/min. But from 16.2 to 27 mN/m, this entrainment frequency changed just 0.33 times. It seems that comparing to high oil-water interfacial tension, the entrainment frequency increases faster in low interfacial tension.

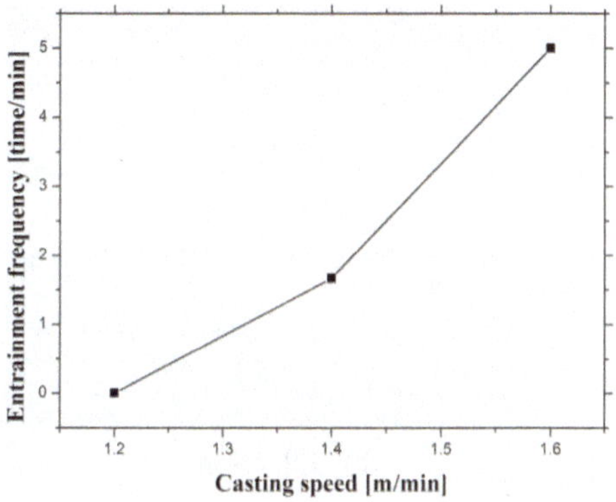

Figure 3. Relation between the casting speed and entrainment frequency.

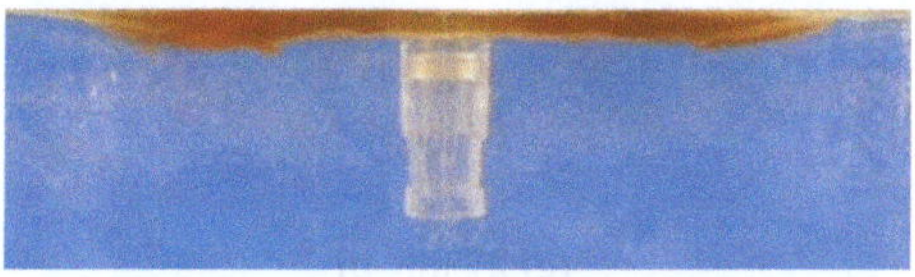

(a) 160 mm

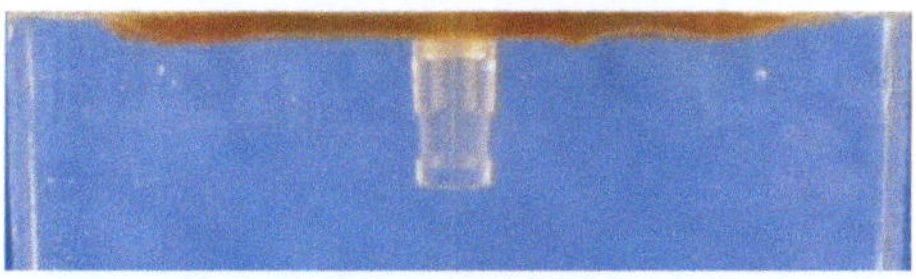

(b) 140 mm

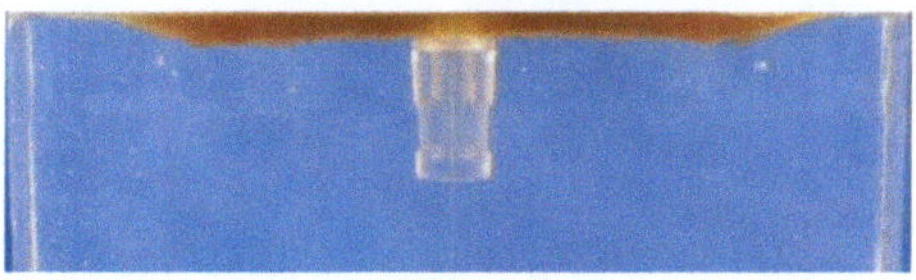

(c) 120 mm

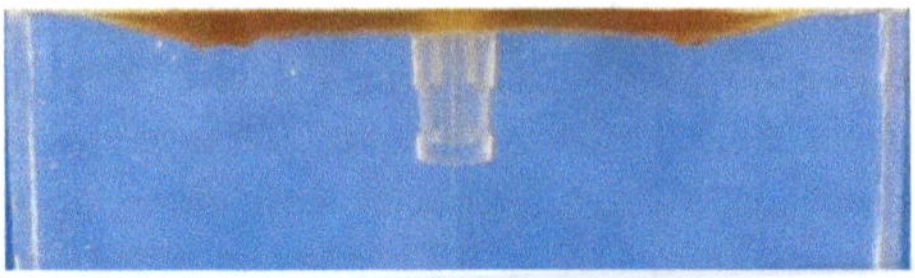

(d) 100 mm

Figure 4. Effect of submergence depth on the interfacial profiles

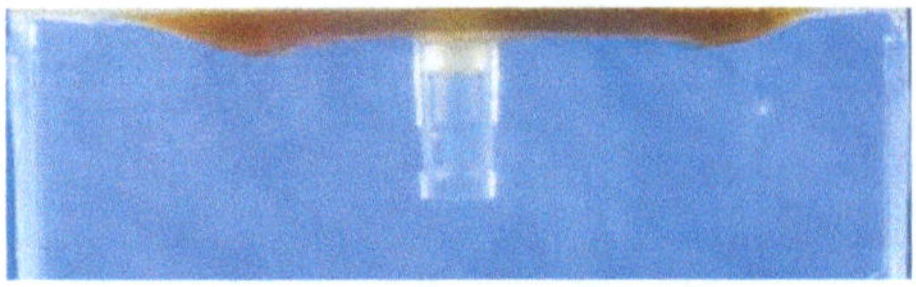

(a) 3.9 mN/m

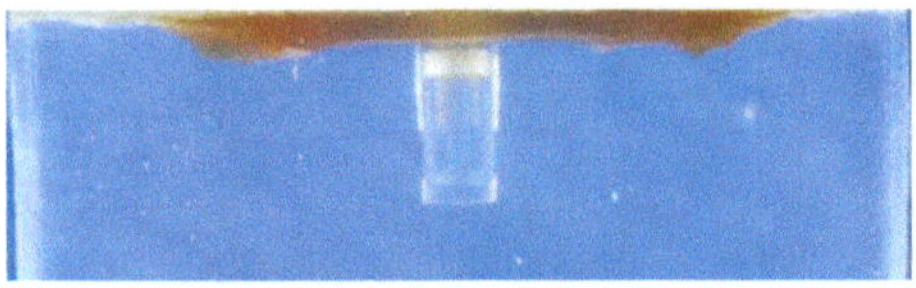

(b) 11.9 mN/m

(c) 27 mN/m

Figure 5. Effect of water-oil interfacial tension on the interfacial profiles

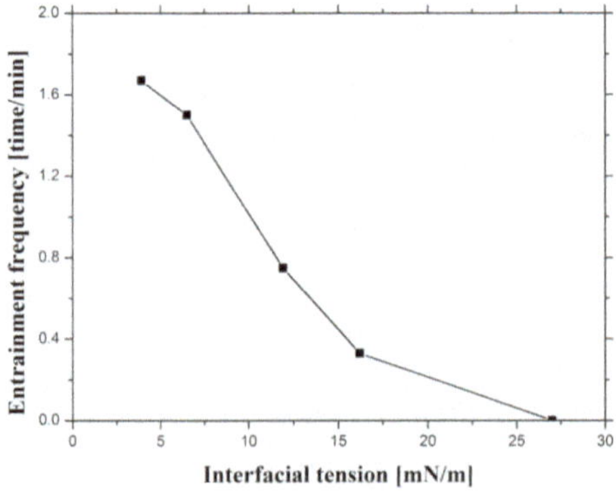

Figure 6. Relation between the water-oil interfacial tension and the entrainment frequency.

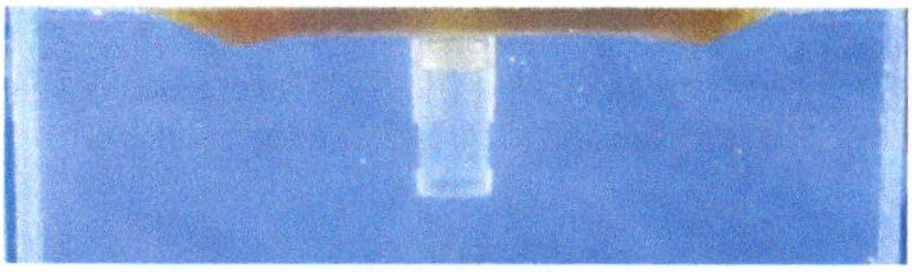

(a) 0.347 Pa.s

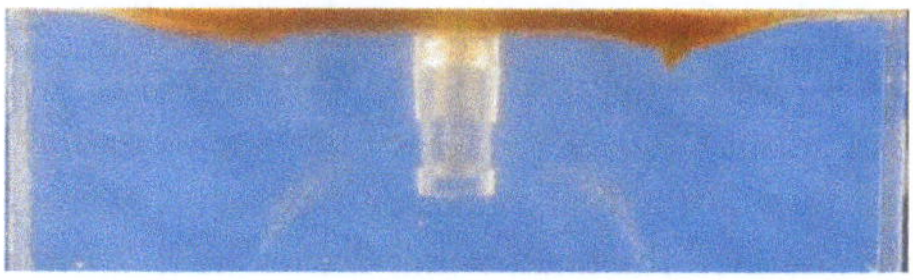

(b) 0.227 Pa.s

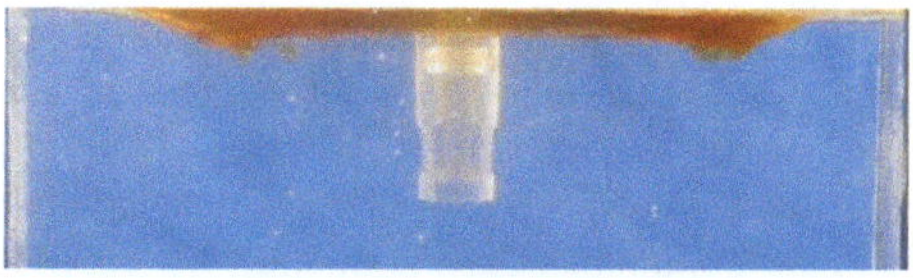

(c) 0.167 Pa.s

Figure 7. Effect of viscosity of oil on the interfacial profiles

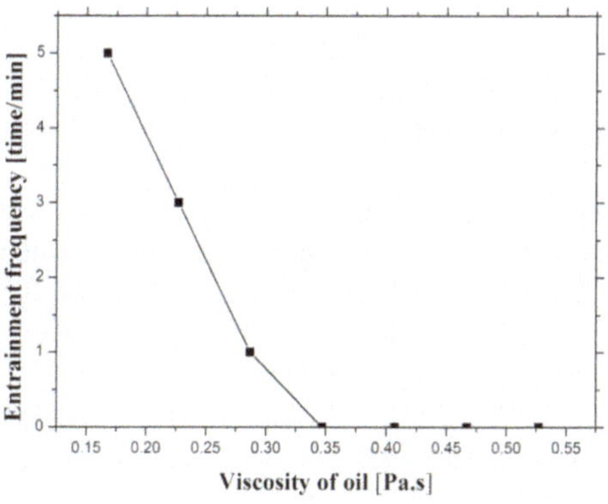

Figure 8. Relation between the viscosity of oil and the entrainment frequency.

Figure 7 shows the interfacial profiles in the situation of different viscosities of oil. The variation of interfacial fluctuation was obvious of different viscosities of oils. For the viscosity of 0.347 Pa.s, the water-oil interface was stable. But when the viscosity dropped down, the interface began to unstable and entrainment formed. As shown in Figure 8, the entrainment frequency decreased rapidly when the viscosity varied from o.167 to 0.347 Pa.s. It is mainly because of that, dropping down of viscosity will lead to increase of the velocity of the oil lay nearby water, which will enhance the momentum of the oil drop and make entrainment forming.

SUMMARY

Using the water model, the effect of various technological parameters on mold powder entrainment was investigated. This water model was established by the similarity theory which was 1/2 scale of the actual casting mold. It was found that the increasing casting speed increased the fluctuation of water-oil interface and entrainment frequency. Moreover, viscosity of oil also had a great impact on the interface fluctuation and entrainment. However, water-oil interfacial tension effect the entrainment which seemed not change the interfacial fluctuation. And submergence depth had little impact on the oil drop entrainment and interfacial fluctuation.

REFERENCES

1. L. Zhang, S. Yang, K. Cai, J. Li, X. Wan, and B.G. Thomas, "Investigation of Fluid Flow and Steel Cleanliness in the Continuous Casting Strand," Metall. Mater. Trans. B, 38 (2007), 63-83.
2. P. R. Scheller, "Interfacial phenomena between fluxes for continuous casting and liquid stainless steel," Ironmaking and Steelmaking, 29 (2002), 154-160.
3. J. Savolainen, T. Fabritius, and O. Mattila, "Effect of Fluid Physical Properties on the Emulsification," ISIJ Int., 49 (2009), 26–36.
4. K. Watanabe, K. Tsutsumi, M. Suzuki, M. Nakada, and T. Shiomi, "Effect of Properties of Mold Powder Entrapped into Molten Steel in a Continuous Casting Process," ISIJ Int., 49 (2009), 1161–1166.
5. Zhang Shengjun, Zhu Miaoyong, Zhang Yongliang and Zheng shuguo, "Study on Mechanism of Entrapment in Slab Continuous Casting Mould with High Casting Speed and Argon Blowing," ACTA METALLURGICA SINICA, 42 (2006), 1087-1090.
6. S. Yamashita and M. Iguichi, "Mechanism of Mold Powder Entrapment Caused by Large Argon Bubble in Continuous Casting Mold," ISIJ Int., 41 (2001), 1529–1531.
7. J. Yoshida, T. Ohmi, and M. Iguchi, "Cold Model Study of the Effect of Density Different and Blockage Factor on Mold Powder Entrainment," ISIJ Int., 45 (2005), 1160–1164.

EPD Congress 2015
Edited by: James Yurko, Antoine Allanore, Laura Bartlett, Jonghyun Lee, Lifeng Zhang, Gabriella Tranell, Yulia Meteleva-Fischer, Shadia Ikhmayies, Arief Suriadi Budiman, Prabhat Tripathy, and Guy Fredrickson
TMS (The Minerals, Metals & Materials Society), 2015

EQUIVALENCY COMPARISON OF HEAT TRANSFER COEFFICIENT IN LIQUID AND GAS QUENCHES

Yuan Lu[1], Yiming Rong[1], Richard D. Sisson Jr[1]

[1]Center for Heat Treating Excellence (CHTE), Worcester Polytechnic Institute
100 Institute Rd, Worcester, MA, USA

Keywords: Gas quench, Liquid quench, HTC, Equivalency

Abstract

With the development of quench technology, there is a trend of using gas quench to replace liquid quench for less distortion and residual stress. The fundamental difference between the liquid and gas quench is the heat transfer coefficient, not only the values but also the shape of the curve as a function of temperature. The equivalent heat transfer coefficient for the liquid and gas quench is analyzed by simulations and experiments. Even it may result in the same hardness in the liquid and gas quench, the steel microstructure may be different because of the different cooling processes, and therefore other steel properties, such as toughness, may be different. The cooling process, microstructures and properties such as hardness and toughness should be examined when designing the liquid or gas quench processes.

Introduction

Gas quench is becoming popular to replace water or oil quench [1] for medium and high hardenability steels, such as 4140, 4340. It has many advantages such as less distortion, less stress, safer and environmental friendly [2]. With gas quench process, mechanical and physical properties can be significantly improved and obtain near shape of metal components [2].

However, the uniformity of gas quench process is an issue compared with liquid quench. According to the work of Jing Wang [3], Elkatatny [2] Bowang Xiao [4] and Cosentino [5], the gas pressure and velocity changes dramatically in the furnace. Current studies on gas quench are focused on the gas flow in the furnace.

Considering the complex of gas pressure and velocity, the gas quench heat transfer coefficient (HTC) is noted, since the HTC has direct influence on cooling curves [6]. In this paper, the equivalent HTC concept is proposed for liquid and gas quench. After verifying the gas quench model based on Dante [7] by experiment, the gas quench model will be used to get equivalent HTC for liquid and gas quench. The steel thermal properties influence and hardness comparison are also discussed in the paper.

Equivalent HTC between Liquid Quench and Gas Quench

HTC is the only difference between liquid quench and gas quench, since the chemical reaction with the surface of the steel is ignored in this paper.

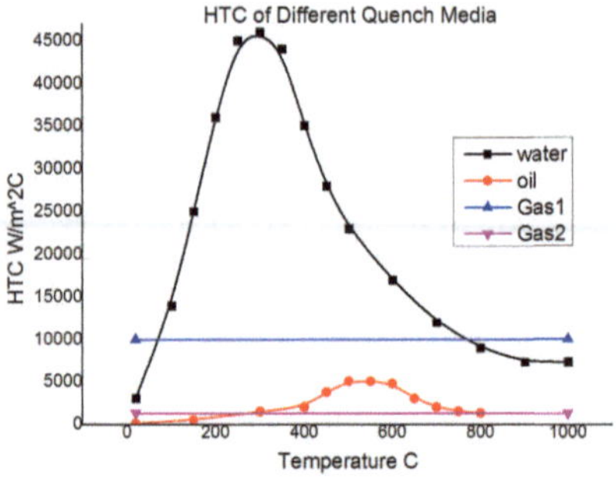

Figure 1. HTC of different quench media (experiment)

Figure 1 is the HTC of different quench media. Liquid quench exhibits three characteristic quenching processes, film boiling, bubble boiling and convection [8]. For gas quench, the single-phase heat transfer process means that the cooling rate is more uniform [6].

In heat treatment, core microstructures and properties are important, because the core cooling rate is the lowest and may form undesired microstructures such as upper bainite and ferrite. The equivalent HTC between liquid and gas quench is defined as the HTC, which has the same cooling curves at the core of the sample. After two different quench processes, if the cooling curves of the core are the same, these two quench HTCs are considered as the equivalent HTC.

Gas Quench Model Verification

The gas quench model sketch is represented in Figure 2. The cylinder sample with 25mm diameter and 100mm length is used. The gas flow is assumed to be the same at the free end of the sample and the sample sides, since the slenderness ratio is large. Gas flow is assumed as laminar flow. In this condition, the gas pressure and velocity are steady during gas quench process.

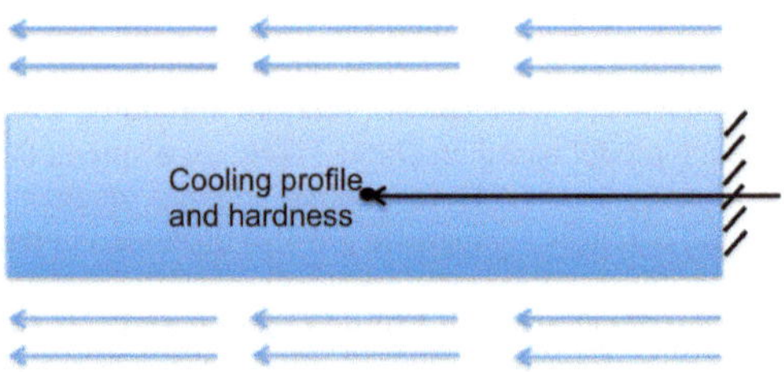

Figure 2. Gas quench model sketch

The experiments are done with the help of Praxair and the gas quench simulation model (based on Abaqus and Dante) is developed with the help of Dante. The steel is 4140 in the experiment.

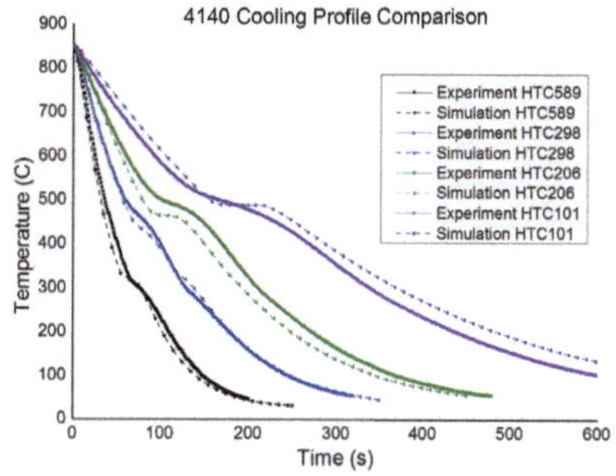

Figure 3. 4140 cooling profile comparison between experiment and simulation

In Figure 3, the cooling curves under different gas quench condition are measured by thermocouple and simulated by gas quench model. The simulation results match the experimental result and it demonstrates the accuracy of gas quench model. To improve the accuracy, the ambient temperature, transfer time from the heating furnace to the quenching chamber and the time required to reach the desired pressure and gas flow speed should be considered.

Equivalent HTC Prediction based on Gas Quench Model

The verified gas quench model is used to simulate the gas quench process and predict the equivalent HTC. Oil quench and gas quench are compared in this paper. The HTC of oil quench is from Figure 1.

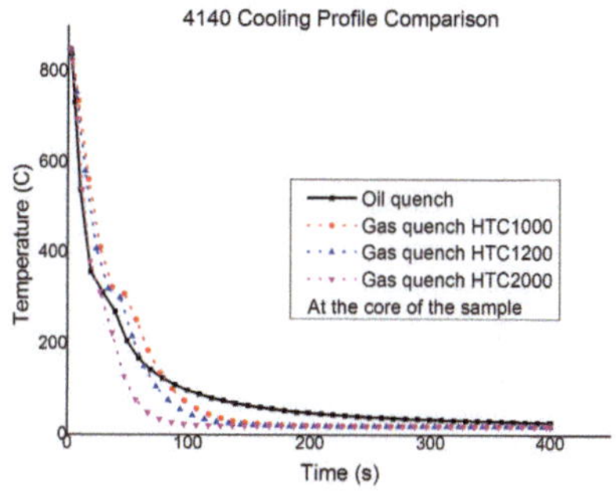

Figure 4. 4140 cooling profile comparison (simulation)

Figure 4 (simulation) is the cooling profile comparison between oil quench and gas quench. The cooling profiles of different gas quench HTCs are simulated to match the cooling profile of oil quench. For gas quench HTC 1000 W/m^{2}C (constant from 20C to 1000C) and HTC 1200 W/m^{2}C (constant from 20C to 1000C), the cooling rates from 850C to 200C is lower than oil quench. In order to increase the cooling rates from 850C to 200C, the gas quench HTC 2000 W/m^{2}C is used. The cooling curves for HTC 2000 W/m^{2}C (constant from 20C to 1000C) matches the oil quench from 850C to 300C. From 300C to 20C, the cooling rates for gas quench 2000 W/m^{2}C is higher than oil quench. No gas quench with constant HTC can become the equivalent HTC compared to oil quench.

One of the advantages of gas quench is great process flexibility that allows to vary cooling rates by adjusting gas pressure and velocity. Gas quench with varying HTCs are considered to find the equivalent HTC compared to oil quench.

The HTC shown in Figure 5 (simulation) is the equivalent HTC for oil quench. From 1000C to 300C, the HTC is 2000 W/m^{2}C. From 300C to 180C, the HTC is 1200 W/m^{2}C. From 180C to 100C, the HTC is 500 W/m^{2}C. From 100C to 20C, the HTC is 100W/m^{2}C. At each stage, the gas quench HTC is the constant. Figure 6 (simulation) are the cooling profiles of oil quench and equivalent gas quench at the core of the sample. Gas quench with varying HTCs is the equivalent HTC compared to liquid quench.

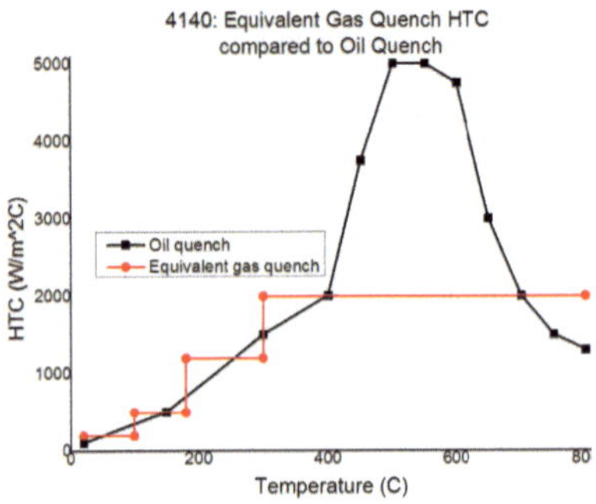

Figure 5. 4140: equivalent gas quench HTC compared to oil quench (simulation)

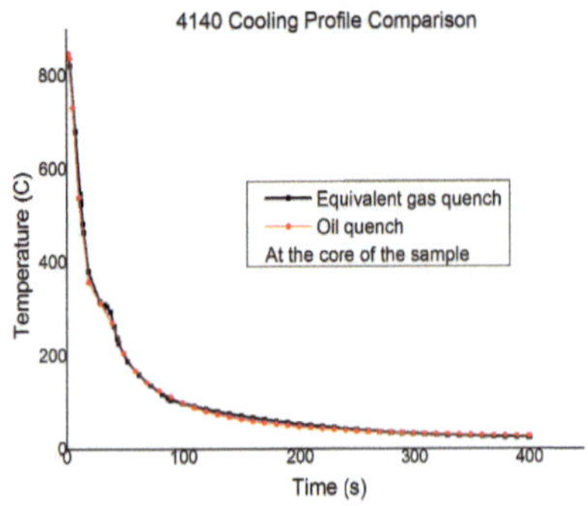

Figure 6. 4140 cooling profile comparison (simulation)

Simulation based on Jominy test is finished to extent the concept of the equivalent HTC. The sketch is shown in Figure 7. The Jominy bar is 25mm diameter and 100mm length. Boundary conditions 2,3 and 4 are air-cooling and boundary condition 1 is oil quench or equivalent gas quench in Figure 5. The temperature profile and the hardenability (along the black line in Figure 7) are compared to verify the equivalency of oil quench and gas quench.

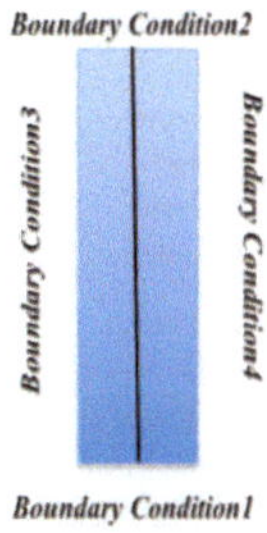

Figure 7. Jominy quench model sketch

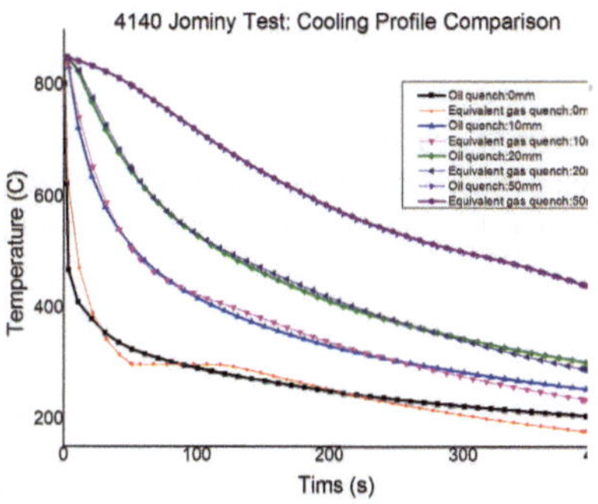

Figure 8. 4140 Jominy test: cooling profile comparison (simulation)

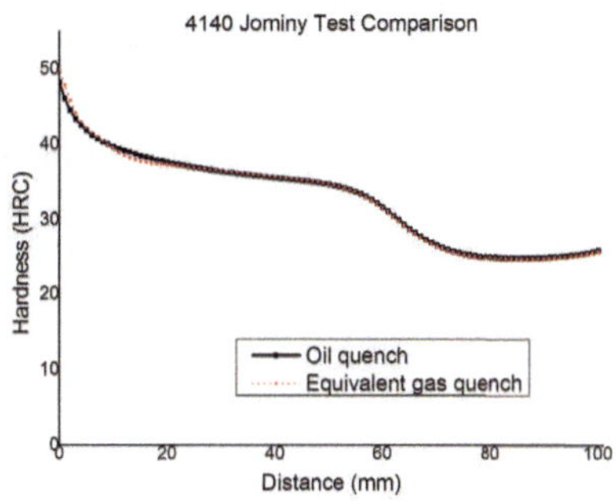

Figure 9. 4140 Jominy test comparison (simulation)

In Figure 8 (simulation), the cooling profiles along the Jominy bar for oil quench and the equivalent gas quench are compared. At 0mm, 10mm, 20mm and 50mm position from the quenched end, the cooling profiles are considered to be the same for oil quench and the equivalent gas quench. In Figure 9 (simulation), the hardenability of 4140 under oil quench and the equivalent gas quench is simulated. Two hardenability curves match perfectly, which demonstrates that the two quench processes generate the same microstructures and properties.

The concept of equivalent HTC should be redefined. After two different quench processes, if the cooling curves, microstructures and properties of all the workpiece are the same, these two quench HTCs are considered as the equivalent HTC.

52100 equivalent gas quench process is simulated as well. The equivalent gas quench HTC is the same as 4140's (in Figure 5). The cooling profile comparison and Jominy hardenability for 52100 are in Figure 10 (simulation) and Figure 11 (simulation).

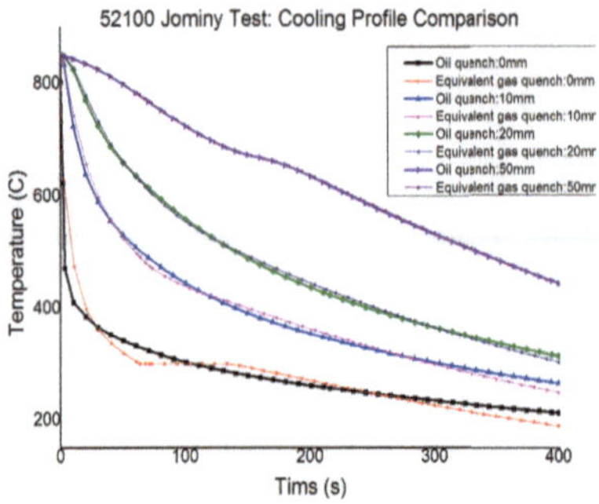

Figure 10. 52100 Jominy test: cooling profile comparison (simulation)

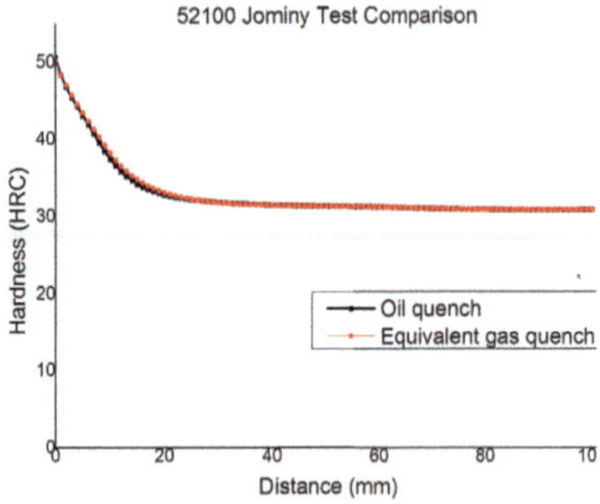

Figure 11. 52100 Jominy test comparison (simulation)

Same Hardness, Different Microstructures

After gas quench, the workpiece may have the same hardness compared with liquid quench. However, the cooling curves are not the same and it leads to different microstructures.

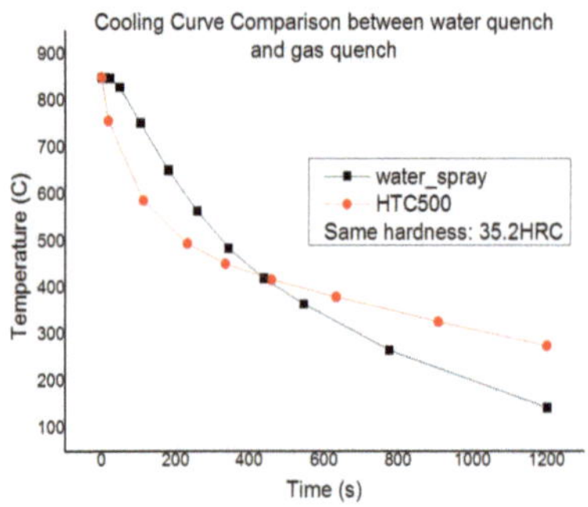

Figure 12. 4140 cooling curve comparison between water and gas quench (simulation)

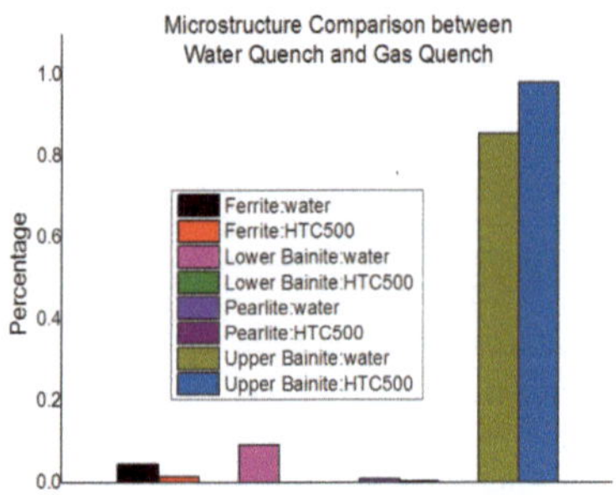

Figure 13. Microstructure comparison between water quench and gas quench (simulation)

In Figure 12 (simulation), the black line is from 50mm distance from the quenched end under water quench condition. The red line is from 5mm distance from the quenched end under HTC500 W/m^{2}C gas quench condition. These two positions have the same hardness, 35.2 HRC with different cooling curves. The microstructure analysis is shown in Figure 13 (simulation). The percentage of lower bainite of water quench is higher than gas quench, while the percentage of upper bainite of water quench is lower than gas quench. Generally, the mechanical properties of lower bainite are better than upper bainite, such as strength, toughness and ductility [9].

Summary and Conclusions

The concept of equivalent HTC, which is the fundamental difference between liquid and gas quench, is proposed. After two different quench processes, if the cooling curves of the core are the same, these two quench HTCs are considered as the equivalent HTC. The equivalent HTC prediction is made based on the verified gas quench model. When compared with oil quench and gas quench, no gas quench with constant HTC can be the equivalent HTC. With the great process flexibility to vary cooling rates, gas quench with varying HTCs are considered to find the equivalent HTC.

After finding the equivalent gas quench HTC, Jominy test is simulated to compare the cooling curves and hardness for the entire workpiece. The concept of equivalent HTC is redefined. After two different quench processes, if the cooling curves, microstructures and properties of all the workpiece are the same, these two quench HTCs are considered as the equivalent HTC.

Even it may result in the same hardness in the liquid and gas quench, the steel microstructure may be different because of the different cooling processes, and therefore other steel properties, such as toughness, may be different.

The cooling process, microstructures and properties such as hardness and toughness should be examined when designing the liquid or gas quench processes.

Acknowledgement

The authors wish to thank Jeffrey C Mocsari from Praxair and Zhichao Li from Dante for the help and the valuable advice. The authors are grateful to the Center for Heat Treating Excellence (CHTE), Worcester Polytechnic Institute for funding under the gas quench project.

References

[1] S.J. Midea, T. Holm, S. Segerberg, J. Bodin, T. Thors, K. Swartström, High pressure gas quenching—technical and economic considerations, in: Proceedings of the Second International Conference on Quenching and the Control of Distortion, ASM International, 1996, p. 157.

[2] Elkatatny, I., Morsi, Y., Blicblau, A.S., Das, S., Doyle, E.D., 2003, Numerical analysis and experimental validation of high pressure gas quenching. Int. J. Therm. Sci. 42, 417–423.

[3] Wang, J., Gu, J., Shan, X., Hao, X., Chen, N., & Zhang, W. (2008), Numerical simulation of high pressure gas quenching of H13 steel. Journal of Materials Processing Technology, 202(1-3), 188–194.

[4] Xiao, B., Wang, G., Wang, Q., Maniruzzaman, M., Sisson, R. D., Jr, & Rong, Y. (2011), An Experimental Study of Heat Transfer During Forced Air Convection. Journal of Materials Engineering and Performance, 20(7), 1264–1270

[5] Cosentino, F., Warnken, N., Gebelin, J.-C., & Reed, R. C. (2013), Numerical and experimental study of post-heat treatment gas quenching and its impact on microstructure and creep in CMSX-10 superalloy. Journal of Materials Processing Technology, 213(12), 2350–2360.

[6] Bozidar Liscic, Hans M.Tensi, Lauralice C.F.Canale, George E.Totten, Quenching Theory and Technology, 2010, ISBN 978-0-8493-9279-5
[7] B. Lynn Ferguson, A. Freborg, G. Petrus, Software simulates quenching, Advanced Materials and Processes (August) (2000) H31–H36.
[8] Herring, D.H. A review of gas quenching from the perspective of the heat transfer coefficient, Industrial Heating 2006, February, 67–72
[9] Gensamer, M., Pearsall, E. B., Pellini, W. S., & Low, J. R., Jr. (2012), The Tensile Properties of Pearlite, Bainite, and Spheroidite. Metallography, Microstructure, and Analysis, 1(3-4), 171–189.

EPD Congress 2015
Edited by: James Yurko, Antoine Allanore, Laura Bartlett, Jonghyun Lee, Lifeng Zhang, Gabriella Tranell, Yulia Meteleva-Fischer, Shadia Ikhmayies, Arief Suriadi Budiman, Prabhat Tripathy, and Guy Fredrickson
TMS (The Minerals, Metals & Materials Society), 2015

DETERMINATION OF CAVITY PROFILE INDUCED BY SUPERSONIC JETS IMPINGING ONTO LIQUIDS SURFACE

Qiang Li[1], Mingming Li[1], Mingxia Feng[2], Zongshu Zou[1]

[1]School of Materials and Metallurgy, Northeastern University
Heping District, Shenyang City, Liaoning Province, 110819, P. R. China

[2]Metallurgy Engineering College, Liaoning Institute of Science and Technology
High-tech development District, Benxi City, Liaoning Province, 117004, P. R. China

Keywords: BOF, Cavity profile, Penetration depth, Impacting diameter, Experiment.

Abstract

The impingement onto the surface of bath by top-blown jets is a significant process characteristic in BOF and EAF steelmaking process. The cavity is one of the most important outcomes of the interaction and plays an important role on reaction kinetics and reactors performance. But to date, the understanding to cavity is still so limited. In present study, a water model for a BOF converter is established and the dimension of cavity profile is investigated. The effects of lance height, gas flow rate, nozzle inclination angle and the amounts of slag are discussed. The results show that penetration depth increases with the increase of gas flow rate and the amount of slag, and the decrease of lance height and nozzle inclination angle. Furthermore, a theoretical model of cavity dimension is proposed for multiply jets impinging liquid bath on the basis of energy balance at the stagnation point of cavity.

Introduction

The phenomena of gas jets impinging onto a liquid pool are commonly encountered in many metallurgical practices, such as oxygen steelmaking process involving Basic Oxygen Furnace (BOF) and Electric Arc Furnace (EAF), vacuum degassing as Vacuum Oxygen Decarburization (VOD) and top-blown copper converter. The extremely significant aspect concerning gas jets impinging onto the surface of liquid are profile and dimensions of impact zone where many key refining reactions are accomplished, and for instance more than 55% of the total carbon is removed during the entire blow in oxygen steelmaking converter [1-2]. On the one hand, depth and diameter of depression are important direct dynamic parameters controlling the rate of slag formation and refining reactions and influencing mixing of the bath during the blow in oxygen steelmaking practices. And thereby they have attracted the great interests of investigators.

In the early of 1960s, Banks and Chandrasekhara [3] have contributed to the study on high velocity gas jet impinging on a liquid surface. A relationship between the momentum of a turbulent jet centerline velocity and the penetration depth was proposed:

$$\frac{M}{\rho_1 g H^3} = \frac{\pi}{2K^2}\frac{h}{H}\left(1+\frac{h}{H}\right)^2 \tag{1}$$

Following them, some experiments [4-7] were conducted to investigate the dimension of cavity, and two major points need be underlined: (i) in order to match the analytical model with their experiments, the value of K is always different and range from 5.13 to 11.5. (ii) The significant of influence of surface tension of liquid were neglected in developing the theory model. In addition, it should been noted that the impinging kinetic energy of jet can not fully be transfer into the bath and is also contributed to deflecting gas flow, splashing and the stirring of bath.

The objectives of this work are to experimentally quantify the profiles of cavity induced by impinging gas jets from multi-hole lance. The effects of lance distance to liquid surface, gas flow rate and inclination angle of jets and the amounts of the upper immiscible liquid on the cavity profile are experimentally investigated. Especially, influence of liquid physical properties on penetration depth is identified. Furthermore, through defining an energy utilization index of impinging kinetic energy of jets and explicitly considering the effect of surface tension, an improved theoretical models of predicting penetration depth for the multiply jets impinging bath is proposed and validates through experimental measures.

Experiments

Experimental Apparatus

Cold model experiments are conducted in a 1/10th scaled-down model of a 150 t BOF steelmaking converter, which are made of acrylic resin. Water and compressed air are used to simulate the hot metal and the oxygen, respectively. Transparent transformer oil with viscosity 0.03 Pa·s, density 880 kg/m^3 and surface tension 0.029 N/m was used to simulate molten slag. The lances with six-hole and single-hole nozzle are fixed above the surface of liquid, and nozzles of different inclination angles (0°, 12°, 14° and 17.5°) are considered. The schematic diagram of experimental set up is shown in Figure 1. Model dimensions and experimental conditions are listed in Table I and II respectively.

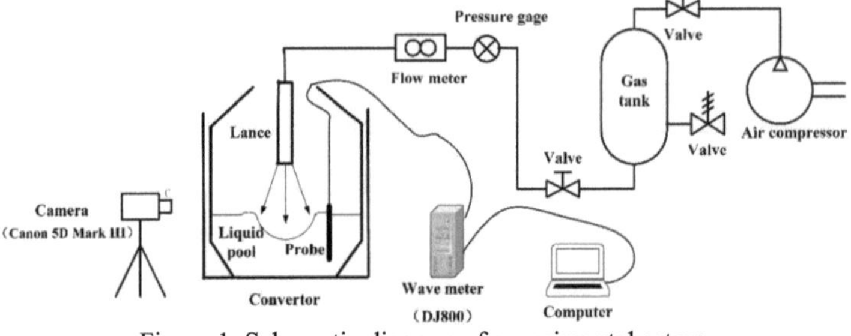

Figure 1. Schematic diagram of experimental set up

Table I. Dimensions of Prototype and Model

Dimension	Prototype	Water model
Bath diameter, D_l (m)	5.8	0.58
Bath depth, H_l (m)	1.545	0.154
Nozzle throat diameter, d_t (m)	0.03	0.003/0.0073
Nozzle exit diameter, d_e (m)	0.0434	0.0043/0.011
Nozzle number, N (-)	6	6/1
Nozzle inclination angle, θ (°)	14	0 (Single-nozzle), 12, 14, 17.5

Table II. Experimental Conditions of Water Model

Parameters	Water model
Top-blown gas flow rate (Nm^3/h)	30.2, 32.7, 35.2, 37.7, 40.2, 42.7
Lance height (m)	0.09, 0.12, 0.15, 0.18, 0.21
The height ratio of slag to metal (-)	0/154, 10/154, 15/154, 20/154

Considerations of Similarity

Experiments are designed based on the geometric and dynamic similarity with the commercial BOF converter. For the geometric similarity the ratios between the nozzle throat diameter (d_t,) and the bath diameter (D_l) and also the bath height (H_l) to bath diameter are considered. The ratio of nozzle throat diameter to bath diameter determines the distribution of jet momentum flow rate on the bath [8]. For the dynamic similarity, four kinds of forces are presented: the gravitational force, the inertial force, the surface tension, and the viscous force. Meidani et al. [6] suggested a final dimensionless function for the top-blown steelmaking converter as following:

$$\frac{h}{H} = f(\mathrm{Mm}, \mathrm{Re}', \mathrm{We}') \qquad (2)$$

where h is the penetration depth; H is the lance height, Mm is the dimensionless momentum flow rate number [8]; Re' and We' are modified Reynolds and Weber numbers. These dimensional numbers reflect the ratio of the jets inertial to gravity force, viscous and surface tension forces of liquid, respectively, and are expressed [6].

Table III shows the comparison of the dimensionless numbers between the water model and the commercial BOF. Since it is not possible to attain complete similarity between the model and the prototype [9], and for the impingement of jets onto liquid surface the inertial forces and gravity force dominates the physical process, and several researchers proved that momentum similarity is the most appropriate similarity criteria for this system [8, 10], so the dimensionless momentum flow rate number was emphasized in current study.

Table III. Comparison of Similarity between Prototype of BOF and Water Model

Similarity	Criterions	BOF	Water model
Geometric	$N^{1/2}d_t/H_l$	0.0476	0.0477
	H_l/D_l	0.2664	0.2655
Dynamic	Mm	2.44×10^{-3}~6.20×10^{-2}	2.44×10^{-3}~6.20×10^{-2}
	Re'	5.88×10^{5}~8.33×10^{5}	1.49×10^{4}~2.10×10^{4}
	We'	4.34×10^{2}~2.03×10^{3}	14.42~67.28

Experimental Results

Cavity Profile

The shapes of cavities induced by impinging jets from single- and multi-nozzle lances are shown in Figures 2(a)~(d). The sectional area of nozzle for single-nozzle lance was equal to the total that of all the nozzles for multi-nozzle lance, and besides the total gas momentum flow rates at nozzles exit for single- and multi-nozzle lances were maintained the same value. Figure 2(a) shows the shape of cavity impinged by single-nozzle jet, typically a deep and narrow one with violent splashing and oscillating, namely the penetrating mode. In contrast, the shape of cavity induced by multi-nozzle jets is a shallower and bigger one, namely the splashing modes. As the nozzle inclination angle increases, cavity becomes shallower, and especially a cavity with scattered depressions at the impinging points of each individual jet and central bulge at the center of cavity is formed at the nozzle inclinational angle 17.5°, as seen in Figure 2(d).

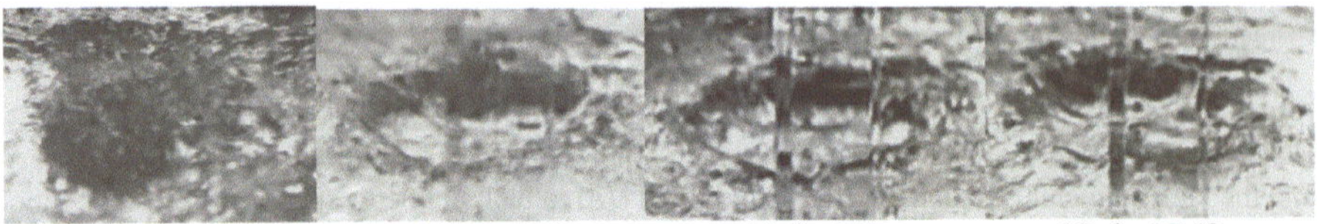

(a) θ=0° (Single nozzle) b) θ=12° (c) θ=14° (d) θ=17.5°

Figure 2. Cavity shapes at different nozzle inclination angles under the gas flow rate of 35.2 Nm^3/h and the lance height of 0.09 m

The differences of cavity shapes impinged by single- and multi-nozzle jets are demonstrated schematically in Figure 3. As the gas flow rate is modified, impinging momentum of jets onto liquid surface along with their interaction structure are changed, thereby leading to different cavity shapes, as shown in Figure 4. Figure 5 shows the cavity shapes at different lance heights and the gas flow rate of 37.7 Nm^3/h. Cavity transits from splashing mode to dimpling one as the lance height increases to 0.18 m. In current considerations, for analogizing the change of slag in a commercial BOF top slag layer with different heights at the beginning of blowing was consider to explore the different penetrating of liquids surface. Figure 6 shows the shape of cavity with different slag height at the same operating condition. Although cavities are splashing mode for all the considering slag heights, increasing cavity depth with increasing slag height are obviously demonstrated.

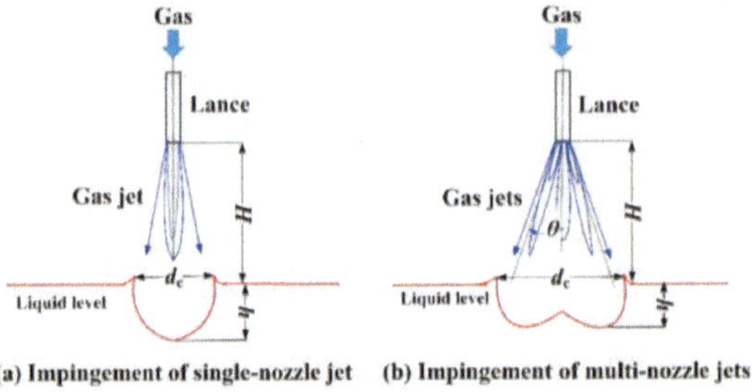

Figure 3. Schematic diagram of impingement on liquid surface of jets from single- and multi-nozzle lance

(a) Q=32.7 Nm^3/h (b) Q=35.2 Nm^3/h (c) Q=37.7 Nm^3/h (d) Q=40.2 Nm^3/h

Figure 4. Cavity profile at different gas flow rates and the lance height of 0.09 m

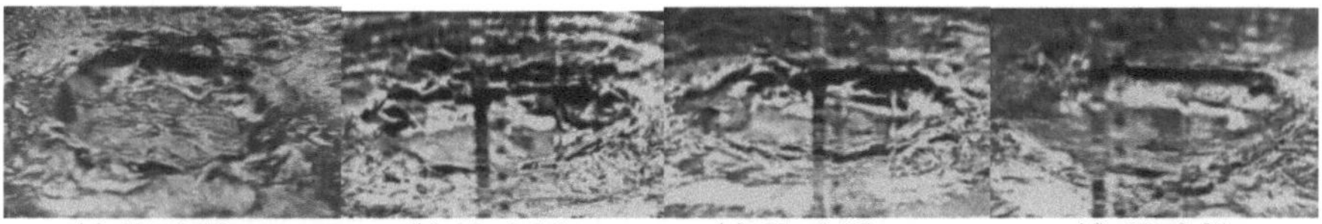

(a) H =0.12 m (b) H =0.15 m (c) H =0.18 m (d) H =0.21 m

Figure 5. Cavity profile at different lance heights and the gas flow rate of 37.7 Nm^3/h

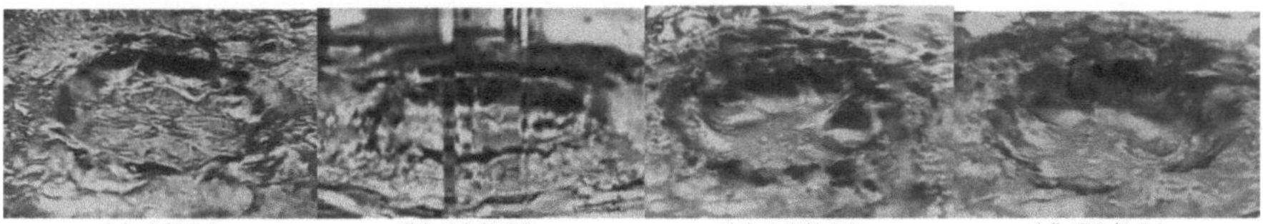

(a) 0/154 (b) 10/154 (c) 15/154 (d) 20/154

Figure 6. Cavity profile for different top phase heights at the gas flow rate of 37.7 Nm^3/h and the lance height of 0.12 m

Cavity Depth

Figure 7 (a) presents the cavity depths at different lance heights and top-blown gas flow rates. It is seen that the cavity depth increases with increasing top-blown gas flow rate and decreasing lance height. The cavity depth for different nozzle inclination angles at a given lance height and gas flow rate is shown in Figure 7 (b), indicating a deeper cavity at the smaller nozzle inclination angle, e.g. cavity depth increases from 0.0203 m to 0.0217 m as nozzle inclination angle decreases from 17.5° to 12°. The cavity depth is defined through the depth of that gas jets penetrate into all the liquid layers. The influence of top-layer liquid (slag) height on cavity depth is presented in Figure 7 (c) as top-layer liquid (slag) is in presence. The figure shows cavity depth becomes deeper with increasing slag height due to smaller kinetic energy consumption on potential energy of top-layer liquid during jets penetrating into liquids.

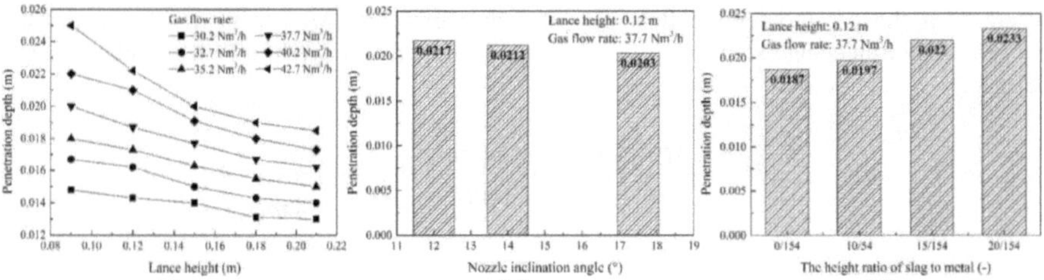

(a) (b) (c)

Figure 7. Cavity depth (a) at different lance heights and gas flow rates, (b) at different nozzle inclination angles, (c) at different slag heights

Discussion

Effect of Liquid Surface Tension

The influencing extent and mechanism of liquid surface tension on cavity depth have not been identified under the normal circumstance. For the following discussion, a general case, namely the impingement on the liquid surface of a single-nozzle jet is analyzed. At the cavity stagnation point fours acting forces work for the formation of cavity, namely inertia force f_s, gravitational force f_g, interfacial tension f_σ, and viscous force τ, as shown in Figure 8.

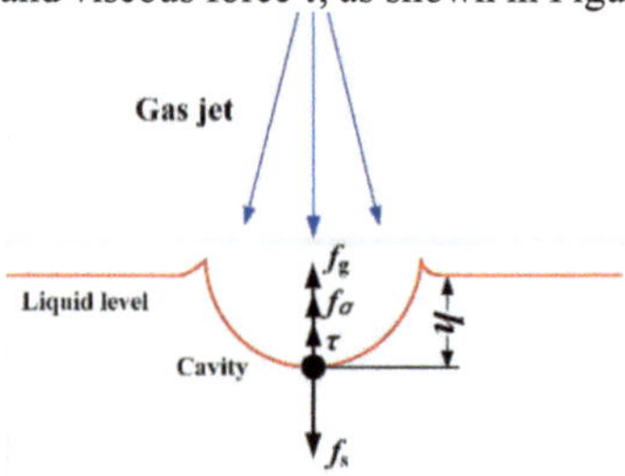

Figure 8. Schematic diagram of acting forces at the cavity stagnation point

At the stagnation point of cavity, following relationship can be obtained by considering the liquid surface tension on the basis of the energy balance.

$$\frac{1}{2}\rho_g V_c^2 = \rho_l gh + \frac{2\sigma}{R_0} \tag{3}$$

where V_c is the impinging velocity of jet; ρ_g and ρ_l are gas and liquid density; h is penetration depth; σ is surface tension of liquid; R_0 is the radius of curvature at the stagnation point, and according to Banks and Chandrasekhara [3], $R_0 = H^2/4K^2h$. In addition, for turbulent gas jet $V_c/V_j = K\, d_e/x$ can be obtained. The x denotes the axial distance from lance tip, and substituting x with lance height H, following predicting model of cavity depth can be obtained:

$$\frac{M}{\rho_l ghH^2} = \frac{\pi}{2K^2}\left(1 + \frac{8K^2\sigma}{\rho_l gH^2}\right) \tag{4}$$

where M is the gas momentum flow rate at nozzle exit, and $M = (\pi/4)\rho_g d_e^2 V_j^2$. In order to consider the influencing extent of surface tension on penetration depth, following relationship is defined:

$$F \equiv \left(\frac{M}{\rho_l ghH^2}\right) \Big/ \left(\frac{\pi}{2K^2}\cdot\frac{8K^2\sigma}{\rho_l gH^2}\right) \equiv \frac{M}{4\pi\sigma h} \equiv \frac{M}{4\pi\sigma H_l \cdot \dfrac{h}{H_l}} \tag{5}$$

where H_l is the bath depth; F represents the ratio of dimensionless momentum consumed in surface tension to the total that of gas jet. h/H_l is the ratio of cavity depth and bath depth. Koria and Lange [11] reported that h/H_l was 11~16%, and Zhou et al. [12] suggested it was less than 30% in their water model study. For current model study, H_l is 0.154 m, and maximum impinging momentum flow rate of each individual jet is 0.3248 kg·m/s^2. Corresponding to the commercial BOF, H_l is 1.545 m, and maximum impinging momentum flow rate of each individual jet is 661.4 kg·m/s^2. The F against the liquid surface tension for present water model and its corresponding prototype of top-blown converter are plotted in Figure 9 (a) and (b) respectively. They show that the influence of liquid surface tension on penetration depth is increased with the increasing liquid surface tension and also the deeper penetration depth.

Present findings also show that the influence of liquid surface tension for the study on penetration depth is significant, especially for the water modeling study where penetration depth could be overestimated over about 10% if liquid surface tension is neglected.

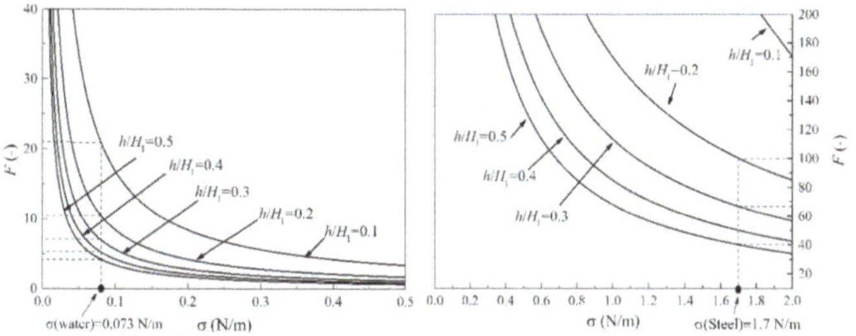

(a) Model (b) Prototype

Figure 9. Effect of surface tension on cavity depth

An Improved Computational Model of Cavity Depth

The impinging kinetic energy of jet at impact point is not only to the formation of cavity, but also contributed to deflecting gas flow, splashing and the stirring of bath, and thereby an energy utilization index η, representing the ratio of impinging kinetic energy acting on the formation of cavity to the total one, was introduced. On the other hand, the impinging momentum flow rate of each jet can be obtained for the multi-nozzle lance, namely $M = M_n \cos\theta / n$, where M_n is the total momentum flow rate of gas jets; θ is the nozzle inclination angle and n is the nozzle number. Thus, a more fundamental penetration depth prediction model is obtained by Eq. (3):

$$\frac{M_n \cos\theta}{\rho_1 g n H^3} = \frac{\pi}{2K^2\eta}\frac{h}{H}\left[(1+\frac{h}{H})^2 + \frac{8K^2\sigma}{\rho_1 g H^2}\right] \tag{6}$$

Figure 10 shows the functional relationship of energy utilization index η with lance heights, which shows the energy utilization index increases with increasing lance heights, namely more impinging kinetic energy is used to the formation of cavity as the lance height increases, and utilization percentage of impinging kinetic energy of jets transforming to bath decreases with increasing lance height.

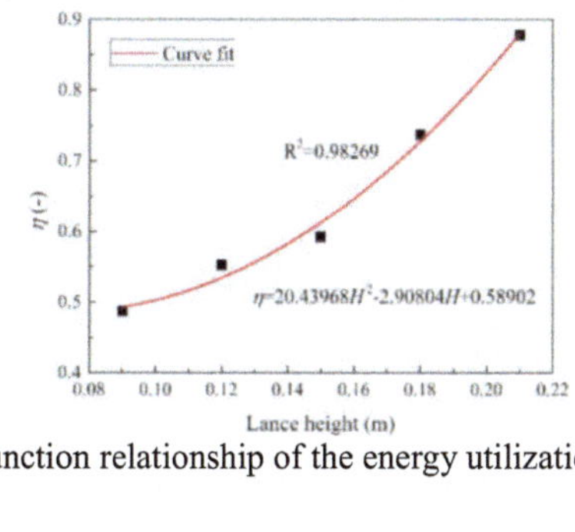

Figure 10. The function relationship of the energy utilization index as lance height

Conclusions

(1) Cavity depth and diameter were experimentally measured at current operating conditions and lance parameters. Cavity depth increases with increasing gas flow rate and decreasing nozzle inclination angle and lance height. Cavity diameter is nearly insensitive to top-blow gas flow rate, but is more dependent on the lance height and nozzle inclination angle.

(2) Influence of liquid surface tension on penetration depth is significant and increases with the increasing liquid surface tension and also the penetration depth. Especially based on the previous predicting model the penetration depth could be overestimated over 10% for the water model study if liquid surface tension is neglected.

(3) An energy utilization index of impinging kinetic energy of jets used for the formation of cavity was defined to eliminate the contributions of jets to deflecting gas flow, splashing and the stirring of bath. The energy utilization index is dependent on the physical properties of fluids, and it increases as lance height increases.

Acknowledgement

The authors are grateful for the financially supported by the National Natural Science Foundation of China (grant no. 51104037, 50774019) and the Fundamental Research Funds for the Central Universities of China (grant no. N120402010).

References

1. D. Price. *Process Engineering of Pyrometallurgy* (London, UK: Institution of Mining and Metallurgy, 1974), 8-15.
2. N. Dogan, G.A. Brooks, and M.A. Rhamdhani, "Comprehensive Model of Oxygen Steelmaking Part 1: Model Development and Validation," *ISIJ Int.*, 51(7) (2011), 1086-1092.
3. R.B. Banks and D.V. Chandrasekhara, "Experimental Investigation of the Penetration of a High Velocity Gas Jet through a Liquid Surface," *J. Fluid Mech.*, 15 (1963), 13-34.
4. F.R. Cheslak, J.A. Nicholls, and M. Sichel, "Cavities Formed on Liquid Surfaces by Impinging Gaseous Jets," *J. Fluid Mech.*, 36 (1969), 55-63.
5. F. Qian, R. Mutharasan, and B. Farouk, "Studies of Interface Deformations in Single- and Multi-Layered Liquid Baths Due to an Impinging Gas Jet," *Metall. Trans. B*, 27(1996), 911-920.
6. A.R.N. Meidani, M. Isac, A. Richardson, et al., "Modelling Shrouded Supersonic Jets in Metallurgical Reactor Vessels," *ISIJ Int.*, 44(10) (2004), 1639-1645.
7. H.Y. Hwang and G.A. Irons, "A Water Model Study of Impinging Gas Jets on Liquid Surfaces," *Metall. Trans. B*, 43 (2012), 302-315.
8. S.C. Koria and K.W. Lange, "A New Approach to Investigate the Drop Size Distribution in Basic Oxygen Steelmaking," *Metall. Trans. B*, 15 (1984), 109-116.
9. M. Lee, S.L. O'Rourke and N. Molloy, "Oscillatory Flow in the Steelmaking Vessel," *Scandinavian Journal of Metallurgy*, 32 (2003), 281-288.
10. M. Lee, V. Whitney and N. Molloy, "Jet-liquid Interaction in a Steelmaking Electric Arc Furnace," *Scandinavian Journal of Metallurgy*, 30 (2001), 330-336.
11. S.C. Koria and K.W. Lange, "Penetrability of Impinging Gas Jets in Molten Steel Bath," *Steel Research Int.*, 58(1987), 421-426.
12. X.B. Zhou, M. Ersson, L.C. Zhong et al., "Mathematical and Physical Simulation of a Top Blown Converter," *Steel Research Int.*, 84(2013), 1-9.

2015 EPD Congress

SYMPOSIUM: MATERIALS PROCESSING FUNDAMENTALS

Materials Processing

Jonghyun Lee,
University of Massachusetts, Amherst

EPD Congress 2015
Edited by: James Yurko, Antoine Allanore, Laura Bartlett, Jonghyun Lee, Lifeng Zhang, Gabriella Tranell, Yulia Meteleva-Fischer, Shadia Ikhmayies, Arief Suriadi Budiman, Prabhat Tripathy, and Guy Fredrickson
TMS (The Minerals, Metals & Materials Society), 2015

IN-SITU GAS MONITORING BY EMISSION SPECTROSCOPY

Thor Anders Aarhaug[1], Alain Marc Ferber[2], Pål Tetlie[1] and Halvor Dalaker[1]

[1]SINTEF Materials and Chemistry, NO-7465 Trondheim, Norway
[2]SINTEF Information and Communication Technology, SINTEF, NO-0373 Oslo, Norway

Keywords: In-situ spectroscopy, emission spectroscopy, high-temperature gas characterization

Abstract

Off-gas monitoring of experimental reactors give little information about intermediate reaction products and reflect chemical equilibria at reduced gas temperatures. A reactor design with optical access was designed and tested in order to be able to qualify and quantify gaseous species from their emission spectra. By use of a sequence of sapphire, UV fused silica and IR transparent sapphire and CaF_2 windows and lenses optical access to a hot, maximum 1800 °C, reaction zone was gained. A cooled area behind the reaction zone was used as a background in order to avoid signal saturation from reactor walls. In addition to molecular emission spectra, atomic emission lines are used to characterize stable gas constituents and radicals. The setup has been designed for the study of methane dehydrogenation experiments in order to investigate the carbon activity of reduction of oxide with methane, but has also been used to monitor gaseous silicon suboxide.

Introduction

Characterization of gases at high temperature is desirable for several reasons. Firstly, the chemical species reacting at a given temperature is not necessarily reflected by the cooler exhaust gas composition which can easily be characterized by commercial gas analyzers. Classification of chemical species in a hot reaction zone is important for the understanding of the chemical reactions taking place. Quantitation of these species is complementary to kinetic modelling of these reactions.

Commercial instrumentation for gas analysis is often limited to temperatures up to 400 degrees. Conventional IR spectroscopy beyond this temperature is limited because molecules will both absorb and emit radiation. At these temperatures, the radiation source can be omitted and the emission from the gas molecules could be analyzed by a spectrometer. Emission spectroscopy is not as established as absorption spectroscopy, and commercial instrumentation is scarce. However, excellent analytical quantitation from emission spectroscopy has been documented [1].

One of the reasons for the lack of commercial emission analyzers are the fact that the optical access to the radiation is not easily standardized, and must be adapted for the application. There is an inherent challenge with thermal blackbody radiation from the spectrometer itself in addition to that of the furnace or gas confinement of elevated temperature. The blackbody radiation is a continuum contributing to noise only. In order to minimize noise, field-of-view has to be limited by cold apertures and using a cold background in the gas confinement in the field-of-view. Emission spectroscopy can also make use of the UV-VIS spectral range. Although electronic transitions are less informative for gas molecules, spectral information from free-electron pairs of intermediate gas radicals can be utilized.

For high temperature furnaces, obtaining optical access to the reaction zone is not trivial. Even if the introduction of optical windows is feasible, their introduction may introduce temperature gradients in the furnace.

One option is to use optical fibers to access the reaction zone. Silica (< 1200 °C) and Sapphire (< 1700 °C) are both materials transparent in the UV-VIS and IR ranges.

Monitoring of carbon activity in reactors using methane as carbon source

Currently, there are many ongoing research projects on the topic of establishing methane as an alternative source of carbon in extractive metallurgy. One of the challenges faced is the decomposition of methane at temperatures above 930 °C. As the cracking of methane produces condensed carbon, the activity of carbon becomes unity. One way to reduce cracking according to (1) is to introduce hydrogen gas. If methane adsorption/decomposition for a situation with active adsorbed carbon is compared to equilibrium between methane, graphite and hydrogen, it can be derived that the activity of the carbon as existing in the thermodynamically unstable methane is determined by the partial pressures of hydrogen and methane in the gas phase [2]. This means that the developed model can be used to prescribe experimental conditions tailored to optimize a certain reaction path. By accurately monitoring the gas composition in situ using the developed gas measurement technique, the carbon activity in the methane can be indirectly measured. Once measured, the potential is opened up for manipulation, and so the carbon activity can be tuned toward the conditions predicted by the model to be favorable for oxide reduction. It is, however, necessary to dehydrogenate methane in order to make use of its carbon:

$$CH_4 = C(s) + 2\ H_2 \tag{1}$$

In order to evaluate the feasibility of methane as a carbon source in the reduction of for Si, Mn and Ti ores, fiber optical emission spectroscopy could contribute significantly to the understanding of these processes. In addition to monitoring the methane concentration in the reactor, the expected radicals from the dehydrogenation of methane, CH_3 (724.6 nm), CH_2 (341.9 nm) and CH (431.42 nm) could be monitored in addition to C and C_2.

Monitoring of formation of SiO gas in silicon metal production

The formation of Silicon suboxide occurs at temperatures higher than 1500 °C by reaction between quartz and silicon carbide:

$$2\ SiO_2\ (s) + SiC\ (s) = 3\ SiO\ (g) + CO\ (g) \tag{2}$$

At temperatures above 1800 °C, the reaction between SiO and SiC could form liquid silicon metal:

$$SiO\ (g) + SiC\ (s) = 2\ Si\ (l) + CO\ (g) \tag{3}$$

In the presence of carbon, however, silicon carbide could be formed:

$$SiO\ (g) + 2C\ (s) = SiC\ (s) + CO\ (g) \quad (4)$$

Silicon suboxide escaping the furnace exhaust impacts the silicon yield of the process as microsilica is formed. The reactivity of the carbon source with SiO (g) is an important parameter for limiting the losses. Although thermochemical calculations indicate low losses [3], kinetics appears to limit the rate of SiC formation [4].

By in-situ monitoring of the SiO formation, improved understanding of the kinetics of the processes of formation and reaction of SiO could be obtained.

Experimental

A tubular graphite furnace was used as a starting point for modifications. The furnace is built around a 457 mm long 60 mm ∅ graphite heating element with radiation shields installed at top and bottom. The heating element was rated for 15 kW. The furnace was designed to have an 80 mm homogeneous temperature zone. This zone was modified for optical access by the installation of a flange on one side of the furnace. A tube with 9 ∅ mm inner diameter was installed perpendicularly through the heating element. The dimensions were carefully chosen in order to allow for differences in thermal expansion between the materials. Slots were machined out of the tube so that obstruction of the gas flux through the furnace was minimal and yet to provide sufficient mechanical integrity for the tube. The tube was electrically insulated from the heating element by using boron nitride sleeves to insulate the probe from the heating element. Macor glass ceramic components were used for fixation to the flange as well as insulation from the furnace outer shell.

Several tube materials were considered. For experiments where monitoring of SiO (g) is the objective, carbon should be avoided in order to prevent formation of SiC. SiC as probe material is promising both with respect to mechanical as well as chemical stability. The cost of the material is substantially higher than graphite. For the initial experiments it was decided to test the design with respect to mechanical stability and evaluate the optical performance of the design. For SiO (g) applications, it is also a possibility of in-situ passivation of the graphite by formation of a SiC layer on the surface that will provide sufficient quantities of SiO (g) in the optical path.

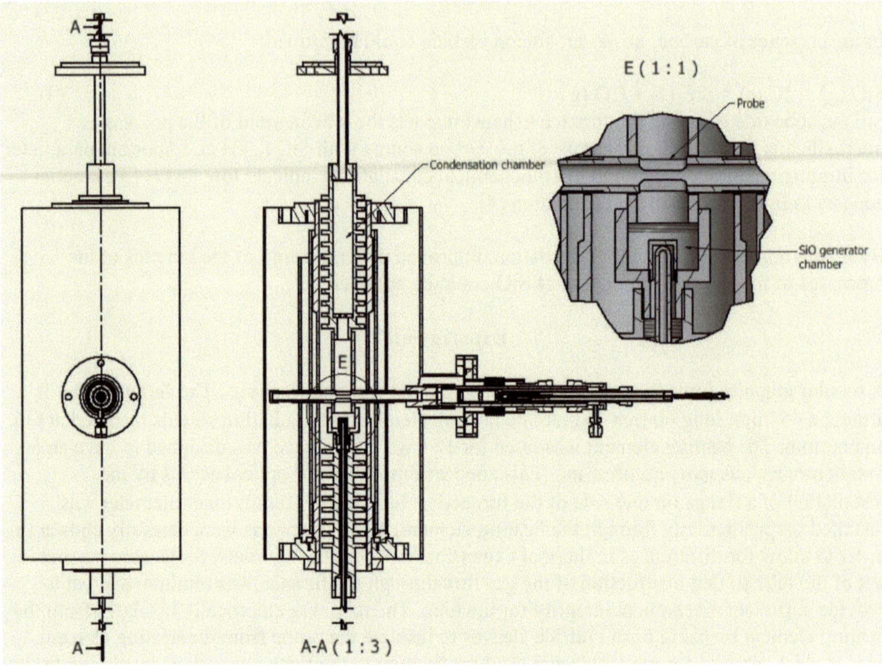

Figure 1. Illustration of graphite tube furnace.

The optical path was modelled using Zemax Premium software. The end of the tube protruding the heating element from the other side of the access was designed as to provide a passively cooled background to the emission from analytes in the optical path.

The light from the emission was focused onto the through two lenses in order to correct for chromatic aberration. An anti-reflective coated sapphire window was used to separate the probe from the furnace gas atmosphere.
The probe was designed with flushing channels, allowing for slow purge argon through the probe. Although purging most likely will dilute the analytical signal, it was considered to be necessary in the event of condensation of SiC and/or SiO_2 in the tube and onto the sapphire window.

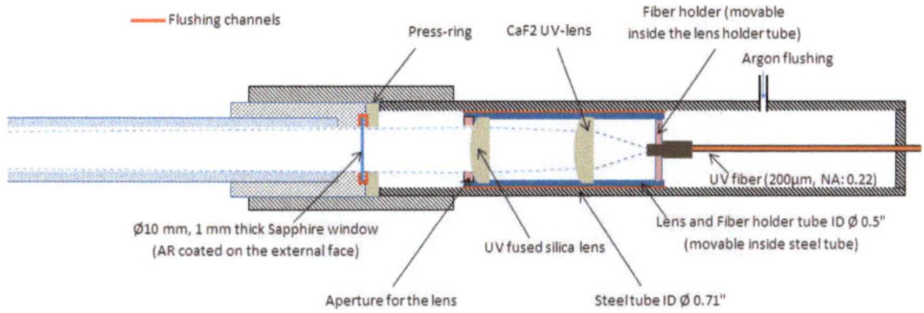

Figure 2. Schematic illustration of the optical probe

In order to continuously monitor a wide spectroscopic range from UltraViolet (200 nm) to Near InfraRed (900 nm), two spectrometers were used: Avantes Multichannel (6 channel) and an Ocean-Optics USV4000-UV-VIS single channel spectrometer. The Ocean-Optics spectrometer was used to scan the optical range from 180 to 875 nm with a resolution of 1.5-2.3 nm Full Width at Half Maximum (FWHM). The Avantes spectrometer provided better resolution: 0.07 to 0.13 nm FWHM. The spectral range from 190 to 920 nm is distributed between the six channels so only one channel can be monitored at a time. An Avantes UV100-SR optical fiber was used. The 200 µm fiber was Solarization Resistant (SR) for applications to wavelengths below 230 nm.

Two experiments were conducted. The setups were almost identical. In the second experiment the The Ar gas flow through the furnace was reduced from 0.5 nLpm to 0.25 nLpm. The purge flow through the optical probe was also reduced from 0.2 to 0.15 nmLpm. A charge consisting of approx. 10 grams of pelletized SiO_2 and SiC in a molar ratio of 1.82:1 was used. In both experiments, the remaining charge after the experiment was found to be approx. 10 % of the original mass.

Results

The design of the optical access appeared to work satisfactory with respect to the passively cooled contrast background. However, the temperature gradient caused condensation that partially blocked the tube on the opposite side of the optical probe. As can be seen from Figure 3, a white condensate of what was by Electron Microsonde analysis verified to be pure SiO_2, was formed. The formation of the web-like structure on this side of the optical access tube only is probably due to the lack of gas purging.

Figure 3. Condensation of SiO_2 inside optical access tube.

On removal of the optical access tube, condensation of green SiC on the white BN sleeves was apparent. This is illustrated in Figure 4. Electron microsonde investigation of the graphite surface of the probe also confirmed formation of SiC, suggesting presence of SiO (g).

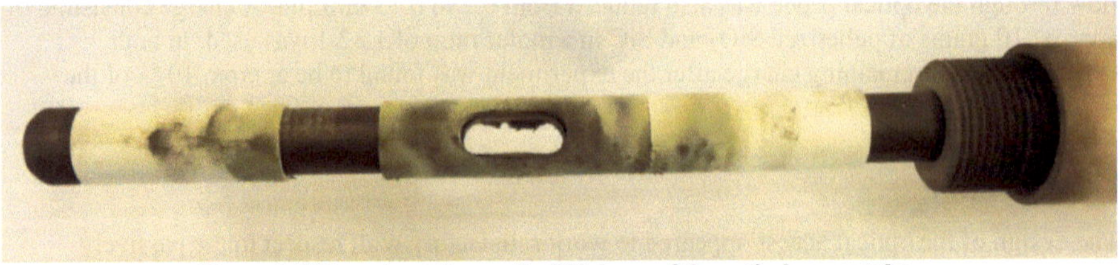

Figure 4. SiC condensation on the BN sleeve of the optical access tube.

Results from in-situ emission spectroscopy

During heating of the furnace, the emission from the probe was recorded. The emission from the VIS part of the spectrum is most pronounced and illustrated in Figure 5. Emission peaks at 589 and in the range 766-770 nm were observed. These peaks have not been identified.

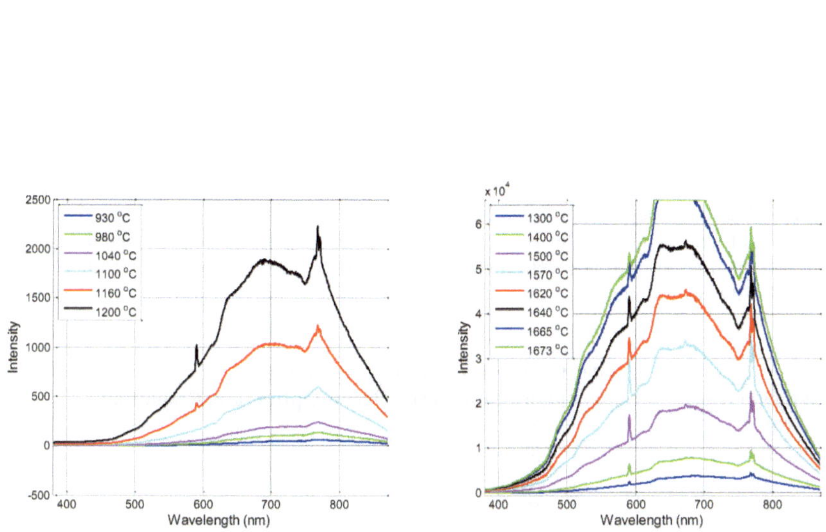

Figure 5. Averaged emission intensities at various temperatures.

For the experiments conducted, the main objective was to document the presence of SiO (g) in the reaction zone. The spectrum of SiO is characterized by band profiles shaded to the red in the range from approx. 200 to 300 nm [5]. In addition, strong emission lines from electronic transitions of atomic silicon are present at 251-253 and 288 nm.

The spectrograms from the high resolution AVANTES spectrometer for the relevant spectral range of SiO is shown in Figure 6. From the data, there is no indication of the presence of SiO (g) in the optical path.

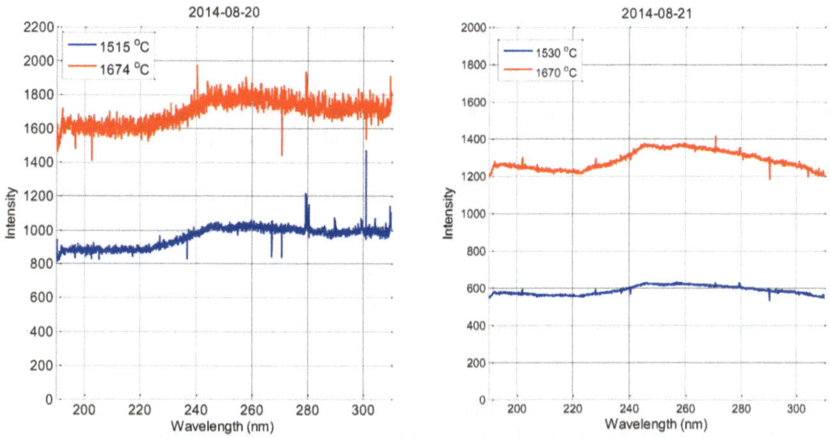

Figure 6. High resolution spectrograms for the 200-300 nm UV-range.

For the two experiments conducted, a lot of spectral information has been recorded. The interpretation of emission spectra from this experiment is currently a work in progress. The

identification of probable compounds is performed by comparison of spectra with spectral libraries as well as the modelling of synthetic spectra.
The major lines found in the spectral data are listed in Table 1. Clearly, more information is obtained with the high resolution spectrometer. It remains to be seen whether this resolution is required once SiO gas is detected.

Table 1. Spectral lines and ranges identified.

Spectral line(s) (nm)	Avantes (high resolution)	Ocean-Optics (low Resolution
280	X	
363-365	X	
403-405	X	
422	X	
589	X	X
671	X	
673		X
766-770	X	X
780	X	

Future work

The main objective of these experiments has been to be able to qualify and quantify SiO (g). This has not been achieved. There are two plausible explanations for the lack of signal. The first is that the purge flow through the probe dilutes the signal too much. The second explanation is that the formed SiO will react with the graphite surfaces inside the furnace, but this formation of SiC on the surfaces does not passivate the graphite surfaces sufficiently over time to produce a measureable quantity of SiO (g) in the light path.

In further experiments, reactions with methane as a source of carbon for reduction of quartz will be conducted. This will produce other gaseous molecules and radicals from the deprotonation of CH_4. New experiments where the gas purging of the optical probe will be systematically changed will be conducted. Eventually, experiments where SiC is used as the optical probe material will be conducted.

Acknowledgements

This work has been conducted as part of the project KISELROX, funded by the Norwegian Research Council.

References

[1] Usseglio, S., Thorshaug, K., Karlsson, A. et. al., App. Spectr., 64 (2010) 2, pp. 141-148.
[2] Ostrovsky, O. and Zhang, G. AIChE Journal, 51 (2006) 1, pp. 225-233.
[3] Myrhaug, Edin, PhD Thesis NTNU 2003:67.
[4] Tangstad, Merete (ed), Metal production in Norway, Akademia Forlag, 2013.
[5] Wooldridge, MS., Danczyk, SA. And Wu, J., J. Quant. Spectroscopy & Radiative Transfer, 64 (2000), pp. 573-584.

2015 EPD Congress

SYMPOSIUM: MATERIALS PROCESSING FUNDAMENTALS

Poster Session

EPD Congress 2015
Edited by: James Yurko, Antoine Allanore, Laura Bartlett, Jonghyun Lee, Lifeng Zhang, Gabriella Tranell, Yulia Meteleva-Fischer, Shadia Ikhmayies, Arief Suriadi Budiman, Prabhat Tripathy, and Guy Fredrickson
TMS (The Minerals, Metals & Materials Society), 2015

Thermodynamic Study on Vanadium Extraction with CO_2 and O_2 Mixed Blowing

Wei-Tong Du[1], Yu Wang[1]*, Gang Wen[1]

[1]College of Materials Science and Engineering; Chongqing University;
Chongqing 400044,China
*Corresponding author; e-mail: wangyu@cqu.edu.cn

Keywords: Carbon dioxide, Mixed injection, Vanadium extraction

Abstract

Carbon dioxide (CO_2), a major component of the greenhouse gases, could be comprehensive utilized as a valuable resource to oxidize vanadium during the converter vanadium extraction process. The thermodynamic software, FactSage, was utilized to study the oxidation ability and theoretical cooling ability of using weak oxidant CO_2 as coolant in vanadium-containing hot metal. Besides, experimental research on vanadium extraction in CO_2 and O_2 mixed blowing extracting process had been undertaken. The results indicated that the elements C, Si, V, Mn etc. could be oxidized by weak oxidant CO_2, and the content of V_2O_3 in slag was approached to pure O_2 blowing. Furthermore, CO_2 had a remarkable cooling ability and the molten bath temperature kept on decreasing as CO_2 blowing rate increases. This research has proved that CO_2 and O_2 mixed blowing in vanadium extraction process are effective.

Introduction

With international communities having increasingly recognized the severity of climate change, the pressure for reduction of CO_2 emissions has become more prominent[1][2]. For this reason, how to reduce the emission of carbon dioxide and take use of carbon dioxide is one of the main problems that has attracted many steel engineers' attention[3][4].

In order to take the utilization of carbon dioxide as a resource in steelmaking process, many analyzed about the application of CO_2 as stirring gas, reaction media and protective gas have been carried out[5][6]. However, vanadium extraction with CO_2 and O_2 mixed gas has not been reported yet.

The traditional technique to extract vanadium during converter steelmaking process was to blow pure O_2 to oxidize V into V_2O_3[7]. However, this process has several problems including strong oxidizer caused overheating and energy consuming. According to the industry vanadium extraction process, the temperature rises quickly at the oxidation period and in this juncture some coolant should be put into the converter to control the temperature of the converter at a proper level. Traditional coolant such as cold bonded pellet, pig iron, iron ore are hard to control and the contained elements P, S, Si can aggravate subsequent metallurgy process burden. In the CO_2 and O_2 mixed blowing technology, CO_2 can blow into bath of the converter as coolant to replace the solid coolants.

Therefore, it is necessary to open up the possibilities for using CO_2-containing oxidants in vanadium extraction. In present work, introduction of CO_2 in suitable amounts to replace O_2

during this process could thus offer a unique solution in view of the fact that an environmentally undesired gas is being used for coolant and oxidizer. This work was supported by National Natural Science Foundation of China (project No.51334001) and Sharing Fund of Large Scale Equipment, Chongqing University (project No. 201406150044).

Experiment

400g metal was prepared by mixing the industrial pure iron and ferrovanadium and the composition was shown in Table I.

Table I. Composition of vanadium containing hot metal

Composition	C	V	Si	Mn	S
wt%	3.53	0.73	0.079	0.40	0.05

The metal was loaded in a corundum crucible with a graphite crucible outside heated in a $MoSi_2$ electric resistance furnace. The experimental apparatus is shown in Fig. 1.

The blowing was began when the temperature of the melt increased to an aimed temperature (1340℃), and lasted for 10min. After confirming the complete melting and homogenization of the iron sample, the corundum lance was immersed into the melt to a depth of half the total height of the bath. Total flow of the CO_2 and O_2 mixed blowing was controlled at 0.6L/min. and with varied proportions of CO_2 as 0%, 20%, 40%, 60%, 80% in the mixed gas, the rest of which is pure oxygen.

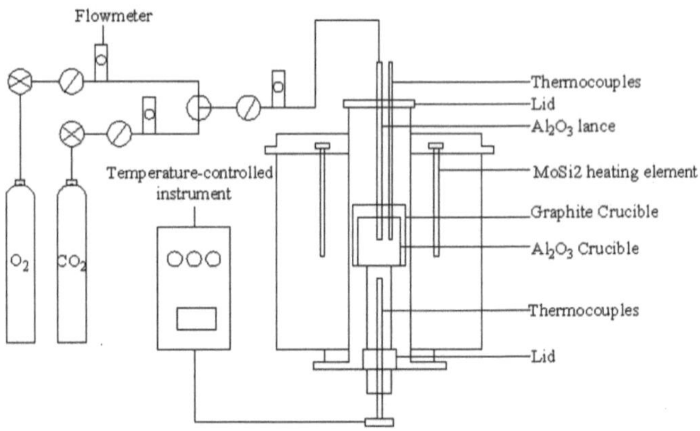

Fig. 1. Schematic drawing of experimental apparatus

Results and Discussion

Chemical Analysis

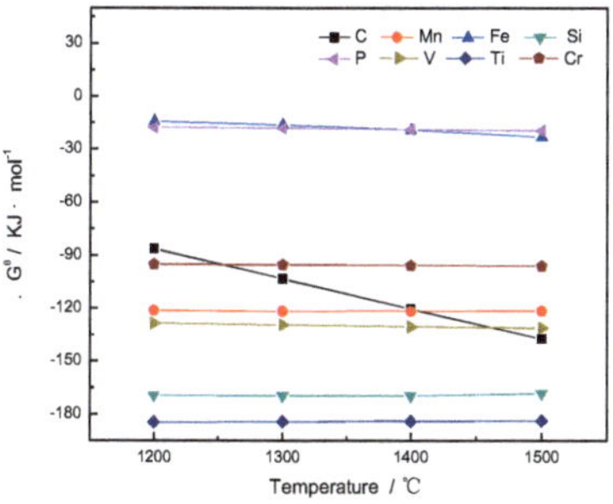

Fig. 2. ΔG^0-T curve of the reactions of elements and CO_2

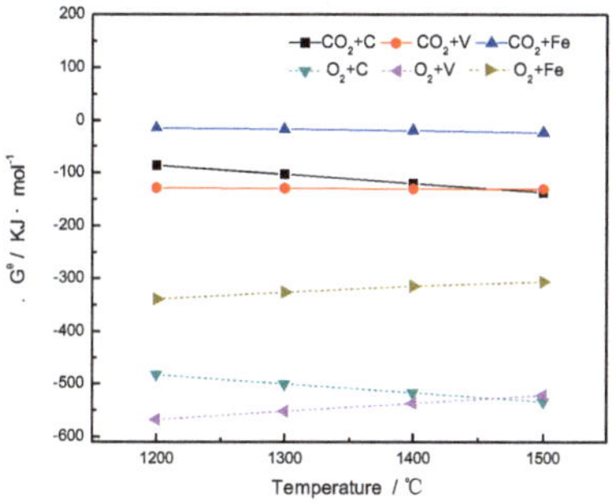

Fig. 3. ΔG^0-T curve of the reactions of elements and O_2 or CO_2

As shown in Fig. 2, from thermodynamic calculation by FactSage, it is possible to use CO_2 as an oxidizer for metallurgy process because the elements' ΔG^0 value in vanadium extraction temperature are all less than zero. It is notable that C and Fe will be oxidized by CO_2 and the reactions are endothermic from this figure.

Fig. 3 shows the primary elements (C, V, Fe) of vanadium containing hot metal. It can be seen in this figure that CO_2 is a weaker oxidizing agent compared to O_2. With the varied proportions of CO_2 in the mixed gas, the oxygen partial pressure in the gas injected for vanadium extraction is different which can be further confirmed by chemical analysis.

As shown in Table II, by lab-scale experiment blowing with different CO_2 proportions as 0%, 20%, 40%, 60%, 80% in the mixed gas, the rest of which is pure oxygen, the elements in hot metal are analyzed by chemical analysis. Because mixed gas blowing with high CO_2 proportions has low oxidize efficiency and it is not proper for steel making[8], we only carry out long time blowing experiment for pure oxygen and 20% CO_2 in the mixed gas, moreover, the rest of experiments are blowing 540s to focus on early and middle stage of vanadium extraction process.

Table II. Results of chemical analysis for iron samples (mass%)

CO_2 proportion (%)	C	V	Si	Mn	S	time(s)
0	2.31	0.023	0.013	0.18	0.049	1200
20	2.23	0.049	0.014	0.18	0.049	1200
40	3.10	0.43	0.041	0.32	0.040	540
60	2.93	0.57	0.054	0.30	0.050	540
80	2.93	0.53	0.055	0.33	0.047	540

Fig. 4 shows the carbon in the molten iron decrease continuously with blowing time. It can be seen that the more proportion of V_{CO2}/V_{CO2+O2} blowing, the lower carbon is exist in the molten iron.

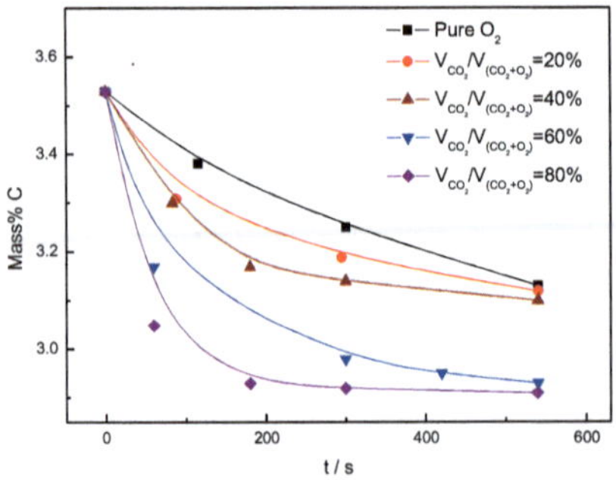

Fig. 4. The carbon in molten iron changes with blowing time.

According to other researchers study[9], decarburization reaction of hot metal is influenced synthetically by the following reactions especially the amount of CO_2 in the mixed gas is large. So the carbon in the molten iron is decreased with CO_2 proportions increases.

$$[C]+CO_2=2CO \qquad \Delta G^0=34580-30.95T(J/mol) \tag{1}$$

$$[C]+[O]=CO \qquad \Delta G^0=-17166-42.5T(J/mol) \tag{2}$$

$$[O]+CO=CO_2 \qquad \Delta G^0=-131945+87T(J/mol) \tag{3}$$

Fig. 5 shows the vanadium in the molten iron decrease continuously with blowing time in pure oxygen or 20% CO_2 in the mixed gas blowing. Compared with conventional vanadium extraction process, blowing 20% CO_2 mixed gas removed pretty much the same [V] in molten steel. Therefore, it can meet the demand of industry standard which demands the content change rate of V is more than 90%.

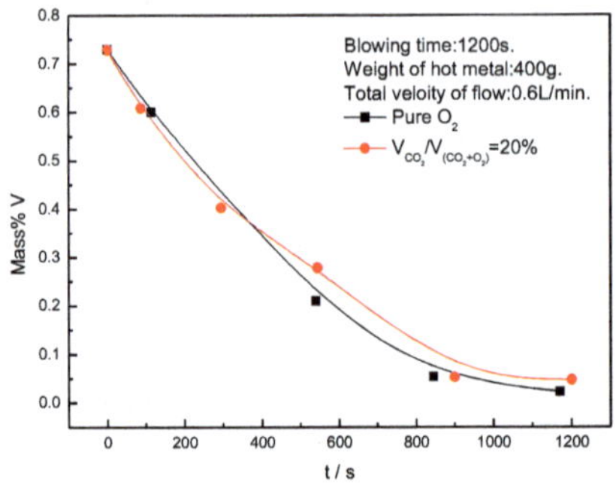

Fig. 5. The vanadium in the molten iron change with blowing time.

CO_2 Cooling Effect

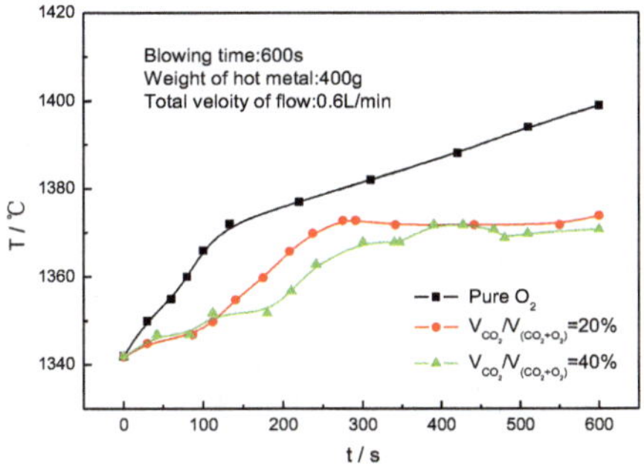

Fig. 6. Changes of molten iron temperature with blowing time

Fig. 6 shows how the temperature of molten iron changes with the blowing time in different CO_2 proportion. From the figure, it can be seen that the bath temperature with pure O_2 blowing keep on increasing with blowing time while the mixed gas contained 20% CO_2 and 40% CO_2 increased slowly and tend to be constant after blowing for 300s.

The temperature of the iron bath can be influenced by CO_2 which is mainly because reactions below are endothermic reactions.

$$[C]+CO_2=2CO \quad \Delta G^0=34580-30.95T \quad \Delta H_C^0=11600\ kJ/kg \tag{4}$$

$$[Fe]+CO_2=(FeO)+CO \quad \Delta G^0=30120-30T \quad \Delta H_{Fe}^0=720\ kJ/kg \tag{5}$$

$$(FeO)=[Fe]+[O] \quad \Delta G^0=459400-87.45T \quad \Delta H_{FeO}^0=4247\ kJ/kg \tag{6}$$

The calculation on heat absorption on introduction of CO_2 can be expressed through the following equation:

$$Q_{CO_2} = Cp_{CO_2} \times \Delta T + (FeO\%) \times \Delta H_{Fe}^0 + \frac{1}{2(CO\%)} \times \Delta H_C^0 + \omega(FeO\%) \times (\frac{56}{72}) \times \Delta H_{FeO}^0 \tag{7}$$

Where Q_{CO2} is cooling capacity of CO_2; and C_{CO2} is specific heat capacity of CO_2 predicted by FactSage; and the item $(FeO\%) \times \Delta H_{Fe}^0$, $\frac{1}{2(CO\%)} \times \Delta H_C^0$, and $\omega(FeO\%) \times (\frac{56}{72}) \times \Delta H_{FeO}^0$ are chemical heat caused by reactions above respectively.

Compared with the conventional coolant, cold-bonded pellet and the waste slag, used in Pan steel which average composition is shown in Table III. And the calculation on heat absorption is shown in equation below.

Table III. Results of average composition for Pan steel coolant (mass%)

Coolant	V_2O_5	TiO_2	CaO	SiO_2	FeO	Fe_2O_3	MnO	P_2O_5	Cr_2O_3	MgO	Al_2O_3
Cold-bonded pellet	0.278	2.805	0.602	7.965	35.79	39.45	0.506	0.134	0.065	0.981	3.275
The waste slag	17.29	10.86	0.561	13.69	41.94	0.50	6.110	0.183	3.070	2.980	3.290

$$Q_{coolant} = Cp_{coolant} \times \Delta T + \omega(Fe_2O_3\%) \times (\frac{112}{160}) \times \Delta H_{Fe_2O_3}^0 + \omega(FeO\%) \times (\frac{56}{72}) \times \Delta H_{FeO}^0 \tag{8}$$

By theoretical analysis above, it is known that CO_2 as coolant posses remarkable cooling ability from Fig. 7 Besides, by use of CO_2, the temperature and compositions of the liquid steel in the converter can be easily controlled. Since gas coolant can be continuously added, the process for converter vanadium extraction can be simplified. On the other hand, the reactant of CO_2 as coolant is CO which can raise the quality of converter gas and reduce the composite energy consumption of the converter steelmaking.

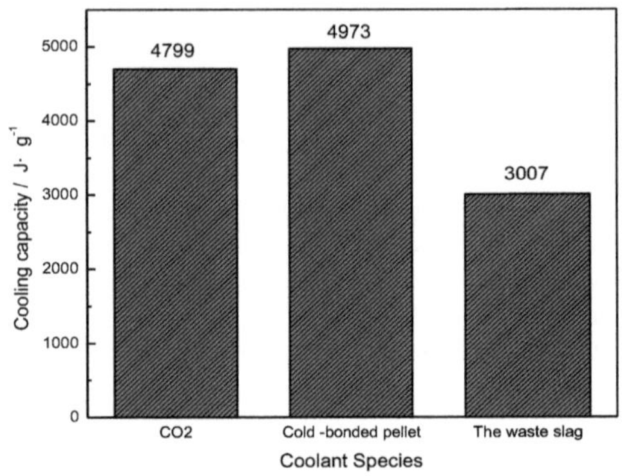

Fig. 7. Comparison with different coolants cooling capacity

Conclusion

Based on the theoretical analysis and lab-scale experimental study above, the following conclusions can be obtained.

(1)The carbon in the molten iron decrease continuously with increasing the proportion of V_{CO2}/V_{CO2+O2} in mixed blowing process. Compared with pure oxygen vanadium extraction process, blowing V_{CO2}/V_{CO2+O2}=20% mixed gas removed pretty much the same [V] in molten steel and can meet the demand of industry standard.

(2)The test of temperature control in CO_2 and O_2 mixed blowing vanadium extraction process shows that the bath temperature has decreased as increasing the proportion of V_{CO2}/V_{CO2+O2}. The cooling capacity of CO_2 is 4799kJ/kg.

References

1. Grimston M. C. et al., “The European and Global Potential of Carbon Dioxide Sequestration in Tackling Climate Change,” *Climate policy*, 2 (2001), 155-171.

2. Wang K. et al., “Scenario Analysis on CO2 Emissions Reduction Potential in China's Iron and Steel Industry,” *Energy Policy*, 35(4) (2006), 2320–2335.

3. Hashimoto K. et al., “Materials for Global Carbon Dioxide Recycling,” *Corrosion Science*, 2 (2002), 371-386.

4. Worrell, Ernst, Lynn Price, and Nathan Martin. "Energy Efficiency and Carbon Dioxide Emissions Reduction Opportunities in the US Iron and Steel Sector," *Energy*, 26(5) (2001): 513-536.
5. S. Hornby Anderson and D.R. Urban. "Cost and Quality Effectiveness of Carbon Dioxide in Steel Mills", *Electric Furnace Conference Proceedings*, (1989), 125-142.
6. Toshie T. et al., "Exploration and Research of Removal of Molten Silicon and Carbon by Blowing Weak Oxidation Gas," *Tetsu-to-Hagane*, 23 (2008), 23-27.
7. Moskalyk, R.R., and A.M. Alfantazi. "Processing of Vanadium: A Review," *Minerals Engineering*, 16(9) (2003), 793-805.
8. Yi C. et al., "Experimental Research on Reducing the Dust of BOF in CO_2 and O_2 Mixed Blowing Steelmaking Process," *ISIJ international*, 49(11) (2009), 1694-1699.
9. Nomura H, Mori K. "Kinetics of Decarburization of Liquid Iron with High Concentration of Carbon," *Tetsu-to-hagané*, 57(9) (1971), 1468.

EPD Congress 2015
Edited by: James Yurko, Antoine Allanore, Laura Bartlett, Jonghyun Lee, Lifeng Zhang, Gabriella Tranell, Yulia Meteleva-Fischer, Shadia Ikhmayies, Arief Suriadi Budiman, Prabhat Tripathy, and Guy Fredrickson
TMS (The Minerals, Metals & Materials Society), 2015

EFFECTS OF POWER ULTRASOUND ON PRECIPITATION PROCESS OF SODIUM SILICATE SOLUTIONS

Tiepeng Li[2)], Jilai Xue[1, 2)], Wenbo Luo[2)], Jun Zhu[1, 2)]*

1) State Key Laboratory of Advanced Metallurgy, University of Science and Technology Beijing, Xueyuan Road 30, Beijing 100083, PR China
2) School of Metallurgical and Ecological Engineering, University of Science and Technology Beijing, Xueyuan Road 30, Beijing 100083, PR China
* Corresponding author: Jun Zhu, E-mail: zhujun@metall.ustb.edu.cn

Keywords: Sodium silicate solution, precipitation process, power ultrasound

Abstract

Sodium silicate solutions for producing silica have been ultrasound treated to enhance the precipitation process in laboratory scale. Higher precipitation rate was found with ultrasound treated solutions than those without ultrasound. The obtained SiO_2 powder products were characterized using XRD, SEM, BET and laser particle size analyzer. The particle size distribution and the specific surface area of the powder products varied with the power input level and the processing time of the ultrasonic treatment. The results indicate that the efficiency of the precipitation process can be improved optimally when the ultrasound is conducted into the solutions before the nucleation stage.

Introduction

Silica is one of the most complex and most abundant families of materials, existing both as several minerals and being produced synthetically. Applications range is from structural materials to microelectronics to components used in the food industry. There are two main processes as gas phase method and liquid phase method currently used for producing silica [1, 2]. In this work, the liquid phase method was adopted to recovery silica and the effect of the ultrasound on the silica products were investigated.

Precipitation process is important for both product quality and productivity in silica production. In recently years, many research focus on increasing the silica precipitation rate and improving its quality. Mu et al [3] investigated the optimum precipitation conditions including reaction temperature, reaction time, molecular ratio of SiO_2/Na_2O, and stirring speed, which result in the SiO_2 extraction rate over 93%. However, the SiO_2 products have the too big particle size which lead to the small specific surface area. Liu et al [4] studied the SiO_2 higee precipitation method to generate silica powder with average particle size 15-20nm, but this method is difficult to industrialize because of great cost and continuous manufacturing.

In order to obtain the high quality SiO_2 by an economical approach, the ultrasound irritation is involved into the precipitation process [5]. The effects of ultrasound are related to ultrasound power level and frequency [6, 7], Na_2SiO_3 concentration and SiO_2/Na_2O molecular ratio of solution, reaction temperature and the seed crystal amount [8].

The aim of this work is to find out the appropriate ultrasound power level and operation time, which positively affect the precipitation process as well as obtaining high quality silica product.

In this paper, the results on ultrasound aided precipitation process of sodium silicate solution are presented, and the precipitation silica product is also characterized.

Experimental

Materials and Chemicals

The sodium silicate solution with constant SiO_2/Na_2O molecule ratio 3.37, SiO_2 27.82% and Na_2O 8.51% was purchased from chemical reagent company in Beijing, whose SiO_2/Na_2O molecule ratio and Na_2SiO_3 concentration was changed by adding sodium hydroxide(CR, Sinopharm Chemical Reagent Co., Ltd) and deionized water (Lab homemade). CO_2 gas (99.5%) used to conduct carbonation precipitation process was provided by University of Science and Technology Beijing.

Analysis Methods

The electronic scales (BS 124S, Sartorius, 0-120 g, Germany) was used to measure the weight of experimental materials and product. The scanning electron microscope (JSM-6510A, JEOL, Japan) analyze morphology of silica powder product. X-ray diffraction analyzer (RigakuD: MAX-RB12KW, Scanning range 10-100°, Scanning rate 0.02°/sec, Cu (40kV, 40mA), RIGAKU, Japan) was used to estimate crystallographic phase characteristics of product. The particle size was measured by laser particle size analyzer (LMS-30, range 0.1-1000 μm, Seishin, Japan). The specific surface area of SiO_2 product was measured by BET (QuadraSorb SI, Quantachrome,).

Experimental Process

300 mL sodium silicate solution with different SiO_2/Na_2O molecule ratio and Na_2SiO_3 concentration in every experiment was added into a beaker in water bath with a certain stirring speed. The ultrasound apparatus is made from Ningbo Scientz biotechnology Co., Ltd.. Ultrasonic operation was conducted with the constant interval of 2 s-on and 3 s-off for various irritation time and solution temperature. The carbon dioxide gas was added into solution by tube for the precipitation process at different temperatures and stirring speed. The pH of the solution was monitored by pH meter every 5 minutes. The carbonation precipitation process was terminated when the pH of the reaction mixture reach to a predetermined value, then the mixture was subjected to filtrate to separate the silicic acid precipitation from the solution, which was calcinated to obtain silica at 120 °C for 24 h. The silica powder was washed several times by deionized water and dried to determine the content and other physical property.

Results and Discussion

Orthogonal Experiment

The orthogonal experiment of carbonation precipitation were investigated, which include 6 parameters as Na_2SiO_3 concentration, molecule ratio of SiO_2/Na_2O, CO_2 flow rate, temperature, stirring speed and solution pH and each parameter has 5 levels as shown in Table I. The result analysis of orthogonal experiment is presented in Table II, which indicates that the optimal operation conditions are Na_2SiO_3 concentration 110g/L, molecule ratio of SiO_2/Na_2O 3, CO_2 flow rate 500 mL/min, temperature 60 °C, stirring speed 500 rpm and solution pH 7.5. The effect of every conditions on SiO_2 precipitation rate in order of decreasing are Na_2SiO_3 concentration, solution pH, stirring speed, CO_2 flow rate, molecule ratio of SiO_2/Na_2O and temperature.

Table I. Factor and Level of the Orthogonal Experiment

Level	Concentration /g·L^{-1}	Molar ratio	Flow rate /ml·min^{-1}	Temperature °C	Stirring speed /rpm	pH
1	10	1	50	25	0	9
2	20	1.5	100	40	100	8.5
3	40	2	200	60	200	8
4	70	2.5	500	80	500	7.8
5	110	3	1000	90	1000	7.5

Table II. Visual Analysis of Orthogonal Experiment

Average 1	0.458	0.649	0.753	0.679	0.591	0.493
Average 2	0.706	0.612	0.56	0.702	0.641	0.594
Average 3	0.778	0.724	0.633	0.814	0.761	0.803
Average 4	0.813	0.792	0.839	0.743	0.875	0.85
Average 5	0.867	0.846	0.838	0.685	0.754	0.877
Range	0.409	0.234	0.279	0.135	0.284	0.384

Precipitation Rate and Efficiency under Ultrasound treatment

The carbonation process was composed of precipitation process and pre precipitation process. The different mode of the ultrasound treatment made different effect on the precipitation rate and time. Therefore, carbonation process was ultrasound treated pre precipitation process, under precipitation process and the whole carbonation process.

Table III. Precipitation Rate and Precipitation Time with Varying Ultrasound Mode

Processing mode	Precipitation rate /%	Precipitation time / min
Without ultrasound	89.34	280
Precipitation with ultrasound	93.33	265
Pre precipitation with ultrasound	89.00	160
Whole process with ultrasound	83.31	295

In Table III, higher precipitation rate was found with ultrasound treated solutions than that without ultrasound. The precipitation time was reduced to 160 min, and the efficiency was apparently improved.

In Figure 1, four curves of pH are given as a function of precipitation time with the ultrasound mode was different under the optimal experimental conditions.

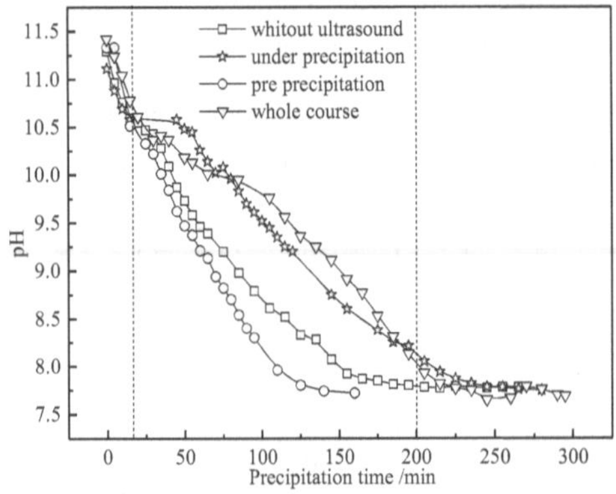

Figure 1. Effects of the ultrasound on the pH values with ultrasound power of 285 W

The possible reactions can be described as follows:

$$CO_2 + H_2O \rightarrow CO_3^{2-} + 2H^+ \tag{1}$$

$$H^+ + OH^- \rightarrow H_2O \tag{2}$$

$$2H^+ + SiO_3^{2-} \rightarrow H_2SiO_3 \tag{3}$$

The value of pH will change with processing of these reactions involving H^+. At the first 20 minutes, the values of pH were the same in the four curves, where the reactions can be (1) and (2); the ultrasound had no effect on these reactions. And then the reaction (3) was occurred and the curves were different because of the different ultrasound mode. And eventually, all of the values of pH reached a constant value abut 7.5. By comparison, when the reaction mixture was ultrasound treated pre precipitation, the precipitation time can reduce about 50 minutes, and the efficiency was improved.

The fact that the ultrasound can improve the precipitation rate means a potential of increasing industrial productivity of SiO_2. With implementation of this technology, the process of precipitation could be shortened with savings in terms of operation costs.

Effects of Ultrasound on SiO_2 Product

After the carbonation process, the residual liquid was removed by filtration with water pump. The product was obtained after further treatments by washing with water, and drying in oven. In the drying process, the SiO_2 appeared:

$$H_2SiO_3 \rightarrow SiO_2 + H_2O \tag{4}$$

The crystalline phases of the product with different ultrasound mode were measured by X-ray diffraction analyzer. In Figure 2, four curves were overlapped. The result shows that the ultrasound has no effect on the structures and the crystal phase of the SiO_2 products.

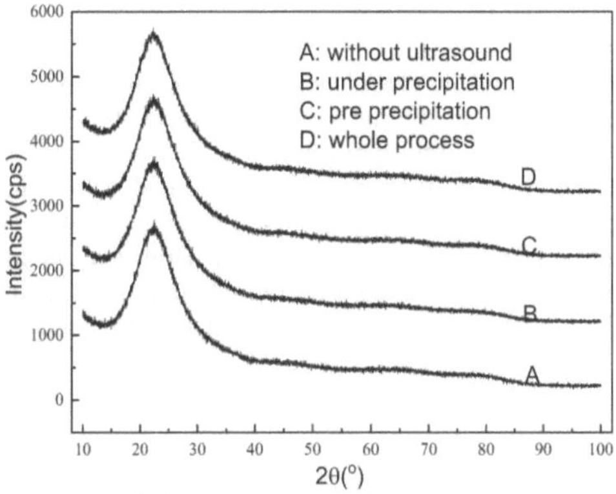

Figure 2. XRD pattern of SiO_2 products with ultrasound power of 285W at 60 °C

Figure 3 is the SEM microphotographs of silica from sodium silicate solution with and without ultrasound, respectively, which both had the identical conditions for sodium silicate solution and precipitation parameters. It is obvious that there are more tiny nuclei on the surface of seeds with ultrasound (b, c, d) than those without ultrasound (a). It means that ultrasound can promote secondary nucleation during the precipitation process.

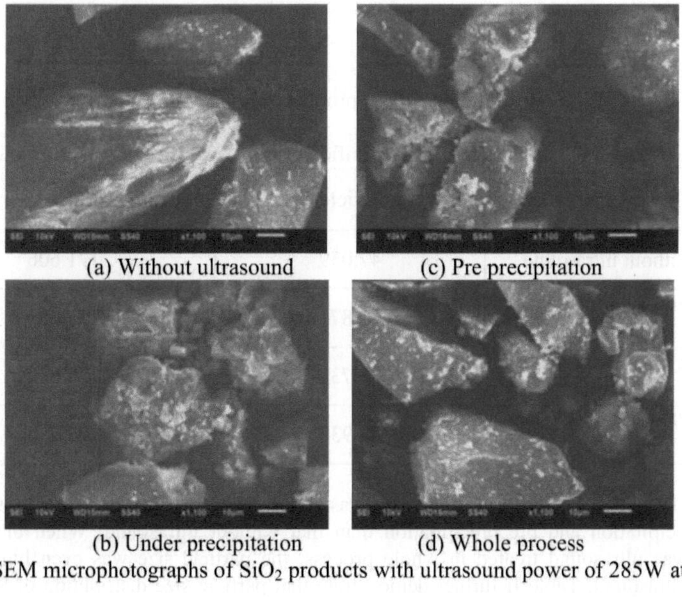
(a) Without ultrasound (c) Pre precipitation

(b) Under precipitation (d) Whole process

Figure 3. SEM microphotographs of SiO_2 products with ultrasound power of 285W at 60 °C

With continuous agitation, the newly formed nuclei could get off the particle surface and enter into the sodium silicate solutions, so that they become secondary nuclei in the solution. The secondary nuclei may act as seeds for further carbonation of the solution and thus increase the precipitation rate.

Figure 4 shows the curves of particle size distribution of the products. Compared with the blank experiment, the volume fraction of the particle size from 100 μm to 350 μm with ultrasound pre precipitation was bigger. Because of the ultrasound treatment, the nucleation ratio of the SiO_2 increased. And the agitation of the ultrasound had an effect on the growth of crystals. On the other hand, the particle size of the product with ultrasound under precipitation was smaller. The largest fraction of the particle under this investigation was one with about 62 μm in size. The agitation of the ultrasound made the particle size smaller. However, the mixture solidified in the carbonation process with ultrasound treatment during whole process, and the ultrasound had no effect on the process. The particle size was even bigger than that without ultrasound.

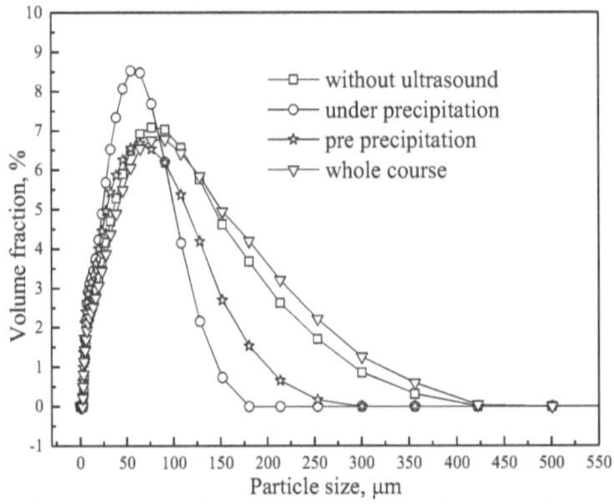

Figure 4. Particle size distribution of SiO_2 products with ultrasound power of 285 W at 60 °C

Table IV. Average Particle Size and Specific Surface Area of Different Ultrasound Mode

Processing mode	Average particle size / μm	Specific surface area /(m^2/g)
Without ultrasound	47.059	171.606
Precipitation with ultrasound	32.873	157.084
Pre precipitation with ultrasound	31.738	172.039
Whole process with ultrasound	47.939	203.035

In Table IV, smaller average particle size was found when the process was ultrasound treated under precipitation and pre precipitation than that without ultrasound. When the carbonation process was ultrasound treated in whole process, the particle size was even bigger than that without ultrasound. Table II further demonstrate that particle size distribution of SiO_2 can vary with ultrasound mode. On the other hand, higher specific surface area was found when the process was ultrasound treated pre precipitation and in whole process than that without ultrasound. The main effect factors of ultrasonic for the higher specific surface area was related to the action of ultrasonic cavitation. Ultrasound inhibited the growth and coalescence of crystals.

To get the optimal ultrasound mode, the precipitation rate, the efficiency, the particle size and the specific surface area should be considered. Therefore, the optimal ultrasound mode was that the carbonation process was ultrasound treated pre precipitation.

Optimal Ultrasound Conditions

The conditions for the ultrasound, such as ultrasound power and treatment time, were of great importance to the carbonation process. The experiment parameters were designed and the resulting data were shown in Table V.

Table V. Experiment for the Optimal Ultrasound Conditions

Ultrasound power / W	Treatment time / min	Precipitation rate / %	Average particle size / μm	Specific surface area /(m2/g)
142.5	5	91.14	40.301	190.240
142.5	10	88.76	32.511	181.836
142.5	15	91.06	39.944	179.245
285	5	92.66	35.486	203.212
285	10	90.93	40.510	210.859
285	15	89.00	42.211	172.039
427.5	**5**	**95.63**	**33.993**	**214.017**
427.5	10	92.93	48.094	188.496
427.5	15	90.11	46.458	197.890

The optional conditions of the ultrasound could be found through comparing the precipitation rate, average particle size and the specific surface area of the product. When the ultrasound power was 427.5 W and the treatment time was 5 min, the smallest average particle size was 33.993 μm, the highest precipitation rate was 95.63%, and the highest specific surface area of the product was 214.07 m^2/g.

Figure 5 illustrates patterns of particle size distribution against precipitation time and ultrasound power. The reaction mixture under the optimal conditions was ultrasound treated pre precipitation process. The ultrasound power and the treatment time were different.

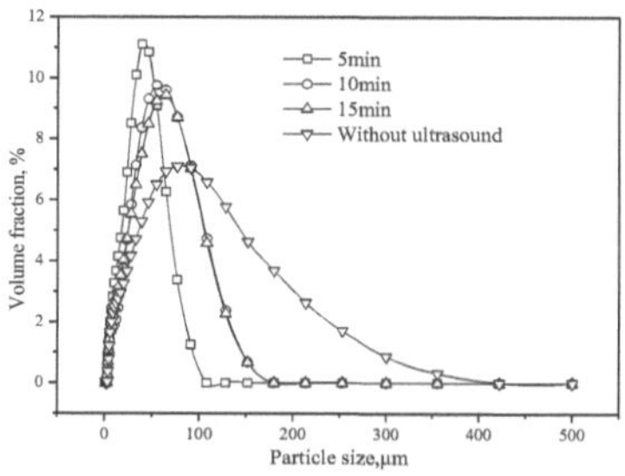

(a) Ultrasound input power of 427.5 W

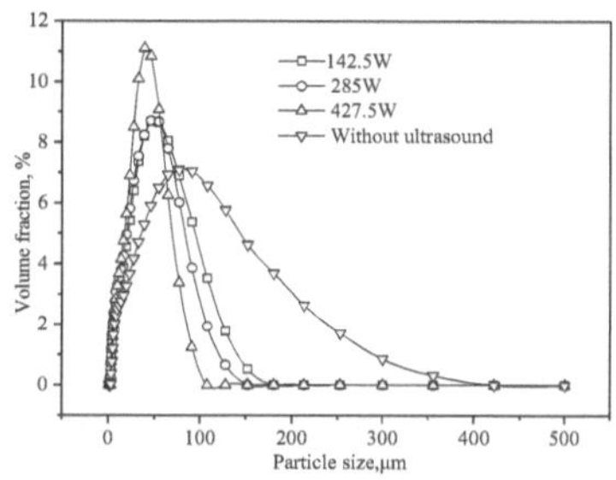

(b) Treatment time of 5 min

Figure 5. Particle size distribution of the products with ultrasound pre precipitation at 60 °C

Figure 5 further demonstrate that particle size distribution of SiO_2 can vary with ultrasound power level and treatment time. The largest fraction of the particle with ultrasound is smaller in size than that without ultrasound. The optional ultrasound power was 427.5W and the optional treatment time was 5 minutes.

Conclusions

1. Precipitation rate is improved in the process with ultrasound, and the highest precipitation rate is 95.63%, which means a potential of increasing industrial productivity of SiO_2 and savings in terms of operation costs.

2. More tiny secondary nuclei were found on the surface of silica with ultrasound, and the major part of the particle smaller in size and the specific surface area increases with ultrasound; the optional ultrasound conditions are with ultrasound of input power of 427.5 W and the treatment time of 5 min.

Acknowledgements

Support from Research Fund of State Key Laboratory of Advanced Metallurgy (No. 41603011) is gratefully acknowledged.

References

[1] Li Suying, Qian Haiyan, "Preparation methods and application actuality of silica white," *Inorganic Chemicals Industry*, 2008, 40 (1): 8-11.

[2] Z.J. Zhang, "research on extraction of alumina and other useful resources from high aluminum fly ash," (Ph.D. thesis, northwest university, 2007.5).

[3] Mu Wenning, Zhai Yuchun, Liu Yan, "Extraction of silicon from laterite-nickel ore by molten alkali," *The Chinese Journal of Nonferrous Metals*, 2009, 19 (3): 570-575.

[4] Liu Hai-di, Jia Hong, Guo Fen, et al, "Preparation of silica SiO_2 utral-fine powder by higee technology," *Inorganic Chemical Industry*,2003,35(1): 13-15.

[5] Jilai Xue, Shaohua Li, Jun Zhu, Baoping Song, "effects of power ultrasound on precipitation process of sodium aluminate solutions," *Light Metals 2006*, 169-172

[6] J. Liu et al, "Effect of Ultrasound Frequency on the Precipitation Process of Supersaturated Sodium Aluminate Solution," *The Chinese Journal of Process Engineering*, 4(2) (2004), 130-135.

[7] G. Ruecroft, D. Hipkiss and M. Fennell, "Improving the Bayer Process by Power Ultrasound Induced Crystallization of Key impurities," *Light Metals* 2005, 163-166.

[8] J. Zhao et al, "Effect of the Reaction Conditions on the Enhancement of $Al(OH)_3$ Precipitation Seeded from Sodium Aluminate Solution under Ultrasound," *Acta metallurgical Sinica*, 38(2)(2002), 166-170.

EPD Congress 2015
Edited by: James Yurko, Antoine Allanore, Laura Bartlett, Jonghyun Lee, Lifeng Zhang, Gabriella Tranell, Yulia Meteleva-Fischer, Shadia Ikhmayies, Arief Suriadi Budiman, Prabhat Tripathy, and Guy Fredrickson
TMS (The Minerals, Metals & Materials Society), 2015

CO_2 CAPTURE AND CONVERSION USING A COBALT(III) SCHIFF BASE COMPLEX AS A CATALYST AT AMBIENT CONDITIONS

Jun Miao, Jilai Xue, Jun Zhu, Kang Liu

Department of Nonferrous Metallurgy, School of Metallurgical and Ecological Engineering, University of Science and Technology Beijing, 100083 Beijing, China

Keywords: CO_2 Fixation, Cobalt(III) Schiff Base, Zeolite 13X, Cyclic Carbonates

Abstract

CO_2 emissions must be reduced by at least 50 % by 2050, and hence the technical solutions to capture and convert CO_2 into value–added products should be considered. A Cobalt(III) Schiff base complex (Salen–Co(III)) has been investigated as a catalyst for synthesis of cyclic carbonate from CO_2 and epichlorohydrin(ECH) with tetrabutylammonium bromide (TBAB) as co–catalyst. To recover Salen–Co(III) for the next cycling operation, it was immobilized onto zeolite 13X through excessive impregnation method. The immobilized catalyst was characterized using XRD, SEM, BET and ICP–AES techniques. Catalytic tests showed that yield of cyclic carbonate reached 90 % using 0.5 mol % Salen–Co(III) at ambient conditions. The immobilized Salen–Co(III) exhibited better catalytic activity than the homogenous one when used at the first cycle, but the yield decreased by some 20 % after five cycles.

Introduction

CO_2 is considered the major cause for global climate warming, which is often found as common emissions in metallurgical and chemical industries. Great efforts have been made in reducing CO_2 emissions through various technologies such as physical adsorption, chemical fixation, biological transformation, etc. However, CO_2 is also the most inexpensive and renewable carbon resource which can be transformed to value–added products through chemical conversion, for instance, cyclic carbonates formed in the coupling reaction of CO_2 with epoxides. The cyclic carbonates have a wide range of technical and engineering applications including: electrolytes for lithium–ion batteries, monomers for synthesizing polycarbonates, degreasers, paint strippers, lacquers and chemical ingredients and so on [1, 2]. From the viewpoint of green chemistry, the processes for such cyclic carbonates need to be energy–minimum to avoid re–generating more CO_2 than the reduced. Therefore, the technologies enabling the CO_2 transformation at atmospheric pressure and room temperature may satisfy this purpose.

In the past few decades, many catalytic systems, such as alkali metal salts [3, 4], metal oxides [5, 6], quaternary ammonium salts [7, 8], ionic liquids [9, 10], transition–metal complexes [11–15] have been developed for the insertion of carbon dioxide into epoxides. However, these catalysts suffer from low catalytic activity and/or selectivity, low stability, separation difficulty, and requirements for high pressure and/or high temperatures. Recently, the combination of Salen–metal complexes and tetrabutylammonium bromide (TBAB) has demonstrated high catalytic activity at ambient conditions [16, 17]. The cost–effective immobilization of Salen–metal on solid supports demonstrate great potential for application in metallurgical and chemical industries because the solid–phase catalysts are relatively easier to recover and to reuse. For further technical evaluation, more details is needed on the preparation processes of Salen–metal complexes and the CO_2 conversion behaviors of the catalyst complex operating at ambient conditions.

In this work, a Salen–Co(III) complex was chosen as a catalyst in the reactions of coupling CO_2 and epichlorohydrin(ECH) with tetrabutylammonium bromide (TBAB) as co–catalyst, and the complex was immobilized onto zeolite 13X through excessive impregnation method. The reaction time, optimal amount of Salen–Co(III) and TBAB, catalytic reactions, recyclability of immobilized catalyst have been investigated under room temperatures and ordinary atmosphere. The results obtained will be evaluated for a potential technical routine towards developing a CO_2 conversion technology for easy preparation and low–cost applications.

Experimental

Preparation of Catalyst

Material Salen–Co(II)(Purity≥98 %) was purchased from Linyi Zhengxin Fine Chemical Co., LTD, Zeolite 13X (LC) was from Tianjin Guangfu Fine Chemical Research Institute. CO_2 (purity≥99.9 %) was supplied by Beijing HuaNeng Special Gas Co., LTD. TBAB (AR) was from Sinopharm Chemical Reagent Co., LTD. The other chemicals used were of AR quality. Salen=(R,R)–(–)–N,N–Bis(3,5–di–tert–butylsalicylidene)–1,2–cyclohexanediamine.

Synthesis of Homogenous Salen–Co(III) Catalyst 1.00 g of Salen–Co(II) was dissolved in 10 ml of CH_2Cl_2, and 0.18 g of HOAc was added into the solution. Afterwards, air was bubbled into the mixed solution under stirring at 50 °C until solvent was removed (Figure 1). Finally, the solid product was dried in oven at 140 °C overnight and became brown powders (0.89 g).

Synthesis of Heterogeneous Salen–Co(III)/Zeolite 13X Catalyst 1.00 g of Salen–Co(III) was dissolved in 5 ml of CH_2Cl_2. Afterwards, 2.00 g of zeolite 13X was added into solution under stirring at ambient conditions for 3 h. Subsequently, the solvent was filtrated under reduced pressure. Finally, the solid product was dried in oven at 140 °C overnight. The resulting product was a yellow powder (2.27 g). The Salen–Co(III) content was determined to be 14.9 % w/w by ICP–AES.

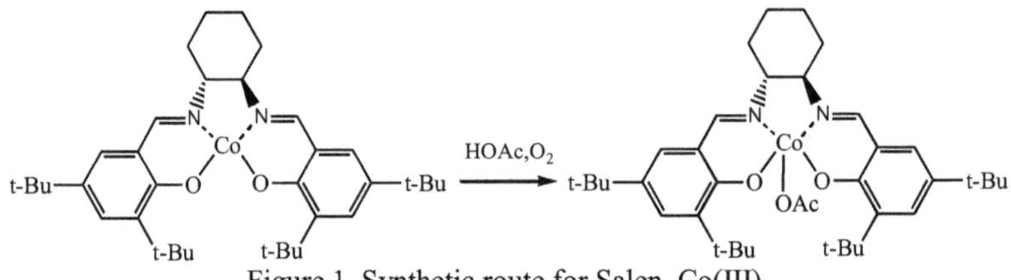

Figure 1. Synthetic route for Salen–Co(III).

Analysis and Characterization

X–ray diffraction (XRD) was operated at Rigaku D/Max–RB 12 kW diffractometer (1.2°/min) using Cu Ka radiation (40 kV, 150 mA). Scanning electron microscopy (SEM) was performed using a JEOL JSM-6480LV scanning electron microscope equipped with an Energy Dispersive Spectrometer (EDS). N_2 adsorption–desorption isotherms were measured with an Autosorb–1C instrument at liquid N_2 temperature. Before measurements, the sample was outgassed at 350 °C for 6 h. The surface area was calculated using the Brunauer–Emmett–Teller (BET) method and the pore size distributions were obtained from the desorption branch of the isotherms, using

Barrett–Joyner–Halenda (BJH) analysis. Salen–metal content was estimated by inductively coupled plasma atomic emission spectroscopy (ICP–AES) analysis using a Perkin–Elmer Optima 7000DV emission spectrometer.

Catalytic Tests

A mixture of epichlorohydrin (25 mmol), tetrabutylammonium bromide (0–2 mol %) and Salen–Co(III) catalyst (0–2 mol %) was vigorously stirred in a gas absorption tube, then CO_2 (99.9 %) was bubbled from the bottom of the mixture without mechanical stirring at atmospheric pressure (Figure 2). After a specified time, a sample of the reaction was analyzed by ^1H NMR spectroscopy to determine the conversion. ^1H NMR spectra were recorded at room temperature on Bruker 400 MHz spectrometers using $CDCl_3$ as the solvents.

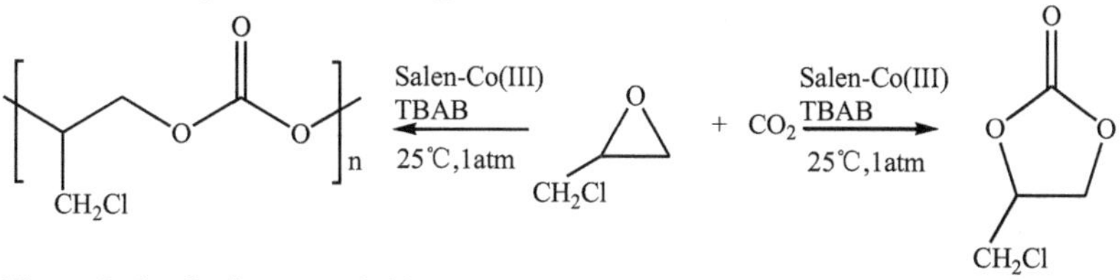

Figure 2. Synthesis route of chloropropene carbonate from epichlorohydrin and CO_2.

Results and Discussion

Microstructure of Catalyst Materials

Figure 3 shows the powder X–ray diffraction patterns of zeolite 13X and immobilized Salen–Co(III). The XRD spectrum of immobilized Salen–Co(III) exhibit a similar pattern to that of zeolite 13X except for slight changes in the intensity of the peaks. These confirm that the framework of zeolite 13X has not destroyed after the preparation process. The changes of relative intensity of the peaks might be correlated to the varied locations of cations which altered by ion–exchange between Co^{2+} and Na^+ during preparation process [18, 19].

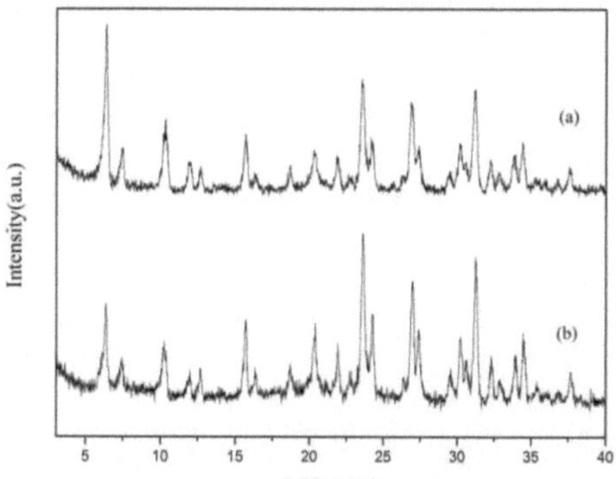

Figure 3. XRD spectra of (a) zeolite 13X, (b) Salen–Co(III)/zeolite 13X.

In Figure 4, SEM image of the Salen–Co(III) powders synthesized alone demonstrate a particle structure in irregular forms. While Salen–Co(III)/zeolite 13X complex material show a microstructure with uniform spheroidal particles, which is similar to the normal structure of zeolite 13X.

Figure 4. SEM images of (1) Salen–Co(III), (2) Salen–Co(III)/zeolite 13X.

N_2 Adsorption/Desorption of Porous Structure

N_2 adsorption/desorption isotherms and pore size distributions for zeolite 13X and Salen–Co(III)/zeolite 13X samples are plotted in Figure 5 and 6, respectively. The isotherm of Zeolite 13X shows higher values than Salen–Co(III)/zeolite 13X within the whole range of relative pressure and a rapid increase in nitrogen uptake with $P/P_0>0.8$. The BET surface area, pore diameter and pore volume of zeolite 13X and Salen–Co(III)/zeolite 13X are presented in Table I. In comparison, a dramatic decrease in surface area and pore volume is observed on Salen–Co(III)/zeolite 13X samples, which can be attributed to part of some open pores being closed after immobilization of the catalyst into the zeolite pores [20].

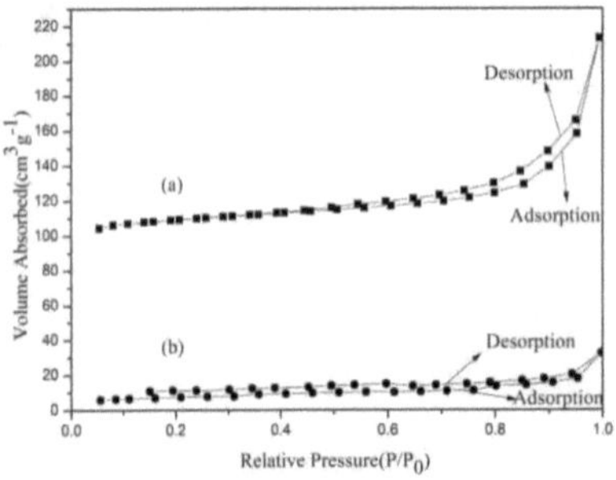

Figure 5. N_2 adsorption/desorption isotherms of (a) zeolite 13X, (b) Salen–Co(III)/zeolite 13X.

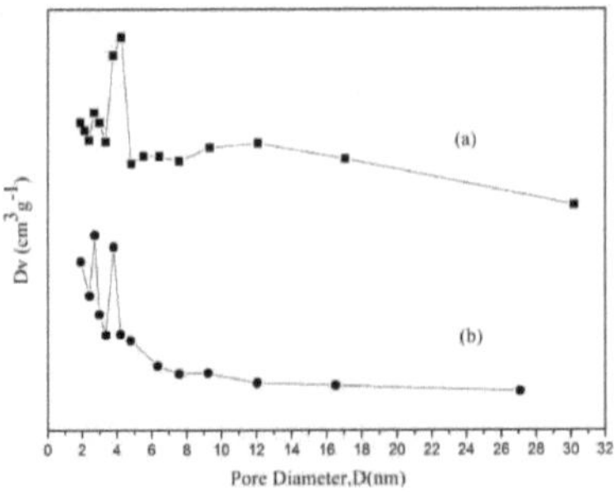

Figure 6. Pore size distributions of (a) zeolite 13X, (b) Salen–Co(III)/zeolite 13X.

Table I. Structural and Textural Parameters of Synthesized Catalyst Samples

Sample	S_{BET} (m^2g^{-1})	Dp (nm)	Vp (cm^3g^{-1})
Zeolite 13X	330	4.25	0.328
Salen–Co/zeolite 13X	24	2.72	0.050

S_{BET} surface area, Dp average pore diameter, Vp total pore volume

Catalytic Properties

For Homogenous Salen–Co(III) Catalyst The catalytic performance of homogeneous catalysts Salen–Co(III) and TBAB were evaluated for the coupling of CO_2 and epichlorohydrin without any solvent. The data of Test 1, 2, 5–7 in Table II indicates that TBAB amount plays an important role in combining with Salen–Co(III). Based on the data of conversion and TOF, the combination of 0.5 mol % Salen–Co(III) and 2 mol % TBAB is the optimal ratio of both catalysts in this study. Test 3, 4, and 5 show that 0.5 mol % of Salen–Co(III) can make obvious increase in conversion rate at a short period together with 2% TBAB addition, while Test 8 and 9 demonstrate that either Salen–Co(III) or TBAB alone may give little or even no catalytic effects recorded under the identical testing conditions. Moreover, Test 4 highlights the very strong synergistic effect happened with a right combination of Salen–Co(III) and TBAB together. It also shows that Salen–Co(III) and TBAB play quite different roles in the catalytic process, as described in literature [21, 22].

Table II. Synthesis of Chloropropene Carbonate Using Different Amount of Catalysts

Tests [a]	Salen–Co(III) (mol %)	TBAB (mol %)	Time (h)	Conversion [b] (%)	Yield [b] (%)	TOF [c] (h^{-1})
1	2	2	24	96	93	1.9
2	1	2	24	92	90	3.8
3	0.5	2	3	32	32	21.3
4	0.5	2	6	62	62	20.6
5	0.5	2	24	93	90	7.5
6	1	1	24	85	82	3.4
7	0.5	1	24	85	82	6.8
8	0	2	6	10	10	-
9	0.5	0	6	0	0	0

[a]All reactions were carried out in the absence of solvent at 25 °C and 0.1 MPa CO_2 pressure.
[b] Conversion is the consumed ECH amount/total ECH amount×100%; selectivity is the amount of chloropropene carbonate/the amount of all product×100%; yield=conversion×selectivity. Conversion and yield were determined by ^1H NMR spectroscopy of the reaction mixture.
[c] TOF is the moles of product/(moles of catalyst •time).

For Heterogeneous Salen–Co(III) Catalyst There are several possibilities of fixation of Salen–Co(III) onto zeolite 13X. Firstly, Salen–Co(III) can be bound to the surface via imine groups which can be protonated and hydrogen bonded to the surface of zeolite. Secondly, the central cobalt ion of the complex can be cationic bonded at ion exchange sites. Finally, complexation of the cobalt by oxygen atoms of zeolite 13X might occur [23]. To test the lifetime and the reusability of the heterogeneous catalyst, a series of experiments were carried out for the model reaction under the following conditions. The solid catalyst was easily recovered by filtration after each reaction cycle and it was directly reused with the same amount of fresh TBAB for subsequent cycles.

As shown in Table III, the original Salen–Co(III) content is 0.9 mol % of epichlorohydrin, where its catalytic activity is as good as 2 mol % homogeneous Salen–Co(III) catalyst. However, the reused catalyst exhibited a decrease in the catalytic activity over five consecutive runs. It could be attributed to the loss of active component immobilized on the surface of zeolite 13X Salen–Co(III) content dropped to be 6 % w/w determined by ICP–AES, and the loss of solid catalyst occurred after each cycle in filtration.

Table III. Recycling Efficiency of Heterogeneous Salen–Co(III) Catalysts

Recycle times	0	1	3	5
Conversion (%)	96	91	73	77
Selectivity (%)	97	95	97	98

Reaction conditions: Salen–Co(III)/zeolite 13X: 1.00 g; Salen–Co(III):0.9 mol %; TBAB: 0.16 g, 2 mol %; time: 24 h; pressure: 0.1 MPa; temperature: 25 °C.

Catalytic Processing Reactions

In summary, Figure 7 presents a series of schematic drawings, step by step, to illustrate the catalytic reactions involved in a cycle of use and reuse of the catalytic complex material prepared in this study. Anion exchange of OAc^- and Br^- provided by tetrabutylammonium bromide proceeds rapidly to form a chemical compound [24] in Step 1, where Salen–Co(III) acts as a Lewis acid coordinating to the epoxide and activating it. Meanwhile, OAc^- or Br^- acts as a nucleophilic reagent to ring–open the epoxide to generate some chemical complex in Step 2. Then, TBAB transforms to tributylamine which reacts reversibly with carbon dioxide to form carbamate salt [25] in Step 3. The activated CO_2 inserting into the chemical complex 2 to produce another chemical complex in Step 4. Subsequently, tributylammonium group displaces from the chemical complex 4 to generate the metal coordinated carbonate in Step 5, and ring-closure forms the cyclic carbonate in Step 6. Finally, Salen–Co(III) separates from the carbonate to regenerate both Salen–Co(III) and TBAB, which can be recovered for reuse.

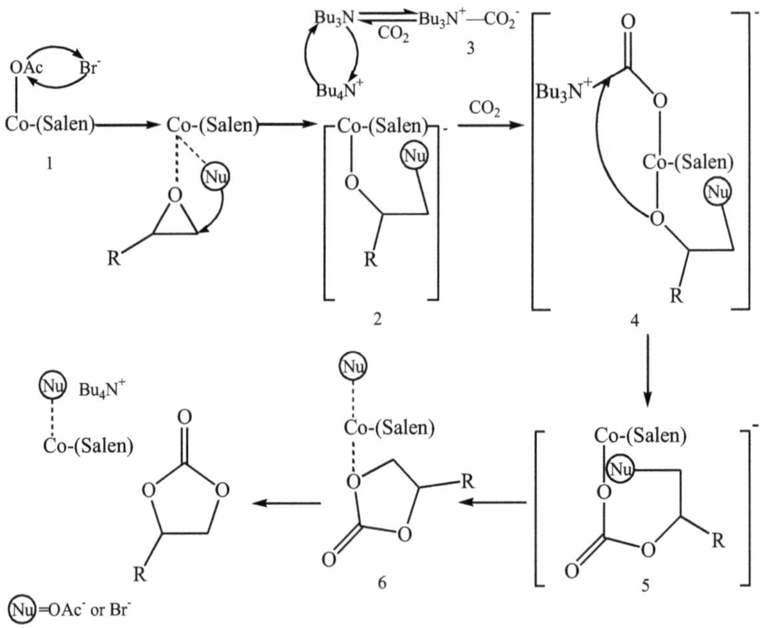

Figure 7. Schematic drawings for catalytic processing reactions of cyclic carbonate synthesis.

Conclusion

1. Cobalt(III)–Schiff based complex has been synthesized and immoblized onto zeolite 13X to form Salen–Co(III)/zeolite 13X catalyst materials. Both homogenous and heterogeneous catalyst materials exhibit high activity for the coupling reactions of CO_2 and epichlorohydrin with tetrabutylammonium bromide (TBAB) as co–catalyst at ambient conditions.

2. The immobilized heterogeneous Salen–Co(III) catalyst is more active than the homogeneous Salen–Co(III). However, the immobilized one shows an activity decrease in the recycling experiments which may be caused by the loss of active component.

3. A right combination of Salen–Co(III) and TBAB together can generate a strong synergistic effect on CO_2 capture and conversion, where both catalysts may play different roles. A better optimization of both catalytic materials and operating parameters with decreasing loss of catalytic materials need further investigation.

Acknowledgements

Part of support from aluminum and carbon industries in China is gratefully acknowledged. Most of analysis work was carried out in Central Laboratory of Metallurgical School at USTB.

Reference

1. A. A. G. Shaikh, and S. Sivaram, "Organic carbonates," *Chemical reviews*, 96 (1996), 951-976.
2. B. Sch. ffner et al., "Organic carbonates as solvents in synthesis and catalysis," *Chemical reviews*, 110 (8) (2010), 4554-4581.
3. N. Kihara, N. Hara, and T. Endo, "Catalytic activity of various salts in the reaction of 2, 3-epoxypropyl phenyl ether and carbon dioxide under atmospheric pressure," *The Journal of Organic Chemistry*, 58 (23) (1993), 6198-6202.
4. P. Ramidi et al., "Synergistic effect of alkali halide and Lewis base on the catalytic synthesis of cyclic carbonate from CO_2 and epoxide," *Chemical Physics Letters*, 512 (2011), 273-277.
5. K. Yamaguchi et al., "Mg-Al mixed oxides as highly active acid-base catalysts for cycloaddition of carbon dioxide to epoxides," *Journal of the American Chemical Society*, 121 (18) (1999), 4526-4527.
6. B. M. Bhanage et al., "Synthesis of dimethyl carbonate and glycols from carbon dioxide, epoxides, and methanol using heterogeneous basic metal oxide catalysts with high activity and selectivity," *Applied Catalysis A: General*, 219 (1) (2001), 259-266.
7. V. Caló et al., "Cyclic carbonate formation from carbon dioxide and oxiranes in tetrabutylammonium halides as solvents and catalysts," *Organic letters*, 4 (15) (2002), 2561-2563.
8. J. Q. Wang et al., "Synthesis of cyclic carbonates from epoxides and carbon dioxide over silica-supported quaternary ammonium salts under supercritical conditions," *Journal of Molecular Catalysis A: Chemical*, 249 (1) (2006), 143-148.
9. J. Sun, S. Fujita, and M. Arai, "Development in the green synthesis of cyclic carbonate from carbon dioxide using ionic liquids," *Journal of Organometallic Chemistry*, 690 (15) (2005), 3490-3497.

10. A. L. Girard et al., "Insights on recyclable catalytic system composed of task-specific ionic liquids for the chemical fixation of carbon dioxide," *Green Chemistry*, 16 (5) (2014), 2815-2825.
11. R. L. Paddock, and S. B. T. Nguyen, "Chemical CO_2 fixation: Cr (III) salen complexes as highly efficient catalysts for the coupling of CO_2 and epoxides," *Journal of the American Chemical Society*, 123 (46) (2001), 11498-11499.
12. C. X. Miao, J. Q. Wang, and L. N. He, "Catalytic processes for chemical conversion of carbon dioxide into cyclic carbonates and polycarbonates," *The Open Organic Chemistry Journal*, 2 (2008), 68-82.
13. T. Chang, L. Jin, and H. Jing, "Bifunctional chiral catalyst for the synthesis of chiral cyclic carbonates from carbon dioxide and epoxides," *ChemCatChem*, 1 (3) (2009), 379-383.
14. A. Ghosh et al., "Cycloaddition of CO_2 to epoxides using a highly active Co (III) complex of tetraamidomacrocyclic ligand," *Catalysis letters*, 137 (1-2) (2010), 1-7.
15. P. Ramidi et al., "Synthesis and characterization of Co (III) amidoamine complexes: influence of substituents of the ligand on catalytic cyclic carbonate synthesis from epoxide and carbon dioxide," *Dalton Transactions*, 42 (36) (2013), 13151-13160.
16. J. Meléndez, M. North, and R. Pasquale, "Synthesis of cyclic carbonates from atmospheric pressure carbon dioxide using exceptionally active aluminium (salen) complexes as catalysts," *European journal of inorganic chemistry*, 2007 (21) (2007), 3323-3326.
17. Y. Xie et al., "Capture and conversion of CO_2 at ambient conditions by a conjugated microporous polymer," *Nature communications*, 2013, 4.
18. K. Li et al., "Highly efficient catalytic oxidation of cyclohexanol with TBHP over Cr-13X catalysts in a solvent-free system," *Journal of Molecular Catalysis A: Chemical*, 387 (2014), 31-37.
19. A. Mobinikhaledi, M. Zendehdel, and P. Safari, "Effect of substituents and encapsulation on the catalytic activity of copper (II) complexes of two tridentate Schiff base ligands based on thiophene: benzyl alcohol and phenol oxidation reactions," *Transition Metal Chemistry*, 39 (2014), 431-442.
20. J. Q. Guan, and J. Liu, "A Copper(II) Schiff base complex immobilized onto SBA-15 silica for selective oxidation of benzyl alcohol," *Transition Metal Chemistry*, 39 (2014), 233-238.
21. M. North, R. Pasquale, and C. Young, "Synthesis of cyclic carbonates from epoxides and CO_2," *Green Chemistry*, 12 (9) (2010), 1514-1539.
22. M. North, P. Villuendas, and C. Young, "Inter-and intramolecular phosphonium salt cocatalysis in cyclic carbonate synthesis catalysed by a bimetallic aluminium (salen) complex," *Tetrahedron Letters*, 53 (22) (2012), 2736-2740.
23. L. Frunza et al., "Host/guest interactions in nanoporous materials I. The embedding of chiral salen manganese (III) complex into mesoporous silicates," *Journal of Molecular Catalysis A: Chemical*, 123 (2) (1997), 179-187.
24. W. M. Ren et al., "Role of the co-catalyst in the asymmetric coupling of racemic epoxides with CO_2 using multichiral Co(III) complexes: product selectivity and enantioselectivity," *Chemical Science*, 3 (2012), 2094-2102.
25. M. North, and R. Pasquale, "Mechanism of Cyclic Carbonate Synthesis from Epoxides and CO_2," *Angewandte Chemie*, 121 (2009), 2990-2992.

EPD Congress 2015
Edited by: James Yurko, Antoine Allanore, Laura Bartlett, Jonghyun Lee, Lifeng Zhang, Gabriella Tranell, Yulia Meteleva-Fischer, Shadia Ikhmayies, Arief Suriadi Budiman, Prabhat Tripathy, and Guy Fredrickson
TMS (The Minerals, Metals & Materials Society), 2015

EFFECT OF LAMINAR COOLING PARAMETERS ON MARTENSITE VOLUME FRACTION AND MECHANICAL PROPERTIES OF HOT ROLLED DUAL PHASE STEEL

Sibel Daglılar, Isıl Kerti, Sinem Yıldırım

Yildiz Technical University, Department of Metallurgical and Materials Engineering, Faculty of Chemistry & Metallurgy, Davutpaşa Campus, 34210, Esenler, Istanbul, Turkey

Keywords: dual phase steel, finishing rolling cooling rate, coiling temperature, laminar cooling system

Abstract

In this study, effect of rolling and coiling temperature in thermo mechanical controlled process (TMCP), on the microstructure and mechanical properties of dual phase (DP) steel according to martensite volume fraction (MVF) rate was investigated. By using various finishing mill exit temperature and cooling conditions that is controlled laminar cooling system, the microstructural evolution of coil improves the mechanical properties with lower costs and higher productivities, in comparison to the heat treatment after rolling. The main part of the paper contains the result of tensile, yield and microstructure properties of DP600 steel produced on hot strip rolling plant and run- out table line. The results show that a higher MVF and an increasing cooling rate after the last deformation raised strength of the DP600 steel. The yield strength and ultimate tensile strength increase by decreasing coiling temperature.

Introduction

Low alloyed multi-phase steels are subject to extensive research efforts especially with regard to automotive applications due to their attractive combinations of mechanical properties. For the automotive/heavy transportation industry, these steels are primarily used for underbody components, which require good press formability and ductility as well as acceptable surface appearance despite in most cases being hidden away. In addition to developing steels that satisfy customer requirements, the additional challenge facing strip steel producers of today is in the production of a hot rolled coil with optimum mechanical properties that satisfy the demands of the final application (i.e. fitness for purpose), whilst causing minimal disruption to rolling practices or increasing costs [1]. Dual-phase steel are low-carbon micro-alloyed steels, characterized by a ferritic multiphase structure (with residual austenite) in which martensite is dispersed. Martensite volume fraction (MVF) is adjusted in the range of 5-20 vol. %. 37 Depending on martensite volume fraction, the tensile strength (Rm) ranges from 500 to about 1000 MPa, and total elongation amounts to 12–34 %. The dual-phase structure of these steels results in adequately low yield stress (Re) and high ultimate tensile strength (Rm), allowing for obtaining sufficiently low Re/Rm ratio. Soft ferrite facilitates the beginning of plastic deformation, while hard martensite increases the strength of steel. The dual-phase structure depends on the chemical composition of the steel, and on thermo-mechanical treatment realized with lower rolling temperatures [2, 3, 4, 5].

Recently, dual phase steel sheet production has been carried out with a thermomechanical treatment during the rolling operations. The required microstructure is obtained in this case with a fine tuning of the process parameters, through a process-design activity and a subsequent engineering process like slab heating temperature, chemical composition, the temperature at exit of finishing, run-out table cooling and coiling temperature. The process-design activity needs to settle the appropriate thermomechanical cycle as a function of the chemical composition of the steel. Controlled rolling (CR) followed by controlled cooling (CC) has been developed during the last decade as one of the thermomechanical-controlled processes for obtaining a good combination of strength, elongation and weldability of dual phase steels in the hot rolled condition [6].

Materials and Experimental Methods

The dual phase steel used in this study was produced industrially by İSDEMİR (Erdemir Group, İskenderun Iron and Steel Co.) as slabs (including melting and casting) with a thickness of 225 mm. Chemical composition of the steels is give in the Table 1. Hot rolled ferrite martensite dual phase steel was produced using simple C-Mn-Si chemistry without adding expensive alloys such as Nb or Mo etc. DP 600 grade steel was aimed to use by automotive wheel manufacturers and chemical composition determined according to manufacturer's specs.

Table 1 - Chemical compositions of the investigated dual phase steel (wt. %).

Steel	C	Mn	P	S	Si	Cu	Ni	Cr	V	Mo	Co	Ti	Al	Nb
DP 600	0,0675	0,9203	0,0354	0,0062	0,2507	0,0433	0,0471	0,658	0,0019	0,0035	0,0071	0,0041	0,0459	0,0007

The experiments were carried out on hot strip mill (HSM) line of İsdemir in Turkey. Slabs were reheated to 1250° C and hot rolled to a thickness 3,7 mm. The rolling and cooling schedule was schematically shown in fig. 1. Stepped cooling processes were employed after the strip was finished rolling at 880° C. After the last deformation step, the strips were cooled below the Ms temperature. First, they were cooled at different cooling rates to the first cooling zone entry temperature (1st CET) in a $\gamma \rightarrow \alpha$ transformation region, until the required volume fraction of ferrite and austenite was obtained. In the first cooling step, laminar cooling water was sprayed on the strip surface, different cooling rate (φ), was applied for strip cooling down from finishing mill delivery temperature (FDT) to ferrite transformation temperature (Ar3). Different cooling rates were used to see the effect on the mechanical properties.

Holding process with air cooling was conducted after first step cooling. Formed ferrite fraction could be controlled by air cooling time at this stage. Temperature decreases almost 30 ° C. During this time a significant change in temperature wasn't observed. Its main purpose is to give time required for the phase transformation.Then second cooling step was applied with higher speed than first step was conducted when adequate ferrite fraction was obtained. Strips was quenched down to martensite formation temperature (Ms) with dense water and coiled at 6 different temperatures. Schematic view of cooling section of ISDEMIR hot mill and cooling curve to plan to obtain are given below (Figure 1).

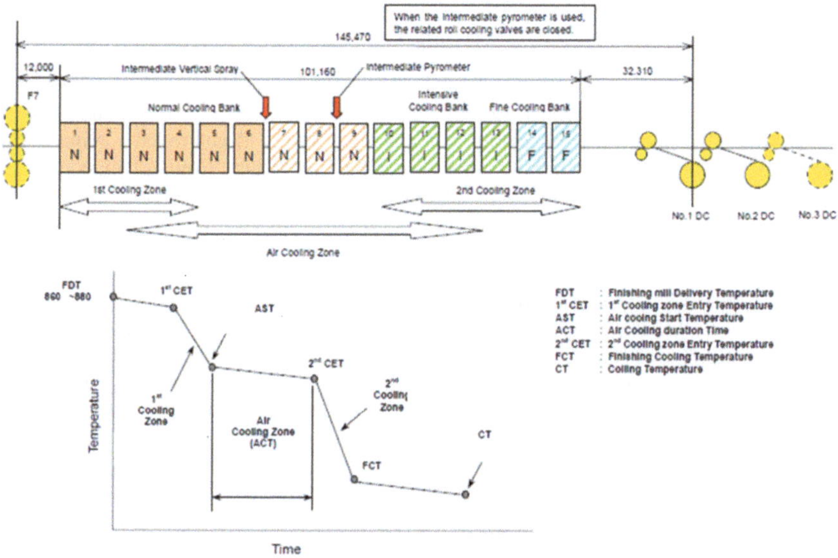

Figure 1. Schematic diagram of rolling and coiling path

As shown in the figure 2. samples taken middle of the coil at the 6 different location and obtained by optical microscope and determined the volume fraction of martesite. Clemex software was used for analysis and calculations by painting method martensitic grains. Tensile test were carried out at room temperature on a Zwick universal tensile testing machine according to EN 10002 -1 standard by 6 samples middle of coil and results were avareged.

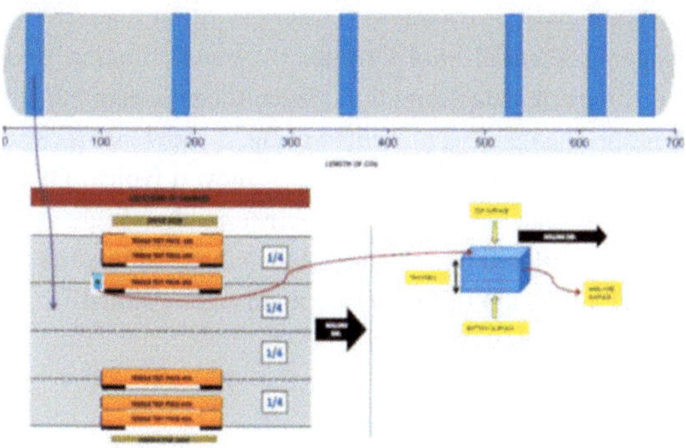

Figure 2. Schematic diagram of sample locations

Results and Discussion

The complexity of processes taking place in hot rolling conditions creates a wide range of possibilities of controlling the microstructure and mechanical properties of thermomechanically treated strips [3]. The principal deciding factor is the ensuring of controlled cooling rate, from finish-rolling temperature in austenite range to coiling temperature. Coiling temperature are the main important parameters that affect the mechanical properties and microstructural evolution of the hot rolled products [6, 7]. The chemical composition of the steel used in this investigation is given in table 1 and all the TMCP parameters were kept constant from reheating of slabs to holding time. The result presented in table 3 shows that coiling temperatures range from 290 °C to 165 °C after the same rolling conditions. High strength and more martensite volume fraction was observed by lower coiling temperature. In order to obtain DP steels with specific volume fractions of ferrite and martensite the appropriate 1st CET temperatures must be determined. There is a certain amount of ferrite (80 – 90 %) and austenite at this temperature. The accelerated cooling of the specimen from 1st CET to below MS results in the transformation of a specific volume fraction of intercritical austenite to martensite. The adjustment of the MVF is possible by changing the 1st CET temperature. Variations of the cooling rate during the $\gamma \rightarrow \alpha$ transformation and the volume fraction of martensite in DP 600 steel are shown in table 3.

Table 3. Effect of coiling temperature on mechanical properties

F6 Temperature °C	Air Cooling Start Temperature °C	Air Cooling Duration Time s	2nd Cooling Start Temperature °C	Coiling Temperature °C	MVF [%]	YS [Mpa]	TS [Mpa]	Elognation %
880	600	5,5	5,5	165	17,3	456	634	25,3
880	650	5,5	5,5	185	15	436	624	26,8
880	680	5,5	5,5	200	12	412	620	27,7
880	690	5,5	5,5	220	10	392	563	28,7
880	710	5,5	5,5	252	7,8	381	554	28,8
880	720	5,5	5,5	290	4,5	362	542	29

Lower 1st CET means longer isothermal hold time for γ transformation, and lesser amount of austenite will be retained for martensite formation. Since there is enougth time, transformation have not been in desired ratio. The result is presented in fig. 2 show that for each steel at constant holding time increasing 1st CET temperatures. All images show a typical DP microstructure with relatively globular martensite islands embedded in a ferrite matrix. Light- colored parts refer to the martensitic phase. The light-colored parts painted by clemex software as shown fig. 2 (e) and percentage were determined by the ratio of martensite in all areas painted.

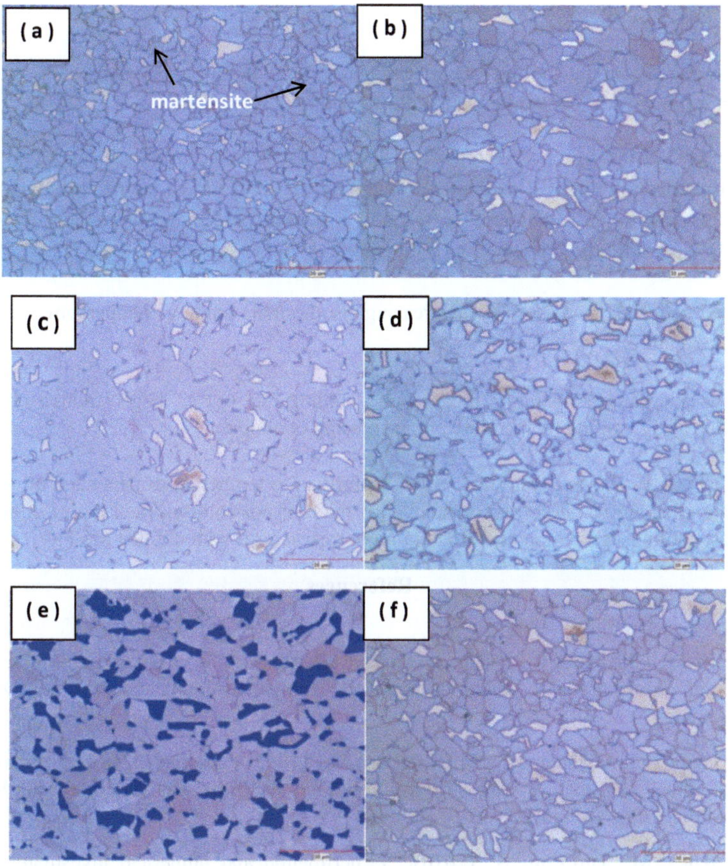

Figure 3. Microstructure of DP steels showing different volume fractions of martensite obtained after second cooling

(a) = 4,5% , (b)= 7,8% , (c) = 10 , (d) = 12 % , (e)=15 % ,(f)= 17,3 %

The results show that a higher MVF and an increasing 1st CET after the last deformation raised strength of the DP600 steel. In the DP steels with high MVF, the amount of dislocations coming in ferrite phase because of volume changes resulted from austenite to martensite transformation is more [6]. The dislocation density in the ferrite and at the F/M interfaces increased with increasing MVF, resulting in a higher fraction of ferrite affected by the martensitic phase transformation. In this way, steel has continuous yield behavior beside strong tensile properties.

Conclusions

1. An increase in the cooling rate after the last TMCP deformation step to TFC during the $\gamma \rightarrow \alpha$ transformation resulted in a higher strength after TMCP.

2. The mechanical properties after TMCP and were affected by the MVF. The mechanical properties after TMCP and were affected by the MVF. Each temperature under the MS according to TTT diagrams correspond a certain percentage of martensite formation. MVF increases by lower coiling temperature

3. YS and TS increases by decreasing isothermal holding temperature and coiling temperature.

4. Lower coiling temperature means rapid cooling rate ($\varphi 1$). High strength obtains with decreasing coiling temperature and increasing cooling rate

5. Test results showed that the $\gamma \rightarrow \alpha$ transformation temperatures depended on the cooling rate after the last deformation step. This made it possible to select the most appropriate fast cooling start temperature to obtain microstructures with defined MVFs.

References

[1] Patel, J., Klinkenberg, C., Hulka, K., "Hot Rolled HSLA Strip Steels For Automotive and Construction Applications", Niobium Products Company GmbH, SteinSrt:28, Düsseldorf, 40210, Germany

[2] De Cosmo, M., Galantucci, L.M., Tricarico, L., (1999). "Design of process parameters for dual phase steel production with strip rolling using the finite-element method", Journal of Materials Processing Technology, 92-93:486-493

[3] Dziedzic, M., Turczyn, S., (2010). "Experimental and numerical investigation of strip rolling from dual phase steel", Archives Of Civil And Mechanıcal Engineering, Vol: 4

[4] Hayat, F., Demir, B., (2009). "The effect of the weld time on dept intensity factor and strength at RSW junctions of commercial DP600 sheet steel" 5. Uluslararası İleri Teknolojiler Sempozyumu (IATS'09), 13-15 Mayıs 2009, Karabuk, Türkiye

[5] Tan, W., et al., "Effects of TMCP Parameters on Microstructure and Mechanical Properties of Hot Rolled Economical Dual Phase Steel in CSP",(2012)., Journal Of Iron And Steel Research, International. 2012, 19(6): 37-41

[6] Niakana, H., Najafizadehb, A., (2010). "Effect of niobium and rolling parameters on the mechanical properties and microstructure of dual phase steels", Materials Science and Engineering, A 527: 5410–5414

[7] Ning, L., et al., (2010). "Effect of Laminar Cooling on Phase Transformation Evolution in Hot Rolling Process", Journal Of Iron And Steel Research, Internnional,17(10): 28-32. 44

EPD Congress 2015
Edited by: James Yurko, Antoine Allanore, Laura Bartlett, Jonghyun Lee, Lifeng Zhang, Gabriella Tranell, Yulia Meteleva-Fischer, Shadia Ikhmayies, Arief Suriadi Budiman, Prabhat Tripathy, and Guy Fredrickson
TMS (The Minerals, Metals & Materials Society), 2015

THE EFFECT OF GRAPHENE ON THE MICROSTRUCTURE AND MECHANICAL PROPERTIES OF ALUMINUM/GRAPHENE PRODUCED BY HPT

Liyuan Zhao[1], Huimin Lu[1*], Zhijiang Gao[1]

[1]Beihang University, School of Materials Science and Engineering;
37 Xueyuan Road; Beijing 100191, China
[*]Corresponding Author: lhm0862002@aliyun.com

Keywords: aluminum/graphene composite, HPT, grain refinement, load transfer, dislocations

Abstract

The aluminum matrix composite reinforced by graphene was successfully fabricated using high pressure torsion (HPT). With only 0.5wt% graphene, the grain size was reduced to 100nm while that of pure Al was about 500nm. The hardness of the composite increased from the center (64HV) to the edge (120HV) of the sample. Moreover, the tensile strength of this composite reached 197MPa, while that of pure Al was 157MPa. The enhancement in mechanical properties was induced by grain refinement, load transfer from the matrix to the reinforcement and dislocations piled up at the grain boundaries.

Introduction

Graphene, consisting of sp^2-hybridized carbon atoms, has been the focus of material science for its superior mechanical, thermal and electrical properties [1]. The modulus of elasticity and fracture strength of Graphene can reach 1TPa and 125GPa, respectively [2], which make it a good reinforcement for composites. Polymer-matrix composites reinforced by graphene have been researched widely while there are only a few studies about metal-matrix composites reinforced by graphene. It may be due to the difficulties in dispersion, fabrication and the unknown interfacial chemical reactions between graphene and the matrix.

In the few available studies about graphene-MMCs, the fabrication of this composite has been realized by powder metallurgy such as hot press, sintering and hot extrusion [3,4]. All these methods required heating which resulted in a coarse grained structure of the matrix and led to the formation of aluminum carbides by a reaction of Al matrix with graphene. Graphene-aluminum nanocomposite showed decreased strength and hardness because of the carbides acting as brittle weakness [5,6]. Moreover, the grain sizes of metal-graphene composites were not specifically mentioned in the literature.

High pressure torsion (HPT) is usually utilized as a typical process of severe plastic deformation (SPD) for ultrafine-grained (UFG) microstructure with grain sizes clearly below 1um. Nowadays, HPT has been adopted to consolidate powders in fabricating single metals and metal-matrix composites without sintering at elevated temperature [7,8]. It may be possible to synthesize a metal-matrix composite reinforced by graphene through HPT without carbides forming.

In this paper, we successfully fabricated an aluminum matrix composite reinforced by graphene platelets using ball-milling and HPT at room temperature. Microstructure and mechanical properties of Al/graphene composite were characterized.

Experimental

Materials used in this study are Al powders (99.7%) with 75 um in diameter and graphene platelets with 0.55-1.2 nm thickness and 1-5 um length. Firstly, graphene platelets (100 mg) were ultra sonicated in ethanol (100 ml) for 2 h. Subsequently, aluminum powders (19.9 g) were added slowly. After ultra sonication, they were mixed homogeneously by the planetary mill (QM-QX2) for 5 h. The mass ratio of ball to powder was 10:1 and the speed was 300 rpm. Then the slurry was filtered and dried for 12 h at 100 °C to get the powder mixture. Ultra sonication is used to disperse graphene platelets. Different from other studies, we dried the mixture after ball milling. Therefore, ethanol can act as a process control agent to impede agglomerations during the milling. More importantly, this approach can reduce the damage to graphene platelets. And then the powder mixture was compacted into 30 mm diameter disc by the tablet press machine at a pressure of 60 Mpa. Finally, 15 mm diameter discs were cut off and consolidated by HPT under 3 GPa for 10 revolutions at room temperature. A rotation speed of 2 rpm during HPT process was adopted. The HPT-processed discs were shown in Fig 1(a) and the final thickness of each disc was about 0.6 mm. For comparison, HPT was also conducted on Al powders without graphene platelets.

The microstructure characterization was examined by scanning electron microscopy (SEM), electron probe X-ray micro-analyzer (EPMA), transmission electron microscopy (TEM) equipped with energy dispersive X-ray spectroscopy (EDX) and X-ray diffraction analysis (XRD). Densities of synthesized materials were measured using electronic density meter (DAHOMETER MH-300). Samples for mechanical properties' tests and TEM were prepared as Fig. 1(b). Akashi MVK-E3 was used to measure Vickers microhardness from the center to the edge of discs, with a load of 50 g for 15 s. Tensile tests were performed using SANS 5504 under a crosshead speed of 0.5 mm min^{-1}. Dog-bone shaped specimens were cut from the HPT–processed discs using electrical spark discharge machine with a gauge length of 5 mm and 1.0 mm in width.

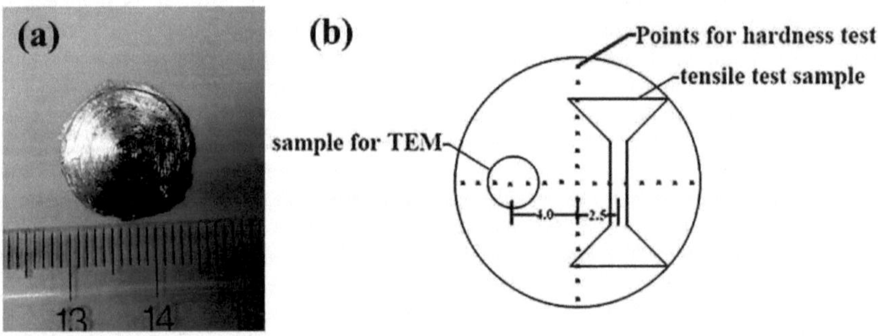

Fig 1. (a) The HPT-processed sample and (b) positions for TEM observation, hardness and tensile test.

Results and discussion

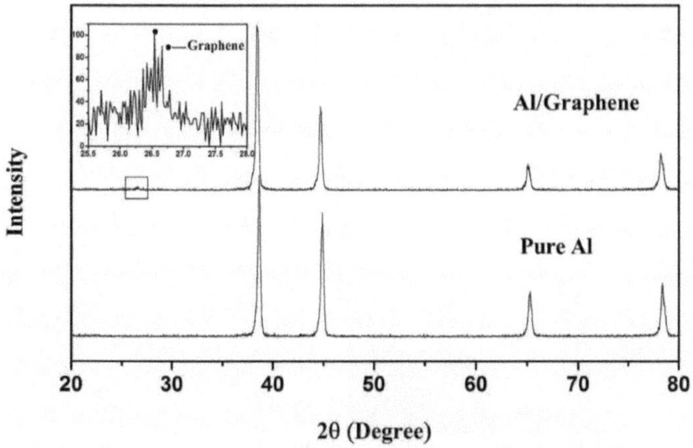

Fig 2. XRD pattern of pure Al and Al/graphene composite.

Fig. 2 presents the XRD pattern of the samples with and without graphene platelets. The peaks for pure aluminum appear at 38.5°, 44.74°, 65.13° and 78.2°. Peaks at 26.54° confirm the presence of graphene. The peaks for Al_4C_3 are not detected. Therefore, HPT can be considered as an efficient way to avoid the formation of possible detrimental phases during the fabrication process of MMCs.

Table I Measured and theoretical densities of HPT-processed samples

	Measured ($g\ cm^{-1}$)	Theoretical ($g\ cm^{-1}$)	Relative density
Al/graphene	2.662	2.698	98.67%
Pure Al	2.669	2.70	98.95%

The densities measured using electronic density meter are listed in Table I for HPT-processed samples with and without graphene platelets. The theoretical density of Al/graphene composite is calculated using the rule of mixture. The measured densities for the samples are very close to the theoretical values: the packing rate is 98.95% for pure Al powders and 98.67% for the composite.

The detailed microstructure analysis for the HPT processed samples is conducted by TEM, as presented in Fig. 3. 100 grains sizes of each sample are measured and it is found that the average grain size of pure Al is ~500nm and ~100nm with graphene platelets. Ultra-fine grained structure is obtained through HPT. The principle of producing UFG structure by HPT is to form subgrain boundaries via dislocation accumulation and rearrangements under severe torsional deformation.

Graphene platelets act as obstacles to pin dislocations and impede their move. That means the dislocation accumulation can be enhanced by graphene. Then subgrain boundaries form. Because of the intense strain produced in the HPT process, large numbers of dislocations are generated and the subgrain boundaries develop into grain boundaries with high-angle misorientation. Consequently, the addition of graphene leads to a further decrease in grain size.

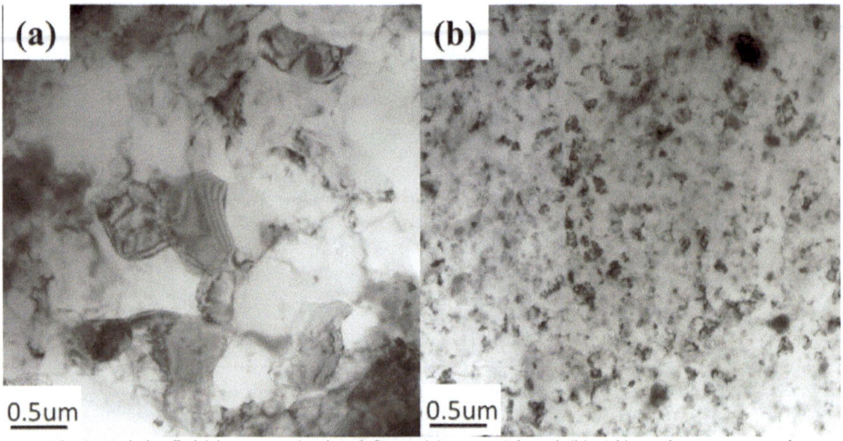

Fig 3. Bright field images obtained from (a) pure Al and (b) Al/graphene composite

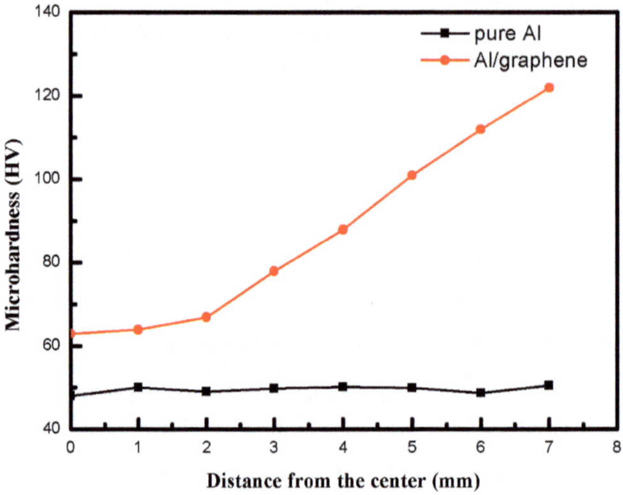

Fig 4. Variations of Vickers microhardness with respect to distance from center of HPT disks for samples with and without graphene

The Vickers microhardness is plotted in Fig. 4 with respect to the distance from the disk center for the samples. The hardness value of Al/graphene composite is 64Hv at the disc center but it increases with the distance to almost twice as 122Hv at 7mm from the center, which is attributed to the inhomogeneous equivalent strain ε_{eq} resulted from torsion during HPT procedure. Mostly, it can be estimated by the relation [9]

$$\varepsilon_{eq} = 2\pi r N / \sqrt{3}h \qquad (1)$$

Where N corresponds to the number of revolutions, r is the distance from the disk center and h is the sample thickness. The equivalent strain increases from the center to the periphery of the disc, followed by more severe plastic deformation of the HPT-processed sample. As a result, the hardness value of Al/graphene composite increases from the center to the edge. However, the hardness value for pure Al is 49Hv at the center and it is constant in hardness with the distance. This trend of constant hardness is because a steady state has been reached as a result of balance between dislocation generation and dislocation absorption. Obviously, graphene platelets play an important role in the hardness enhancement.

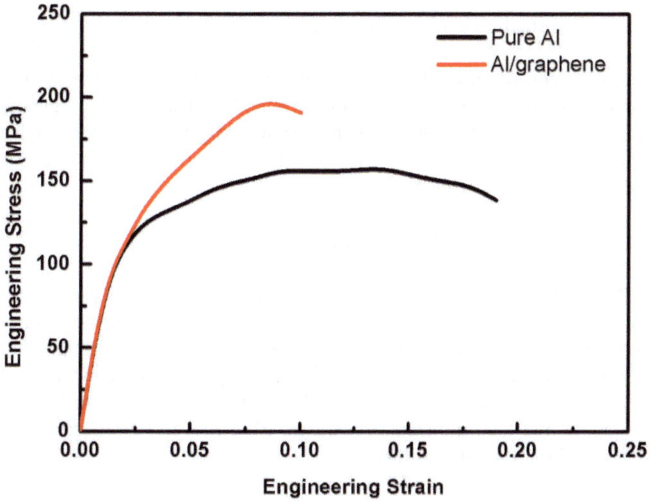

Fig 5. Stress-strain curve plotted for HPT-processed specimens of pure Al and Al/graphene composite

Fig. 5 shows the tensile stress-strain curve of HPT-processed samples with and without graphene. In consistence with the hardness measurement, the sample with graphene shows higher strength than the sample without graphene. The tensile strength is 197 Mpa for Al/graphene and 157 Mpa for pure Al. The specimen for tensile test is performed at position 3.5mm away from the center. As discussed above, Al/graphene composite has not reached a steady state, which means the tensile strength can be much higher at the periphery of the disc.

Grain refinement can make great contributions to the enhanced strength. According to the Hall-Petch relation, the strength increases as the grain size decreases. As shown in Fig. 4, the grain size reduces to ~100nm with 0.5 wt% graphene platelets. In addition, the presence of nanosized graphene platelets leads to the formation of residual dislocation loops around each particle which produces a back stress preventing dislocation motion and resulting in an increase in strength. That's how the Orowan looping mechanism works. Furthermore, load transfer from the matrix to reinforcement is also involved in strengthening.

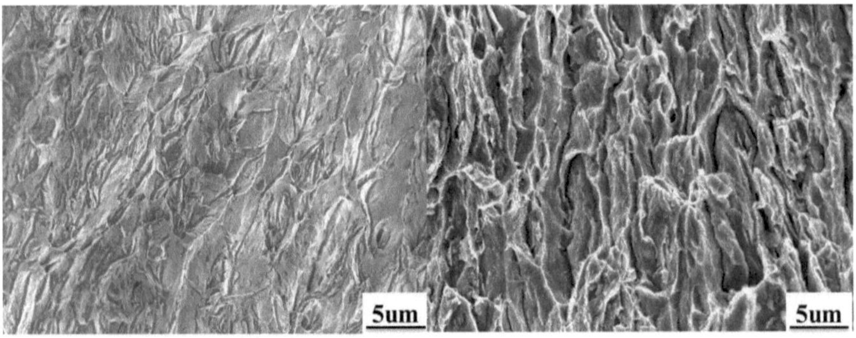

Fig 6. SEM fracture surface of (a) pure Al and (b) Al/graphene

The tensile fracture behavior of pure Al and Al/graphene composite is shown in Fig. 6. It is known that the matrix and the reinforcement have poor wettability, leading to the generation of pores and cavities. It is clear from the SEM fracture images that Al/graphene composite has more pores and cavities than pure Al. These pores and cavities are responsible for the crack and fracture initiation so the composite has lower ductility.

Conclusions

Al-matrix composite reinforced by graphene platelets has been successfully synthesized through high pressure torsion. Neither heating nor sintering is required so carbides do not occur. Ultrafine-grained microstructure is achieved in Al/graphene composite. The composite has not reached a steady state so the hardness value increases from the center to the periphery of the disc. Al/graphene composite exhibits higher tensile strength than pure Al, which is attributed to grain refinement, Orowan looping and load transfer mechanism. It's believed that graphene has great application potential in metal matrix composites.

References

[1] A.K. Geim, et al., "The rise of graphene," Nat. Mater., 6 (2007), 183–191.

[2] C. Lee, et al., "Measurement of the elastic properties and intrinsic strength of monolayer graphene," Science, 321 (2008), 385–388.

[3] J. Wang, et al., "Reinforcement with graphene nanosheets in aluminum matrix Composites," Scr. Mater., 66 (2012), 594–597

[4] M. Bastwros, et al., "Effect of ball milling on graphene reinforced Al6061 composite fabricated by semi-solid sintering," Compos. Part B, 2014, 60, 111–118

[5] S.F. Bartoluccia, et al., "Graphene–aluminum nanocomposites," Mater. Sci. Eng. A, 528 (2011), 7933-7937

[6] Y. Tang, et al., "Thermal expansion of a composite of single-walled carbon nanotubes and nanocrystalline aluminum," Carbon, 42 (2004), 3260–3262.

[7] S.D. Kaloshkin, et al., "Composed phases and microhardness of aluminium-rich aluminium-iron alloys obtained by rapid quenching, mechanical alloying and high pressure torsion deformation," Mater. Trans., 43 (2002), 2031–2038.

[8] V.V. Stolyarov, et al., "Processing nanocrystalline Ti and its nanocomposites from micrometer-sized Ti powder using high pressure torsion," Mater. Sci. Eng. A, 282 (2000), 78–85.

[9] M. Das, "Microstructures and mechanical properties of HPT processed 6063 Al alloy," Mater. Sci. Eng. A, 585 (2012), 525-532

[10] Q.B. Nguyen, et al., "Increasing significantly the failure strain and work of fracture of solidification processed AZ31B using nano-Al_2O_3 particulates," J. Alloys Compd., (2008), 244–250.

EPD Congress 2015
Edited by: James Yurko, Antoine Allanore, Laura Bartlett, Jonghyun Lee, Lifeng Zhang, Gabriella Tranell, Yulia Meteleva-Fischer, Shadia Ikhmayies, Arief Suriadi Budiman, Prabhat Tripathy, and Guy Fredrickson
TMS (The Minerals, Metals & Materials Society), 2015

THE EXTENT OF DOPANT ACTIVATION AFTER MICROWAVE AND RAPID THERMAL ANNEALS USING SIMILAR HEATING PROFILES

T. Gunawansa[1,2], Zhao Zhao[2], N. David Theodore[3], A. R. Lanz[4], T. L. Alford[2]

[1]College of Engineering, Norfolk State University, Norfolk, Virginia 23504
[2]School of Matter, Transport, and Energy, Arizona State University, Tempe, Arizona 85287
[3]CHD-Fab, Freescale Semiconductor Inc., 1300 N. Alma School Rd., Chandler, Arizona 85224
[4]Department of Mathematics, Norfolk State University, Norfolk, Virginia 23504

Keywords: Microwave, Annealing, Silicon

Abstract

Many sustainability issues arise during the manufacturing processes that are currently used for solar cells. Solar energy is a renewable energy source that is independent of the earth's resources, and it is therefore important for the development of more sustainable technologies. Microwave annealing (MW) has been proposed as a technically feasible fabrication scheme for large area silicon solar cells. Apart from that, microwave annealing has been demonstrated to be a promising alternative for repairing damage and electrically activating dopants in ion-implanted semiconductors for integrated circuit manufacturing. A microwave oven is cheaper than conventional furnace systems. In addition, microwave heating is much more efficient than conventional furnace heating, as heating is directly produced inside the material. This minimizes the loss of energy due to heating of the ambient. There is a need for more efficient processing techniques. In this study, microwave annealing is used as an alternative to the current post-implantation processing. Arsenic-doped silicon was microwave annealed (with an alumina-coated silicon carbide susceptor) to activate dopant atoms and to repair damage that was caused by ion implantation. Sheet resistance and Hall effect measurements were used to assess the extent of dopant activation. Rutherford backscattering spectrometry (RBS) with ion channeling was conducted to determine the extent of recrystallization. The dopant activation and recrystallization resulting from microwave annealing is compared with that resulting from conventional rapid thermal annealing (RTA) with the same heating profile. The results show that when compared to RTA, susceptor-assisted microwave annealing results in better dopant activation for shorter anneal times under the same heating conditions.

Introduction

The most commonly used method for the introduction of dopant atoms into silicon is ion implantation. [1]. When high concentrations of dopants are implanted into silicon, the surface layer is damaged. This is very much the case when heavy atoms such as arsenic are implanted. The depth of the damaged layer is directly proportional to the energy of the implanted dopant and can even extend well past the intended junction depth of the device region [1]. Large amounts of lattice damage can cause increased sheet resistance. High temperature anneals are performed to repair the damage that is created during ion implantation and to electrically activate the implanted dopants [2]. The high temperature used for such processing can cause significant vertical and lateral diffusion of dopant atoms. These can degrade device performance [2]. RTA has been used to reduce the diffusion of dopants during annealing. RTA uses a lamp to heat up the wafer to the temperatures that are needed to repair the damaged silicon lattice and to activate the dopant atoms [3]. However, uneven heating can occur because there are differences in the emissivities of the various near-surface device materials. In addition, the photons used in the RTA are not able to penetrate beyond the surface regions of the silicon [3]. An alternative to RTA is microwave heating of the silicon.

Thompson *et al.* have reported benefits of microwave heating for solid state reactions in silicon [4,5]. Microwave annealing causes more volumetric heating of the silicon wafer due to the greater penetration depth of the microwave radiation [6]. Microwave annealing can deliver much higher power in shorter time than RTA. Solar cells based on microwave heating have recently been reported by Herman *et al.* [7]. They used microwave heating to produce nanoparticles and then used microwave annealing to form thin-film solar cells. In the present study, shorter processing times have been achieved using susceptor-assisted microwave heating of the ion implanted silicon. Ceramic composite susceptors, made of alumina and silicon carbide, were used to assist the microwave heating. So that the temperatures that are required for the repair of lattice damage and for the electrical activation of the As^+ dopants can be achieved.

Experiment Details

The starting material was p-type (001) oriented-silicon wafers (ρ = 1-10 Ω-cm, approximately 1 x 10^{15} cm^{-3} boron doped). The wafers were exposed to 180 keV arsenic ion implantation with one of three doses: 1, 2, and 4 × 10^{15} cm^{-2}. The wafers were tilted by 7° off-axis with respect to the incident ion beam during ion implantation. The theoretical projected range, R_P, is 0.12 μm and the transverse straggle, ΔR_P, is 32.6 nm according to TRIM simulation. A single-frequency (2.45 GHz), 2.8 x 10^4 cm^3 cavity applicator microwave system, with a 1300 W magnetron source, was used to anneal the As^+ implanted silicon. A SiC microwave susceptor (which can strongly absorb microwaves and convert their energy into heat) was used to enhance the anneal temperature and to enable more uniform heat transfer. A flat 1.5 cm x 1.5 cm groove was machined into the convex side of a semi-cylindrical piece of the susceptor. 1.4 cm x 1.4 cm Si samples were mounted on top of the groove and then annealed for various times ranging from 20 to 120 seconds. A Raytek Compact MID series pyrometer (mounted through the microwave cavity wall right above the suspector) was used to monitor the near surface temperature.

Figure 1. An image of the susceptor used for microwave annealing.

Each sample was heated using the setup described above. Once the microwaves were switched off, the oven was opened and a high power fan was used to cool the sample down to room temperature. To monitor the dopant activation after microwave annealing, the sheet resistance as measured using an in-line four-point-probe system equipped with a 100 mA Keithley 2700 digital multimeter. Carrier concentration and mobility measurement were performed using an Ecopia HMS-3000 Hall effect measurement system with a 0.98 Tesla magnet.

The heating curve was collected during the microwave annealing. In order to compare the dopant activation resulting from RTA and MW annealing, it was proposed to anneal each sample using a Heat Pulse 610 RTA with a similar heating curve. Therefore, RTA heating recipes were created to intentionally match the microwave temperature-time-rate profiles. The areas under the RTA and microwave anneal heating curves were used to determine if the two curves were closely matched (in addition to the difference in temperature between the two plotted lines). The heating curves for the RTA recipes were plotted and compared with the heating curves of the microwave annealed samples. The best match (between RTA and microwave anneal temperature-time profiles) was for the 2 x 10^{15} cm^{-2} dose annealed for 50 seconds. However, recipe 1 used for the RTA graph was up to 27 Celsius degrees below the microwave annealed sample at certain times. In recipe 1, the steps included ramping up to 400°C for 4 seconds at a rate of 140 °C/s, ramping up again to 586°C for 44 seconds at a rate of 9 °C/s, and then cooling down. In order to decrease the difference in temperature (between the RTA and microwave profiles), a second recipe was created for a new sample. In recipe 2, the steps included ramping up to 400°C for 8 seconds at a rate of 100 °C/s, ramping up again to 586°C for 40 seconds at a rate of 10.5 °C/s, and then cooling down. In the new heating curve (compared to the microwave anneal curve), we can see that the RTA did well in reaching its first ramp up temperature. However, the temperature decreased before ramping up to reach the highest temperature. The difference in temperature between the two graphs (RTA versus microwave) was around 50 degrees Celsius. To avoid the dip in the RTA heating curve after the first ramp up temperature was reached, a hold was inserted in the recipe and an additional ramp-up was set to reach the midpoint in the peaks in the temperatures based on the microwave heating curve. The final recipe, recipe 3, had steps that included ramping up to 400°C for 8 seconds at a rate of 95 °C/s, holding for 1 second, ramping up to 435.2°C for 9 seconds at a rate of 25 °C/s, holding for 1 second, ramping up to 458.9°C for 7 seconds at a rate of 23 °C/s, holding for 1 second, ramping up to 586°C for 19 seconds at a rate of 18 °C/s, and then cooling down. Figure 2 shows the close match between the RTA and microwave heating curves applied to the silicon samples implanted with a dose of 2 x 10^{15} cm^{-2} and for an anneal time of 50 seconds. Each RTA recipe had a purge time of 300 seconds, a ramp up section to reach the initial high heating temperature, a second ramp up section to reach the highest temperature, a ramp down rate of 100 °C/s to reach 275°C, and a finish step to completely cool down the machine to 250°C before the sample could be removed.

Once the recipe was created and all of the samples were annealed, each sample was tested using the four-point-probe method, a Hall Effect measurement system and RBS/ion channeling. Hall measurements were used to determine the carrier concentration, mobility and electrical resistivity of the samples. The samples were mounted onto a circuit board using the Van der Pauw configuration. A copper wire was attached to the board with Leitsilber 200 silver paint. The corners of the samples were scratched to ensure that the copper wire could be attached with indium to the samples. The Ohmic connection of each contact was tested. To measure the Hall voltage, a current was forced through the diagonally opposite pair of contact and the Hall voltage was measured across the remaining pair of contacts. A higher dose sample had a higher carrier concentration with a negative value indicating that it was n-type.

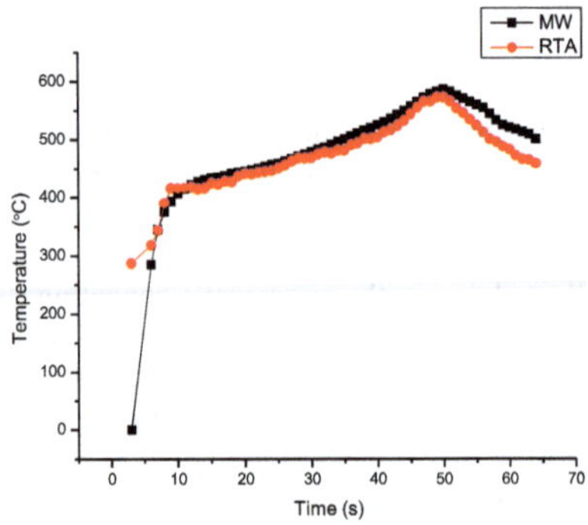

Figure 2. Comparison of the heating curves for microwave and RTA annealed 2 x 10^{15} cm^{-2} dose samples heated for 50s.

Results

The arsenic implanted silicon samples were annealed to study (a) the extent to which the damage caused by ion implantation can be repaired, and (b) the extent of dopant activation. During microwave annealing, the highest temperature reached for all three doses was approximately 695°C when the samples were annealed for 100 seconds. The sheet resistance, R_S, was determined for each sample. The four-point-probe method was used to measure the resistance in Ohms, *V/I.* To determine the proper correction factor, *C.F.*, the diameter, *d*, of the sample was divided by the distance between two specific points on the four-point-probe, *s*, which was set to be 0.1016 cm. Since all of the samples were 1.4 cm x 1.4 cm, the *d/s* ratio was close to 15. In addition, since the samples were square shaped and based on the calculated d/s ratio (of 15), the correction factor was calculated to be 4.3882. The final sheet resistance was calculated by

$$R_S = (V/I) \times C.F. \qquad (1)$$

Figure 3 presents the sheet resistance values for all three doses for each of the different MW and RTA annealing times. For 20s and 40s MV anneals, for each dose, the sheet resistances are so high that they are out of the measurable range. This indicates that there is almost no dopant activation. However, a 50s MW anneal dropped the sheet resistance to a measurable and stable value. This stability indicates that the saturation point for the dopant activation in microwave annealing is 50 seconds. Based on this result, two anneal times were chosen for the RTA: 50s and 100s. The sheet resistances for the 50s RTA anneals were out of range (for all three doses); hence, the single hollow plotted points for the 100s anneals.

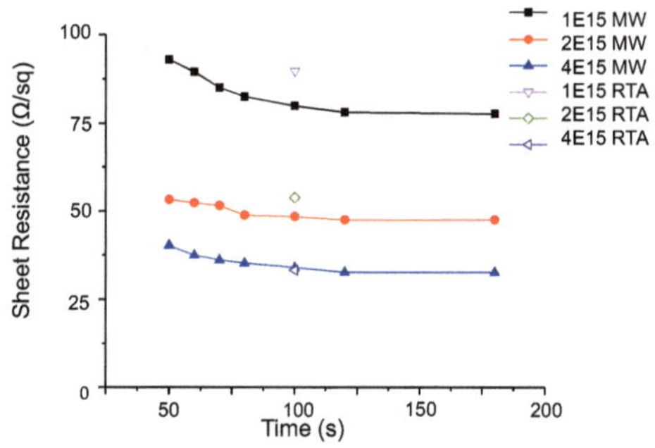

Figure 3. Comparison of the sheet resistance for each dose (MW and RTA samples)

Table I compares the Hall measurements, including the carrier concentration, mobility, and resistivity for each MW dose.

Table I. Hall measurements obtained from the MW samples.

Dose (cm^{-2})	Time (s)	Current (mA)	Carrier Concentration ($\#/cm^3$)	Sheet Resistance (Ω/sq)
1E15	50	1	0.9E20	92.6
1E15	100	1	1.0E20	80.3
2E15	50	1	1.7E20	53.5
2E15	100	1	1.8E20	48.7
4E15	50	1	3.4E20	40.4
4E15	100	1	4.0E20	34.2

Table II shows the Hall measurements for each RTA dose, the measured carrier concentration and the sheet resistance values.

Table II. Hall measurements obtained from the RTA samples.

Dose (cm^{-2})	Time (s)	Current (mA)	Carrier Concentration ($\#/cm^3$)	Sheet Resistance (Ω/sq)
1E15	50	1	-	-
1E15	100	1	1.8E20	90.0
2E15	50	1	-	-
2E15	100	1	1.8E20	54.0
4E15	50	1	-	-
4E15	100	1	4.3E20	33.4

Figure 4 shows the normalized yield of 50s MW and RTA annealed samples from Rutherford backscattering spectrometry (RBS)/ion channeling measurements using the RUMP program. Figure 5 represents the Rutherford backscattering spectrometry measurements obtained from 100s MW and RTA annealed samples. The black solid line in the RBS figures is where the incident alpha-particle beam is incident on the sample in a random direction. The aligned spectra are where the incident beam is aligned along the 001 crystalline direction of the substrate.

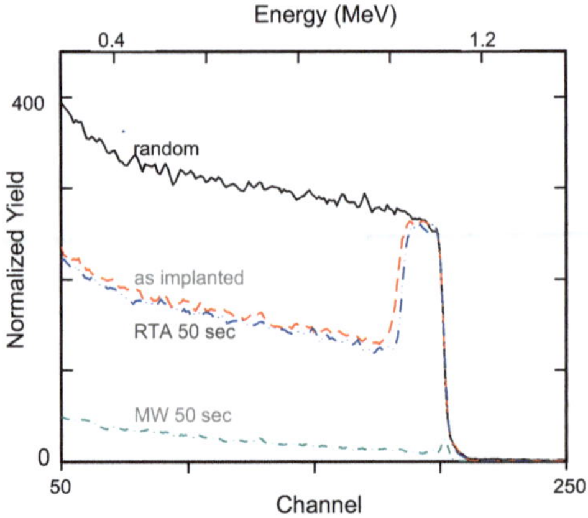

Figure 4. RBS results from 50s annealed 180 keV, 2 x 10^{15} cm^{-2} As^{+} implanted Si.

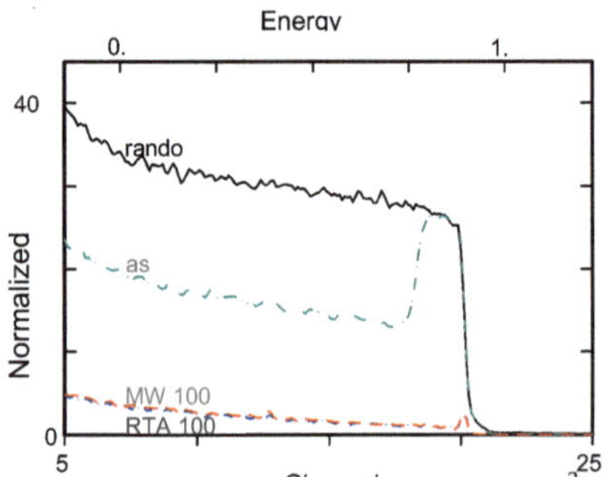

Figure 5. RBS results from 100s annealed 1... cm^{-2} As^{+} implanted Si samples.

Discussion

The sheet resistance was too high to be measured after the sample was RTA-annealed for 50s, indicating no dopant activation. In contrast, the 50s microwave annealing of ion-implanted Si sample with the same heating profile result in in-range measurable sheet resistances. This comparison suggests that dopant activation was able to be achieved quicker and effectively through microwave annealing. The RTA was able to produce 100s annealed samples that had in-range measurable sheet resistance, similar to the 100s microwave annealed samples. The sheet resistance between the RTA and MW samples were not exact, the closest results came from the 4 x 10^{15} cm^{-2} dose. The difference between the two methods of annealing was 0.8 Ω/sq. The largest sheet resistance difference created by these two methods was 10.1 Ω/sq for the 1 x 10^{15} cm^{-2} dose. The data collected can conclude that the higher the dose and anneal time of arsenic in the silicon, the smaller the difference in the sheet resistance between the RTA and MW annealing.

RBS and Hall measurements shows that the MW results in better dopant activation and damage repair for shorter anneal times compared to RTA annealing. This is the case for identical heating profiles (for MW versus RTA). In the RBS spectra, reduced yield is due to the reduced backscattering in the crystalline regions. The lower the back scattering yield, the better the crystalline quality. The high yield of the as-implanted Si in the aligned direction (that comes up to the yield level from the random direction) indicates that the sample is amorphous, or heavily damage after ion implantation. After the 50s RTA anneal, the aligned spectrum still comes up to the random level. This indicates that the Si is still amorphous. Typically, damage repair and recrystallization occur together with dopant activation. This heavy damage of the 50s RTA sample is consistent with its out-of-range high sheet resistance which implies that there is no (or minimal) dopant activation. However, the 50s MW anneal (see Figure 4) has recrystallized the previously amorphous section. This good crystalline quality from the surface to the bulk is achieved with the same heating profile as that of the RTA, indicating that MW annealing is more efficient in terms of dopant activation and damage repair. Figure 5 showed that the 100s RTA and 100s MW anneals resulted in similar reductions of the normalized yields. This indicates that both methods achieve almost the same recrystallization qualityCompared to 50s MV annealed sample, the ion channeling spectra of 100s MV annealed sample shows almost the same backscattering yield, indicating additional 50 s MV annealing does not enable further recrystallize and the full recrystallization is done within 50s MV annealing. Hence, MV annealing can achieve faster recrystallization than RTA. The ion channeling spectra of 50s and 100s RTA treated samples indicate that additional annealing does repair the lattice and improve the crystalline quality further in RTA. Full recrystallization of the ion- implanted silicon has occurred after 100s of RTA.

Conclusion

This paper presents results from microwave and rapid thermal annealed ion-implanted silicon with similar heating profiles. Sheet resistance, carrier concentration, and ion channeling results are discussed. Sheet resistance measurements showed that microwave annealed samples reached the saturation of dopant-activation quicker than RTA samples did. RBS ion-channeling showed that microwave annealing produced recrystallization in a shorter time-frame than rapid thermal annealed did. Further optimization and modeling of the microwave annealing process is needed. Given such optimization and modeling, microwave annealing is likely to be a viable alternative to conventional rapid thermal annealing, especially for large area silicon solar cells.

Acknowledgments

I sincerely thank my PI, Prof. T. L. Alford, and my mentors Ms. Zhao Zhao, Dr. N. David Theodore, and Dr. A. Lanz for their assistance and guidance. I extend my thanks to Dr. Trevor Thornton, the National Nanotechnology Infrastructure Network Research Experience for Undergraduates Program, the National Science Foundation, and the Center for Solid State Electronics Research at Arizona State University for providing the opportunity and funding for this research experience. This work was enabled by use of the NNIN supported NanoFab at Arizona State University

References

1. K.N. Tu, J.W. Mayer, L.C. Feldman, *Electronic Thin Film Science* (Macmillan Publishing Company, New York, NY, 1992).

2. J.D. Plummer, M.D. Deal, P.B. Griffin, *Silicon VLSI Technology: Fundamentals, Practice and Modeling* (Prentice Hall, Upper Saddle River, NJ, 2000).

3. D.C. Thompson, H.C. Kim, T.L. Alford, J.W. Mayer, Appl. Phys. Lett. **83** 3918 (2003).

4. D.C. Thompson, T.L. Alford, J.W. Mayer, T. Hochbauer, M. Nastasi, S.S. Lau, N. David Theodore, K. Henttinen, llkka Suni, Paul K. Chu, Appl. Phys. Lett. **87** 224103 (2005).

5. T.L. Alford, M.J. Madre, R.N.P. Vemuri, N. David Theodore. **520** 4315 (2012).

6. B. Flynn, W. Wang, C.-h. Chang, and G. S. Herman, Microwave assisted synthesis of Cu_2ZnSnS_4 colloidal nanoparticle inks, *Physica Status Solidi* A, **209**, 2186 (2012).

7. T. Sato, Jpn. J. Appl. Phys. **6** 339 (1967).

EPD Congress 2015
Edited by: James Yurko, Antoine Allanore, Laura Bartlett, Jonghyun Lee, Lifeng Zhang, Gabriella Tranell, Yulia Meteleva-Fischer, Shadia Ikhmayies, Arief Suriadi Budiman, Prabhat Tripathy, and Guy Fredrickson
TMS (The Minerals, Metals & Materials Society), 2015

APPLICATION OF COMPUTATIONAL THERMODYNAMICS TO STEEL PROCESSING: THE CASE OF STEEL CLEANNESS

A. Costa e Silva[1], L. Goulart[2], E. Araujo, Jr[2], R. Batista[3] and A. Martins[3]

[1]EEIMVR-UFF Volta Redonda, Brazil
[2]VSBM, Barra Mansa, Brazil
[3]CSN, Volta Redonda, Brazil

Keywords: thermodynamics, steelmaking, non-metallic inclusions, tin foil, long products

Abstract

Steel cleanness is of great importance to the performance of almost all steel products. From very clean steel for bearings to long products used in civil construction, processing variables – in special those in steelmaking operations- must be properly balanced to achieve adequate cleanness for the desired performance. With the present complexity of steels as alloy systems, it is not efficient nor appropriate to develop these processes on empirical basis alone. Computational thermodynamics can greatly improve process development to achieve the desired cleanness level at reasonable costs. In this work, examples of these calculations and their applications to real steel processing are presented and discussed. The advantages, limitations and areas in which improvement in these calculations is desired are highlighted and discussed.

Introduction

Steel cleanness influences a large range of properties: ductility, toughness, fatigue life, processability (rolling, bending, polishing, etc.), corrosion resistance being some examples [1]. In recent decades, a large effort has been dedicated to optimize steel processing via proper process design to achieve less harmful or even beneficial non-metallic inclusions [2,3,4,5]. In many cases the optimal solution defies common sense and intuition. For this reason, the classical empirical approach has given way, since the 1980's to more elaborated techniques. Presently, computational thermodynamics has reached a critical role in understanding and designing steelmaking (primary and secondary) operations, in order to achieve the most desired inclusion population possible. In the following sections, examples of the applications of computational thermodynamics to real steel "cases" are presented and discussed, highlighting advantages, limitations and opportunities for improvement in the use of this technique.

Control of Inclusions in Bearing Steels

High carbon, high chromium steels (such as AISI 52100) are widely used in bearing applications. Several experimental studies on fatigue as well as analysis of fatigue failures [6] have indicated that not only the total amount of inclusions must be very low but also there are types of inclusions that are less detrimental to the fatigue life [7,8,9]. In order to optimize fatigue life, very low total oxygen and aluminum content must be reached and calcium aluminates as well as magnesium spinel inclusions must be avoided [9,10]. This can be achieved via the equilibration of the steel with a slag with proper composition, sufficient rinsing time to eliminate inclusions and judicious control to avoid any possible source of reoxidation [11]. Three key variables in the slag design are basicity (%CaO/%SiO_2) and % Al_2O_3 and %MgO. Figure 1 shows the effect of %MgO in slag on the content of Al, O and Mg dissolved in the steel. One can then tailor the slag to avoid Mg contents that would give rise to spinel (and do the same with respect to basicity versus aluminates). Figure 2 shows the comparison of the measured and calculated Al and O in three heats of this steel. The agreement is good.

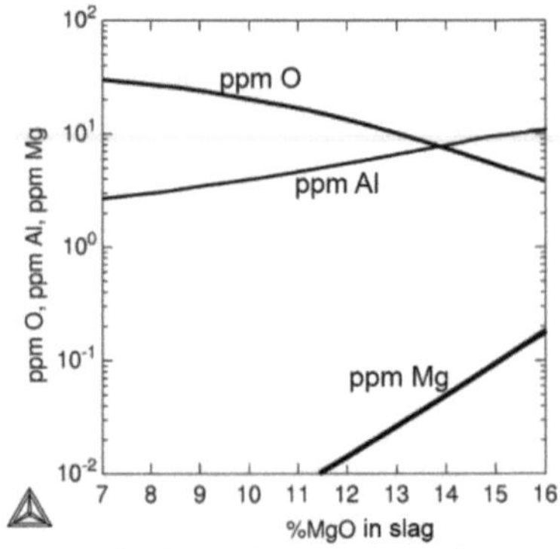

Figure 1. Effect of %MgO in slag with constant $\%Al_2O_3$ and $B=\%CaO/SiO_2$ on Al, O and Mg content in bearing steel in equilibrium with slag. Calculated with Thermo-calc and SLAG2 database [12,13], 1540°C.

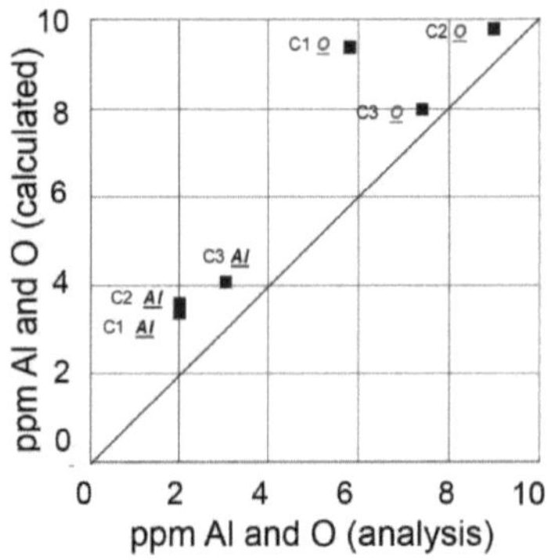

Figure 2. Comparison of calculated and measured Al and O content in bearing steel after ladle furnace processing. Calculated with Thermo-calc and SLAG2 database [12,13].

Calcium Treatment of Alloy Steels

Calcium treatment of steel has been widely used to shape control sulfides and to avoid nozzle clogging in continuous casting [14]. More recently, some results have indicated that calcium treatment to form liquid inclusions can also promote cleanness [15]. Adjusting the Ca addition is a complex task since the optimum amount in solution depends on Al, total O (Ot) and S in the steel, as well as on casting temperature. Measurement of total oxygen content is a time consuming process and there is no analytical method, presently, that can supply results in time for the decision concerning the amount of Ca to be added. Most plants rely on historical series of total O to predict the amount of Ca to add in order to reach the castability window with a majority of liquid inclusions [16,17]. Castability window calculations based on total oxygen are now everyday tasks in many steel plants. In this work an attempt at evaluating the effectiveness of the Ca treatment via oxygen activity sensor is presented, since this measurement gives almost instantaneous results. To do this, the classical castability window graphs (for constant Ot, S and Temperature) were changed into Ca versus dissolved oxygen graphs, as shown in Figure 3. Superimposed on the calculation graphs performed for three levels of Ot, are measurements taken in a series of experimental heats in which Ot was also analysed afterwards. The results show that oxygen activity is a good indicator of correct Ca treatment and can be used to check the efficiency of the treatment before releasing the heat to the caster (Figure 3). When the same calculations are performed for Si-Mn killed steels, the measurement should be even more discriminatory, since the values of oxygen activity are higher and cover a wider range [18].

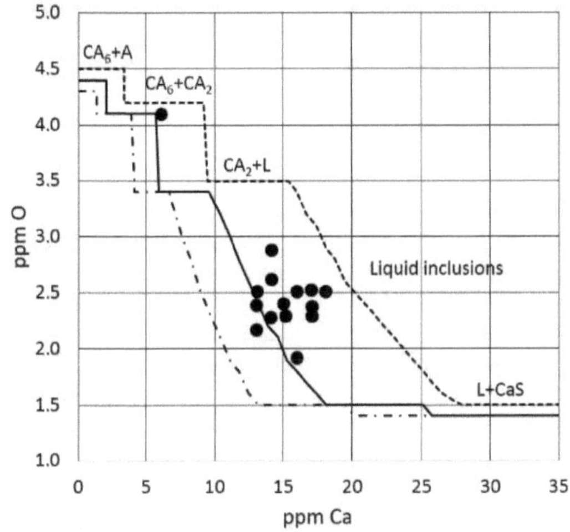

Figure 3. Calculated soluble oxygen in steel containing 0.025%Al, 0.01%S, and various values of Ot (15, 20 and 30ppm, left to right) as a function of Ca content at 1540°C, compared with experimental measurements. Non-metallic phases present indicated for each region of composition. Calculated with Thermo-calc and SLAG3 database [12,19].

Tin Foil Steel Deoxidation Practice – Flat Products

Tin foil is usually cold rolled to very low thickness (0.14mm). Albeit the cleanness requirements are not as stringent as, for instance D&I can steel, large alumina inclusions or alumina clusters can be cause of rejection at the final stages of rolling, after a significant expenditure has already been made in processing. Figure 4 shows the type of defect observed in cold rolled foil for tin coating. Foils are 100% inspected automatically and visually for this rare occurrence. Two deoxidation practices are used for this type of steel. In the first case (called "killed" in this work), enough Al is added to fully deoxidize the steel and reduce the FeO and MnO contained in the slag from tapping from the converter. This is close to 1.7kg Al/t of steel. In the second practice (called "fully killed" in this work) additional aluminum is added to guarantee that the steel will remain fully killed during the complete secondary metallurgy (close to 3.5kg Al/t). The results of experimental heats produced according to both practices were compared. The fully killed heats had a lower rejection index due to chemical composition deviations as well as a final lower cost of added elements (when Al and Mn costs are considered). More important, from the cleanness point of view, total oxygen and soluble oxygen measured during continuous casting were lower when the fully killed practice was used.

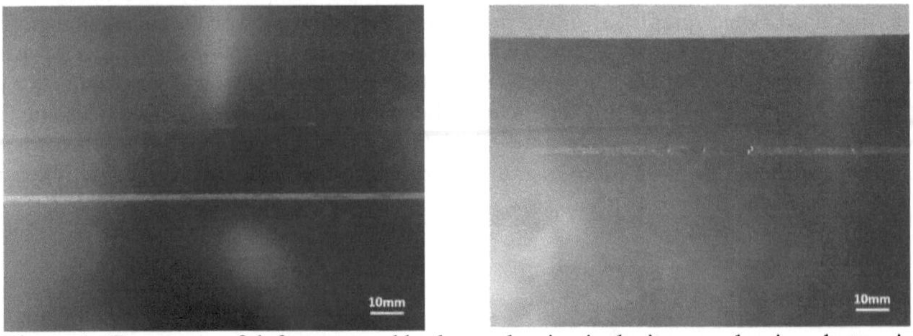

Figure 4. Two types of defects caused by large alumina inclusions or alumina clusters in cold rolled foil.

Furthermore, when evaluating the alumina morphology present in the steels produced according to both practices, the killed steel had alumina with dendritic and plate-like morphologies whereas the fully killed steel had mostly faceted alumina inclusions. This indicates that the remaining alumina inclusions in the fully killed steels were formed earlier in the secondary refining, whereas those in the killed steel formed at a time closer to casting.

An evaluation of the activity of FeO in the slag as well as the use of aluminum in the secondary refining explain these observations, as the fully killed heats maintain lower activities of FeO during the process due to the high amount of Al available to prevent re-oxidation of the steel and hence formation of "fresh" inclusions (Figure 5).

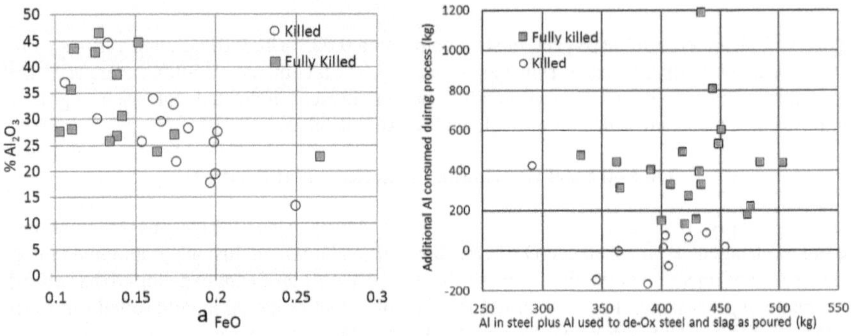

(a) (b)

Figure 5 (a) Alumina content in slag as a function of activity of FeO in secondary refining process. Fully killed steel maintains lower activity of FeO during the processing. (FeO activities calculated with Thermo-calc and SLAG3 database.) (b) Aluminum consumed in addition to that dissolved in steel plus the amount used for deoxidation of steel and slag at pouring in relation to the aluminum used for deoxidation during tapping. Fully killed steel is "protected" against re-oxidation by higher amount of Al added during the process, starting at tapping.

The agreement between the solubility product calculated and measured in both practices is good, as shown in Figure 6.

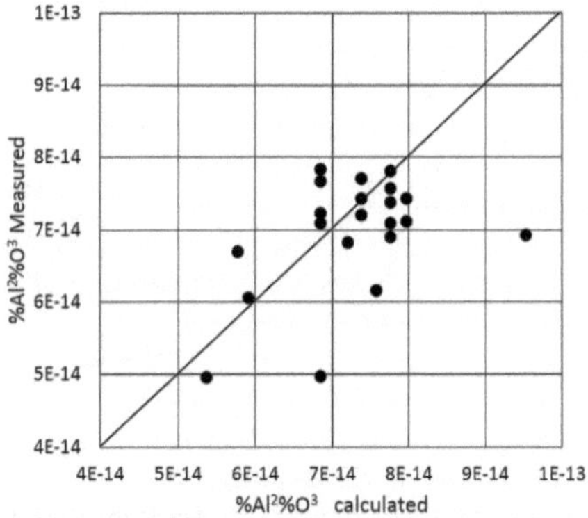

Figure 6. Calculated and measured solubility product of alumina at the end of secondary refining. Deviations are probably due to the fact the activity of alumina was approximated as equal to unity in all calculations. Thermo-calc and SLAG3 database [12,19].

As a result of the adjustment of the deoxidation practice observations over ten months of production indicated that the number of heats in which the defects shown in Figure 4 was observed decreased from 0.066% of the heats when the killed practice was used to 0.027% of the heats when the fully killed deoxidation practice was employed.

Calcium Carbide Additions during Pouring in EAF Steelmaking of Long Products

In order to evaluate the potential improvement in deoxidation of long products (rebars) associated to the addition of CaC_2 during pouring in Si-Mn deoxidized steels, thermodynamic calculations and experimental heats were performed. This practice has already been used successfully in other mills [18]. However, in the present case, in order to try and achieve maximum productivity, ladle furnace time was kept at the minimum needed to adjust temperature and alloying element compositions (less than 15 minutes total). The calculated results in Figure 7 shows that CaC_2 additions have an excellent potential as deoxidizing agent resulting in cleaner steel and higher yield of expensive alloying additions. Besides, sulfur removal can be enhanced. However, the experimental results indicate that the treatment time should be increased, in order to approach equilibrium between slag and metal in the ladle furnace. This practice is now being followed, with a longer time in ladle furnace when better cleanness and lower sulfur is necessary. When increased Si and Mn yield is the single objective, productivity can be increased since the deoxidation level achieved with CaC_2 in very short times is the same as when only Si and Mn are used. This is associated with better yield of these alloying elements in the steel and maximum economic advantage.

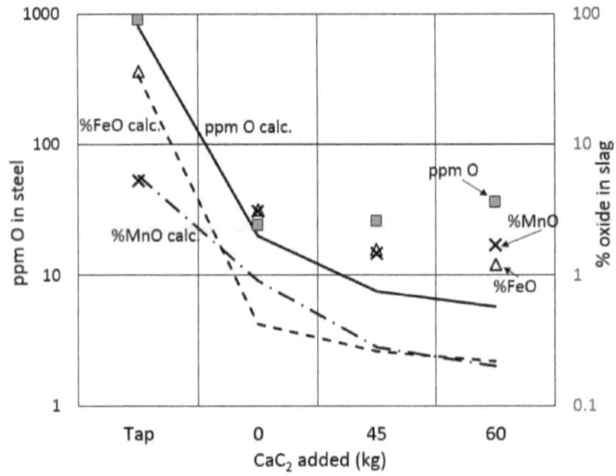

Figure 7. Effect of increasing additions of CaC_2 to a 50t EAF heat during tapping. Lines are calculated values for %FeO and %MnO in slag and dissolved O in steel (ppm O). Points are experimental values after tapping and less than 15 minutes total treatment in ladle. Metal-slag equilibrium is not achieved, albeit the deoxidation with calcium carbide results in savings in Mn and Si (not shown in graph). Potential of this technique is fully explored with longer times in ladle furnace to approach the equilibrium lines. (Calculated with Thermo-calc and SLAG3 database[12,19].

Conclusions

Computational thermodynamics is an important tool for steelmaking process design. Albeit only equilibrium states can be calculated, deviations from equilibrium usually supply an estimate of kinetic problems. With the increase in the requirements for cleanness, strict adherence to narrow composition ranges and the pressure for more efficiency and economy, the use of computational thermodynamics has spread widely to evaluate a large number of steelmaking problems. Currently, there are still some limitations in thermodynamic databases, due to the complexity of the systems involved and the difficulty in performing accurate thermodynamic measurements in these systems. Fe-Mg-O as well as Fe-P-PO_4^{-3} (in CaO-SiO_2-FeO-Al_2O_3-MgO slags) are some examples of systems that would greatly benefit from improved thermodynamic measurements and/or from improved models to describe the solutions (mostly in the case of slags/oxide mixtures).

Acknowledgments

ACS thanks the colleagues in the Brazilian steel industry for the collaboration in many projects, their continuous incentive and interest in solving new problems and their focus on relevant problems to be solved.

References

1. H. Colpaert, A. Costa e Silva, *Metalografia dos produtos siderúrgicos comuns*, *4th ed.* (São Paulo, SP: Editora Blucher; 2008), 119-174.

2. T. Onoe et al., "Shape control of inclusions for steel tire cord (development in ladle arc refining)", *Trans. ISIJ*, 27 (1987), B249.

3. S. Maeda et al., "Shape control of inclusions in wire rods for high tensile tire cord by refining with synthetic slag", *Steelmaking Conference Proceedings*, (1989), 379–85.

4. H. Gaye, P.V. Riboud, J. Welfringer, "Use of slag model to describe slag-metal reactions and precipitation of inclusions", *Ironmaking & Steelmaking*, 15 (6) (1988), 319–22.

5. T. Oshiro et al., "Verbesserung der Dauerhaltbarkeit von Ventilfederdraht", *Stahl und Eisen,* 10 (21) (1989), 1011–5.

6. Y. Murakami, *Metal fatigue: Effects of small defects and nonmetallic inclusions* (Kidlington, England Elsevier Science, 2002).

7. T. Lund, J. Akesson, "Oxygen Content, Oxidic Microinclusions, and Fatigue Properties of Rolling Bearing Steels", *STP987 Effect of Steel Manufacturing Processes on the Quality of Bearing Steels* (West Conshohocken, PA, ASTM, 1988), 308-330.

8. C. Moyer, "Fatigue and life prediction of bearings", ASM Handbook, vol 19 (Materials Park, OH: ASM International, 1996), 355-362.

9. J. Monnot, J.Y. Cogne, B. Heritier, "Inclusion Cleanness and Fatigue Resistance of Rolling Bearing Steels", *Rev. Metall. - CIT,* 83 (10) (1986), 761-770.

10. J.Y. Cogne, B. Heritier, J. Monnot, "Cleanness and Fatigue Life of Bearing Steels" (Clean Steel 3; Balatonfured; Hungary; 2-4 June 1986), 26-31.

11. A. Costa e Silva, "An overview of the use of CALPHAD methods in steelmaking", *J. Min. Metall. B,* 35B (1) (1999), 85–112.

12. B. Sundman, B. Jansson, J.O. Andersson, "The Thermo-Calc databank system", *CALPHAD*, 9 (1985), 153–90.

13. TCAB, "SLAG2- IRSID database" (Stockholm, Sweden, TCAB, 2000).

14. A. Costa e Silva, "Calcium and magnesium thermodynamics in steel and its impacts on secondary steelmaking: a computational thermodynamics approach", *Rev. Metall. - CIT,* 105 (4) (2008), 181–93.

15. M. Mercier, "Adição Dinâmica de CaSi", (MSc Thesis, B. Horizonte, Brazil, UFMG, 2014).

16. Y. Kusano et al., "Calcium treatment technologies for special steel bars and wire rods", *ISIJ Int.,* 36 (1996), S77–80.

17. A. Costa e Silva, "Application of computational thermodynamics to steelmaking and processing- Keynote Lecture". *Proceedings of the 19th IAS Steel Conference.* (Rosario, Santa Fe, Argentina, IAS, 2013), 1–21.

18. M. Scal, private communication with author, GERDAU, December 2013.

19. TCAB, "SLAG3 Database" (Stockholm, Sweden, TCAB, 2012).

EPD Congress 2015
Edited by: James Yurko, Antoine Allanore, Laura Bartlett, Jonghyun Lee, Lifeng Zhang, Gabriella Tranell, Yulia Meteleva-Fischer, Shadia Ikhmayies, Arief Suriadi Budiman, Prabhat Tripathy, and Guy Fredrickson
TMS (The Minerals, Metals & Materials Society), 2015

REACTION BETWEEN MnO-SiO_2 OXIDES WITH LOW FeO CONTENT AND SOLID STEEL DEOXIDIZED BY Si AND Mn DURING HEAT TREATMENT

Chengsong Liu, Jingshe Li, Haiyan Tang*, Xiaojie Gao

School of Metallurgical and Ecological Engineering, University of Science and Technology Beijing; 30 Xueyuan Road, Haidian District; Beijing, 100083, China

Keywords: Solid-state reaction, Heat treatment, Diffusion couple, Inclusion

Abstract

In order to clarify the reaction between MnO-SiO_2 oxides with low FeO content and Fe-Mn-Si alloy, two diffusion couples were produced by the new method using confocal scanning laser microscopy (CSLM). The interface of the alloy and oxide, content of Mn and Si in the alloy near the interface, size distribution and composition of the particles which precipitated in the alloy were observed and analyzed using the electron probe microanalysis (EPMA). Results show that though the FeO content in the oxides decrease to 1% which is lower than the equilibrium with the molten steel at 1873K, the diffusion of oxygen from oxide to alloy still exists during the heat treatment at 1473K and causes shorter Particle Precipitated Zone (PPZ) and Manganese Depleted Zone (MDZ) normally for the reduction of diffusion flux of oxygen.

Introduction

The inclusions in the final steel product directly affect the quality and performance of the steel. However, more and more studies proved that after heat treatment and rolling the final inclusions were often different with those in molten steel. The modification of some kinds of inclusions in the alloy by heat treatment has already been clarified. Choi et al[1] investigated the behavior of non-metallic inclusions in Al-Ti deoxidized steels at 1473K. It was observed that the composition of inclusions might change to Al-Ti-Fe-O, Al-Fe-O, Fe-Ti-O oxide from pure Al_2O_3, Al-Fe-O, TiO_x respectively after the heat treatment at 1473K depending on the contents of Al and Ti. Shibata et al[2-3] investigated the solid-state reaction at the interface between Fe-Cr steel and MnO-SiO2 oxide by the heat treatment at 1473K using diffusion couple method. It was shown that MnO-SiO2-type inclusions have changed to MnO-Cr_2O_3-type inclusions in the case of low Si content in the steel. On the other hand, in the case of high Si content, MnO-SiO_2-type inclusion was stable after heat treatment.

In previous work[4], solid-state reaction between the Fe-Mn-Si alloy and MnO-SiO_2-FeO oxide whose compositions were in equilibrium at 1873K was clarified by heat treatment at 1473K using an improved method to ensure good contact between the alloy and oxide and suppress the influence of element diffusion at 1673K. However, due to the great diffusion flux of oxygen from the oxide to alloy, sulfur effect on the reaction became ambiguous. So, it is necessary to clarify the solid-state reaction between the Fe-Mn-Si alloy and MnO-SiO_2-FeO-S oxide under small diffusion flux of oxygen. In this study, two diffusion couples with low FeO content in the oxide were also produced by the improved method using confocal scanning laser microscopy (CSLM) and studied by the heat treatment at 1473K. The interface of the alloy and oxide, content of Mn and Si in the alloy near the interface, size distribution and composition of the particles which precipitated in the alloy were observed and analyzed using the electron probe microanalysis (EPMA).

Experimental Method

Table 1 shows the compositions of Fe-Mn-Si-S alloy and MnO-SiO_2-FeO-S oxide which were used for producing diffusion couple L1 and L2. According to previous work[4], the oxide with 3 mass% FeO content has an equilibrium relation with the alloy at 1873K including sulfur. So in the case of diffusion couples L1 and L2, the FeO content (1 mass %) in the oxide was lower than the equilibrium content.

Table 1. Initial compositions of the alloy and oxides used for the diffusion couple experiments.

Diffusion couple	Alloy (mass%)				Oxide (mass%)			
	Fe	Mn	Si	S	MnO	SiO2	FeO	S
L1	96.7	3.2	0.1	0.009	67	31	1	0.09
L2					67	31	1	0.28

In order to avoid the intense diffusion during pre-melting of oxide at 1673K, the improved method using CSLM was introduced to produce the diffusion couple and ensure good contact. First, the alloy and oxide with the appropriate compositions were prepared by an arc furnace and an electrical resistance furnace respectively. A hole where the oxide was placed during the heating at 1673K was made in the alloy which was machined into a cubic shape. Al_2O_3 crucible was used to contain the samples. Figure 1 shows the schematic diagram of experimental set-up. Second, Argon gas and Ti foil was used to reduce oxygen partial pressure further after the evacuation of the CSLM chamber up to 5.0×10^{-3} Pa. The experimental temperature increased from room temperature to 1673K by 100K/min. After the oxide melted, the sample was quenched immediately. By this method, the holding time at 1673K was minimized and the element diffusion was suppressed. Then, the sample, a piece of Ti foil and a block of alloy with the same composition were sealed into a quartz tube filled with pure Ar gas. The tube has gone through the heat treatment at 1473K for 10h and 50h. After the heat treatment, the quartz tube was also quenched by water. A vertical cross section of each quenched sample was mirror polished with SiC coated sandpaper and diamond paste. Finally, the interface of the alloy and oxide, content of Mn and Si in the alloy near the interface, size distribution and composition of the particles which precipitated in the alloy were observed and analyzed using EPMA.

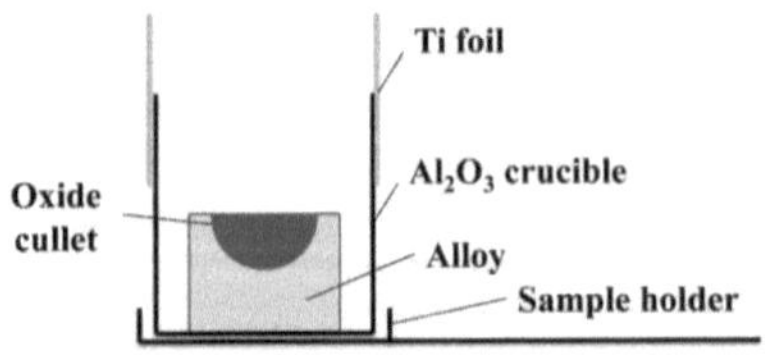

Fig. 1 Experimental set-up to produce diffusion couples using CSLM.

Results

Figure 2 shows the interface between the alloy and the oxide in L1 before and after the heat treatment for 10h and 50h. Good contact between oxide and alloy was confirmed although few particles were formed and narrow particle precipitation zone (PPZ) was observed in the alloy near the interface after the heating at 1673K shown in Figure 2(a). While in the case of L1-10 and L1-50, after the heat treatment for 10h and 50h, both number and size of the particles increased. The width of PPZ also became larger with the increase of heat treatment time. In these figures, several phases were observed in the oxide and the compositions analyzed by EPMA were also shown in Table 2. This result indicated that the composition of gray phases was close to $2MnO \cdot SiO_2$, dark gray phases was close to $MnO \cdot SiO_2$ and white particle was recognized as metallic phase. As shown in Figure 2(a) ~Figure 2(c), the main phase was $2MnO \cdot SiO_2$ but the amount of $MnO \cdot SiO_2$ rose with the increase of heat treatment time. Figure 3 shows the interface between the alloy and the oxide in L2 before and after the heat treatment for 10h and 50h. Similar phenomena were observed in L2 as those in L1. However, compared with the results in L1, the width of PPZ became shorter in all cases of L2.

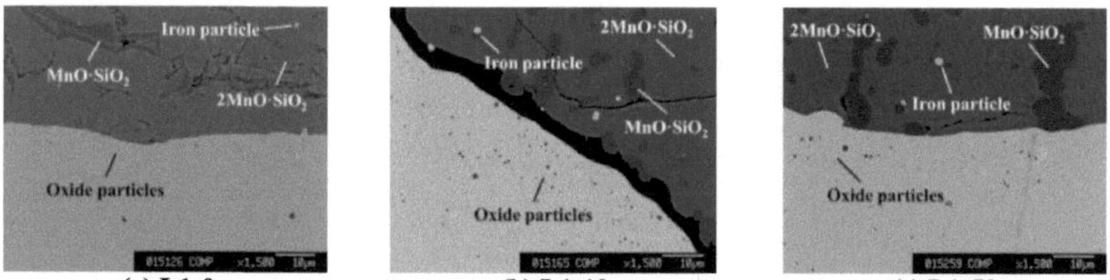

(a) L1-0 **(b) L1-10** **(c) L1-50**

Fig. 2 Interface of the alloy and oxide in the diffusion couple L1 before and after the heat treatment.

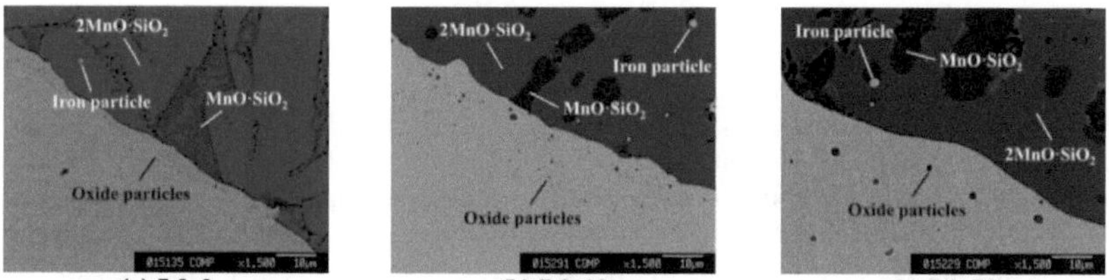

(a) L2-0 **(b) L2-10** **(c) L2-50**

Fig. 3 Interface of the alloy and oxide in the diffusion couple L2 before and after the heat treatment.

In diffusion couple L1 and L2, the MnO content in the oxide increased while the FeO and S content decreased with the increase of heat treatment time. For example in L2, after the heat treatment for 10h and 50h, the MnO content in the dark phase of oxide increased from 51.02% (L2-0) to 51.30% (L2-10) and 52.32% (L2-50). The FeO content decreased from 1.14% (L2-0) to 0.95% (L2-10) and 0.78% (L2-50). The S content decreased from 0.19% (L2-0) to 0.06% (L2-10) and 0.02% (L2-50). While in all cases, there was trace of S content in the gray phase of the oxide though the MnO and FeO contents also increased and decreased as the heat treatment time increased, respectively.

Table 2. Compositions of oxide in diffusion couple L1 and L2 before and after the heat treatment.

Diffusion	Gray phase				Dark phase			
couples	MnO	SiO_2	FeO	S	MnO	SiO_2	FeO	S
L1-0	67.30	31.76	0.92	0.01	51.38	46.69	1.82	0.11
L1-10	68.03	31.26	0.70	0.01	52.29	46.93	0.76	0.02
L1-50	68.32	31.17	0.50	0.01	52.52	47.24	0.22	0.02
L2-0	67.31	31.28	0.91	0.01	51.02	46.94	1.14	0.19
L2-10	67.70	31.66	0.63	0.01	51.30	47.69	0.95	0.06
L2-50	68.21	31.32	0.45	0	52.32	46.88	0.78	0.02

Figure 4 shows the Mn and Si contents in the alloy near the interface of L1 before and after the heat treatment at 1473K. It is clear that there was a little decrease in Mn and Si contents near the interface for the suppression on the reaction between the alloy and oxide at 1673K by the improved pre-melting method. The width of the Mn-depleted zone (MDZ) was only 13μm or less.

After the heat treatment for 10h, Mn and Si contents gradually decreased towards the interface from 3.1% and 0.1% to 1.4% and 0.02%, respectively. As the heat treatment time increased to 50h, the width of MDZ became larger and the minimum value of Mn content was slightly increased to 1.9%. However, after the heat treatment for 10h and 50h, there was no slight increase of Mn content close to the interface which was different with that of diffusion couple A0~A2 in previous work[7]. It was the same for Si content in the alloy in L1-10. While as the heat treatment time elongated to 50h from 10h, a region where the Si content was higher than that in the bulk alloy came out. The maximum value of Si content and the distance from the interface to the position where Si content reached original content was 0.14% and 35μm, respectively.

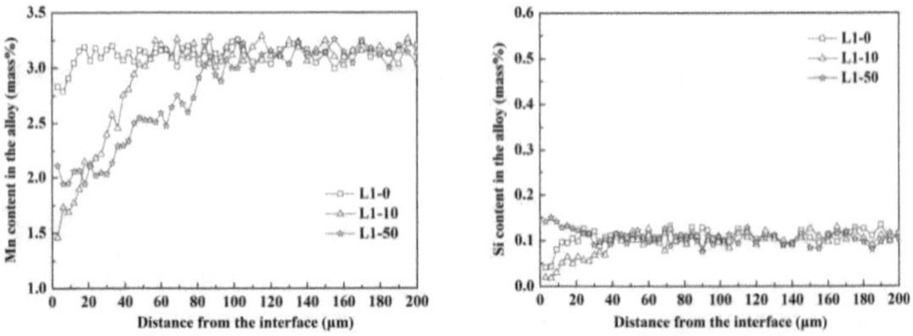

(a) Mn content (b) Si content

Fig. 4 Mn and Si contents in the alloy near the interface of diffusion couple L1 before and after the heat treatment at 1473K.

Figure 5 shows the Mn and Si contents in the alloy near the interface of L2 before and after the heat treatment at 1473K. After the heating at 1673K, the Mn and Si contents in the alloy close to the interface decreased a little respectively which were similar to those in L1-0. After the heat treatment at 1473K for 10h, Mn content also gradually decreased towards the interface from 3.1% to 2.0% while Si content increased towards the interface from 0.1% to 0.13%. Unlike that in L1-10, the Mn content slightly increased in the region close to the interface in about 10μm or less. As the heat treatment time increased to 50h, the width of MDZ also became larger and the minimum value of Mn content was slightly increased to 2.1% and the distance from the interface to the position where Mn content showed minimum value sightly increased to 18μm. As for Si content in the alloy, when the heat treatment time rose to 50h, the maximum value of Si content and the distance from the interface to the position where Si content reached original content was 0.14% and 90μm, respectively.

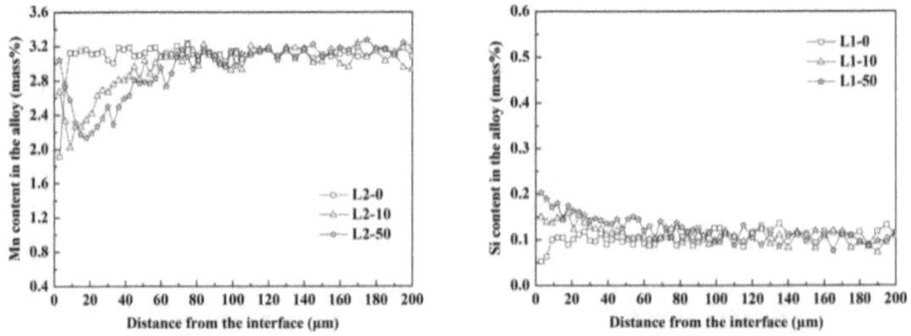

(a) Mn content (b) Si content

Fig. 4 Mn and Si contents in the alloy near the interface of diffusion couple L2 before and after the heat treatment at 1473K.

Figure 6 displays the PPZ and MDZ of L1 and L2 in this study. In this Figure, after the heating at 1673K, the PPZ and MDZ in L1-0 and L2-0 were restrained and less than 15μm. As the time for heat treatment at 1473K increased, the PPZ and MDZ of L1 and L2 all increased and the MDZ was always larger than the PPZ after the heat treatment. In addition, compared with those in L1, the width of PPZ and MDZ of L2 were relatively shorter. For example, with the heat treatment time elongated to 50h from 10h, the PPZ and MDZ of L1 increased from 35μm and 58μm to 55μm and 87μm while those in L2 increased from 20μm and 50μm to 50μm and 72μm.

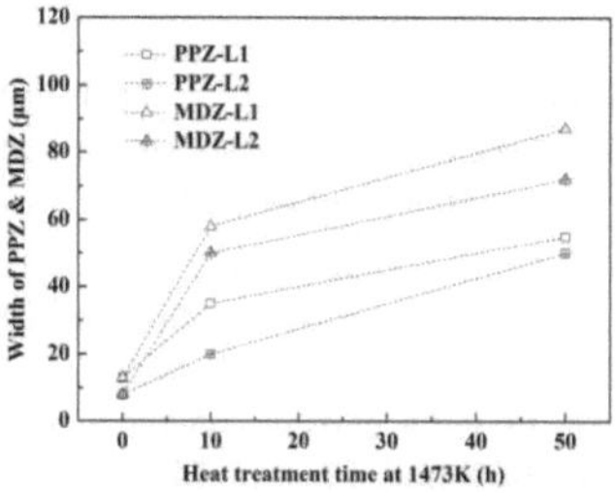

Fig. 6 PPZ and MDZ of the diffusion couple L1 and L2 before and after the heat treatment at 1473K.

Figure 7 presents the size distribution of particles in diffusion couple L1 and L2 before and after the heat treatment. From Figure 7(a), it could be concluded that after heat treatment for 10h and 50h, though the PPZ area increased from 650μm^2 to 1750μm^2 and 2750μm^2, number of the particles smaller than 0.5μm kept decrease. However, after heat treatment for 10h, number of the particles smaller than 1.0μm and larger than 0.5μm increased but it decreased after heat treatment for 50h. Moreover, after heat treatment for 50h, the number of the particles of larger size increased. As for the case of L2, similar size distribution of particles could be observed before and after the heat treatment at 1473K for 10h and 50h. Compared with those in L1, there was a slight decrease in number of the particles with almost all sizes in L2.

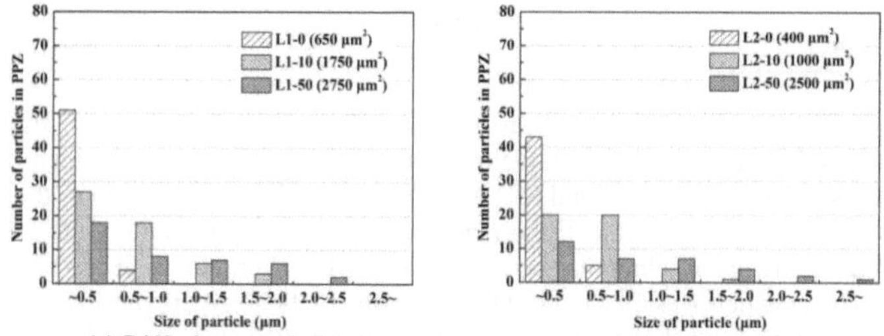

(a) Diffusion couple L1 **(b) Diffusion couple L2**

Fig. 7 Size distribution of the particles in the PPZ of the diffusion couple L1 and L2 before and after the heat treatment at 1473K.

Figure 8 shows the composition of oxide particles precipitating in the PPZ of the alloy near the interface before and after the heat treatment in diffusion couple L1 and L2. The main chemical composition of the particles in L1 and L2 before the heat treatment was close to the $2MnO \cdot SiO_2$ containing 5~30 mass% MnS. After the heat treatment for 10h and 50h, the MnS content increased much even to 95 mass% in some particles although the ratio of Mn and Si was still close to 2 $MnO \cdot SiO_2$.

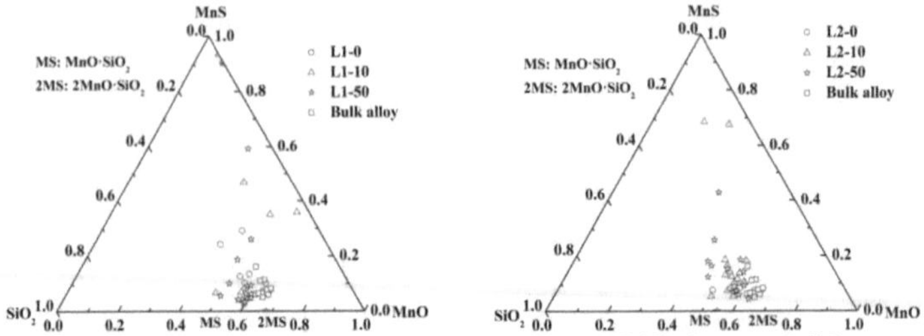

(a) Diffusion couple L1 (b) Diffusion couple L2

Fig. 8 Composition of the particles in the PPZ of diffusion couple L1 and L2 before and after the heat treatment at 1473K.

Discussion

In this study, based on Wagner model[5], regular solution model[6] and relative thermodynamic data[7], the composition of MnO-SiO_2-FeO oxide which was in equilibrium with the Fe-Mn-Si alloy (S<0.009 mass%) at 1873K under a certain oxygen partial pressure was calculated. The activity of oxygen in Fe-Mn-Si alloy and FeO in MnO-SiO_2-FeO oxide was obtained. The thermodynamic data at 1873K were also extrapolated to 1673K and 1473K to calculate the activity of oxygen and FeO due to the lack of basic data in solid phase. The results were shown in Table 3. It could be seen that the activity of FeO at 1673K and 1473K is 0.022 and 0.013, respectively. In this case, though the FeO content in the oxides decrease to 1% which is lower than the equilibrium with the molten steel at 1873K, part of FeO in the oxide is unstable, the diffusion of oxygen from oxide to alloy still exists during the heat treatment at 1473K. Consequently, just as the diffusion couple in previous work, excess FeO is decomposed to metallic iron, and the resulting oxygen diffuses to alloy and reacts with the Mn and Si in the alloy. MnO-SiO_2 particles were formed in the alloy near the interface.

Table 3 Change in the activity of oxygen and FeO with temperature in equilibrium with the alloy.

Temperature/K	Activity of oxygen/-	Activity of FeO/-
1873	0.0065	0.033
1673	0.0018	0.022
1473	0.0004	0.013

In the diffusion couple L1, after the heat treatment for 10h and 50h at 1473K, the slightly increase of Mn and Si content in the alloy close to the interface disappeared which indicated that in this case, cations including Mn^{2+} and Si^{4+} have not diffused from the oxide to the alloy. It was thought that during the heat treatment, the reaction between the alloy and oxide caused oxygen vacancy formed in the oxide which has positive charge. However, due to the low FeO and sulfur content in the oxide which was 1% and 0.09% respectively, the oxygen vacancy was limited and could not bring cations from the oxide to alloy. While in the diffusion couple L2, when the sulfur content in the oxide increased to 0.28%, the oxygen vacancy concentration increased and became excessive which resulted in relatively strong unbalanced electrical charge in the oxide. The diffusion of cations was considered to be caused mainly by the diffusion of anions (such as O^{2-} and S^{2-}) in order to maintain electrical neutrality.

Conclusion

(1) Though the FeO content in the oxides decrease to 1% which is lower than the equilibrium with the molten steel at 1873K, the diffusion of oxygen from oxide to alloy still exists during the heat treatment at 1473K.

(2) The width of PPZ and MDZ become shorter due to the decrease of FeO content in the oxide, though the MDZ was always larger than the PPZ after the heat treatment. The MnS content in the particles increase from 5~30 mass% to even 95 mass% after the heat treatment for 10h and 50h.
(3) The diffusion of cations including Mn^{2+} and Si^{4+} from oxide to alloy is probably determined by the oxygen vacancy concentration which resulted in relatively strong unbalanced electrical charge in the oxide.

Acknowledgement

The authors would like to acknowledge the funding of the National Natural Science Foundation of China (Grant no. 51074021 and no. 51304016) and the funding of State Key Laboratory of Advanced Metallurgy (Grant no. 41603014).

References

[1] W. Choi, H. Matsuura, F. Tsukihashi. Changing behavior of non-metallic inclusions in solid iron deoxidized by Al-Ti addition during heating at 1473K, ISIJ Int., 2011, 51, (12), 1951-1956.
[2] H. Shibata, K. Kimura, T. Tanaka, et al. Mechanism of change in chemical composition of oxide inclusions in Fe-Cr alloys deoxidized with Mn and Si by heat treatment at 1473K, ISIJ Int., 2011, 51, (12), 1944-1950.
[3] H. Shibata, T. Tanaka, K. Kimura, et al. Composition change in oxide inclusions of stainless steel by heat treatment, Ironmaking Steelmaking, 2010, 37, (7), 522-528.
[4] C. S. Liu, K. H. Kim, S. J. Kim, et al. Reaction between MnO-SiO_2-FeO Solid Oxide and Solid Steel Deoxidized by Si and Mn during Heat Treatment at 1473 K, Metall. Trans. B, 2014, Submitted.
[5] M. Hino and K. Ito. Thermodynamic Data for Steelmaking, 167-170; 2011, Sendai, Tohoku University Press.
[6] S. Ban-ya. Mathematical expression of slag-metal reactions in steelmaking process by quadratic formalism based on the regular solution model, ISIJ Int., 1993, 33, (1), 2.
[7] The Japan Institute of Metals. Physical Chemistry of Metals, 198-209; 1996, Tokyo, Maruzen Press.

EPD Congress 2015
Edited by: James Yurko, Antoine Allanore, Laura Bartlett, Jonghyun Lee, Lifeng Zhang, Gabriella Tranell, Yulia Meteleva-Fischer, Shadia Ikhmayies, Arief Suriadi Budiman, Prabhat Tripathy, and Guy Fredrickson
TMS (The Minerals, Metals & Materials Society), 2015

NUCLEATION AND GROWTH IN THE EQUIAXED ZONE OF METAL MATRIX COMPOSITES

Alicia E. Ares[1,2], Carlos E. Schvezov[1,2]

[1]Materials Institute of Misiones, IMAM (CONICET-UNaM), University of Misiones; 1552 Azara Street, Posadas, Misiones, 3300 Argentina.
[2]Member of CIC of the National Research Council (CONICET) of Argentina.

Keywords: Metal Matrix Composites, Solidification, Nucleation, Growth

Abstract

In the present work it was observed the nucleation and growth of grains in the equiaxed zone of metal matrix composites (MMCs) samples directionally solidified wit the presence of columnar-to-equiaxed transition (CET) in the samples. In order to do so, first, it was defined the difference between the real temperature at a given instant, which it is used in the analysis, its variation and furthermore, the variation in the solid fraction with time between the beginning and the end of solidification. It was determined a growth law assuming a lineal variation of equiaxed grain radius as a function of time, and also, it was expressed a law of grain density as a function of time, considering the presence of SiC and Al_2O_3 particles in the matrix. Finally, the results obtained were validated with measured values of grain density at each position of the thermocouples in the equiaxed zone of the solidified samples.

Introduction

In order to predict the microstructure formation in solidification processes, should take into account the basic mechanisms of nucleation and growth on a microscopic scale, along with the equations of macroscopic continuity. An exact solution of growth only (i.e. no nucleation) would require a method for front location on the scale of the whole process but with an incredibly intricate shape of the solid/liquid interface. Since such a task is impossible without a powerful computational tool, are to be searched for coupling approximate microscopic phenomenon of microstructure formation to macroscopic solutions using continuity equations [1, 2]. The microstructural aspects of columnar morphology can be predicted using standard calculations including macroscopic aspect or not [1-5] and the supercooling of solidification front in continuity equations [6-7].

For the equiaxed microstructures, the approach has to be different since the growth rate is not directly related to the velocity of the isotherms and itself depends also on the density of grains. Although the basic modeling concepts introduced in equiaxed solidification works of Oldfield (8) in 1966, only recently this approach has been extended to calculate solidification of gray cast iron (9-11), white cast iron [12-15], spheroidal gray [12 to 15,16] iron, and dendritic alloys [5, 6,11,17-19].

In the present work it was observed the nucleation and growth of grains in the equiaxed zone of metal matrix composites (MMCs) samples directionally solidified. In order to do so, first, it was defined the difference between the real temperature at a given instant, which it is used in the analysis, its variation and furthermore, the variation in the solid fraction with time between the beginning and the end of solidification.

It was determined a growth law assuming a lineal variation of equiaxed grain radius as a function of time, and also, it was expressed a law of grain density as a function of time, considering the presence of SiC and Al_2O_3 particles in the matrix. Finally, the results obtained were validated with measured values of grain density at each position of the thermocouples in the equiaxed zone of the solidified samples.

Experimental Procedure

The directional solidification process was done as was reported elsewhere [23, 24]. As an example of structures obtained by directional solidification process, in Figure 1 is showed the macrostructure and microstructures of one sample of MMCs with columnar-to-equiaxed transition, CET (Zn-27%Al + 8 vol pct SiC).

From the equilibrium diagram data of Zn-Al system polynomial expressions of variation of liquidus and solidus temperatures were obtained with the atomic concentration. Also, the specific heat, density, thermal conductivity and enthalpy were determined as a function of temperature and concentration. The solid fractions were determined considering a volume element solidifying with two faces at different temperatures.

Temperature measurements during the directional solidification were reduced to five positions in the metal in which five thermocouples are located. This requires knowledge of the temperature range as a function of time and position during solidification of the metal matrix. In particular, in order to locate the CET, requires knowledge of the local temperature and the temperature gradients during the movement of the interphase. This information was obtained by developing a heat transfer model for the Zn-Al (ZA) matrix system directionally solidified.

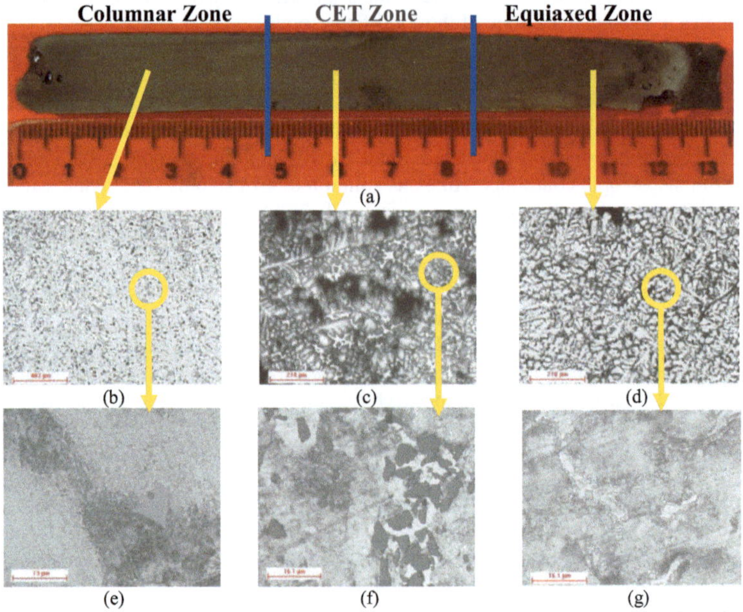

Figure 1. Typical macrostructure of MMCs (Zn-27%Al + 8 vol pct SiC) obtained by directional solidification process. (b) to (g) Microstructures of different zones: (b) and (e) columnar zone; (c) and (f) CET zone; (d) and (g) equiaxed zone.

Results and Discussion

For the analysis presented in this work, first we defined how the actual instantaneous temperature difference was calculated between the onset temperature of directional solidification of MMCs and the temperature of the MMCs during solidification. For this, as shown in Figure 2, it was resolute first, the liquidus temperature ($T_{Liquidus}$) experimentally determined in a given position of the sample. The temperature difference as the difference between the $T_{Liquidus}$ of alloy and the instantaneous temperature (T_i) of the metal matrix of Zinc-Aluminum (ZA) alloy in a given position of the sample was then calculated, which allows determining a temperature difference which is a function of time and that takes a value in each instant between the start and the end of solidification. Expressed as an equation: $\Delta T(t) = T_{Liquidus} - T_i(t)$; where: ΔT (t) is the actual instantaneous temperature difference, $T_{Liquidus}$ is the liquidus temperature determined experimentally for a given position in the alloy and T_i (t) is the instantaneous temperature determined experimentally for a given position for the alloy.

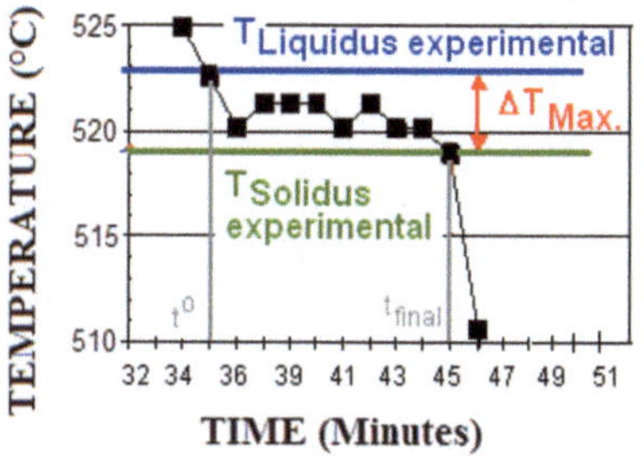

Figure 2. Temperature versus time curve for corresponding to thermocouple T_3 (5 cm from the base of the sample).

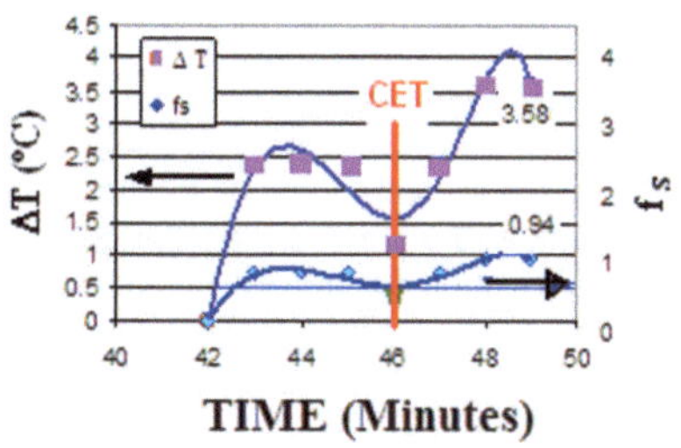

Figure 3. ΔT and f_S versus time curves.

Using the variation curves of temperature versus time, $\Delta T = \Delta T(t)$, and the variation of solid fraction versus time, $f_s = f_s(t)$ from Figure 2, was determined that:

$$\Delta T(t) = \Delta T_{máximo} \cdot f_s(t) \tag{1}$$

This is justified by the symmetry of the Zn-Al phase equilibrium diagram between atomic concentrations. By observing Figure 2 indicates that the initial conditions for the analysis are:

$$\begin{aligned} f_s &= 0 \rightarrow \Delta T = 0 \\ f_s &= 1 \rightarrow \Delta T = \Delta T_{Max.} \end{aligned} \tag{2}$$

In the same figure it can be seen that $\Delta T(t)$ *and* $f_s(t)$ from the beginning to the end of solidification, follow a similar pattern. If Figure 3 is analyzed it is observed that, when solidification is completed, which is detected by a thermocouple in a given fixed position, the instant of the end of solidification (t = 49 min.) The temperature difference reaches the maximum value ($\Delta T = \Delta T_{máx} = 3.58°C$) and this corresponds to a solids content close to unity ($0.94 \approx 1$).

In this figure, the thermocouple is the nearest to the position of the transition from columnar to equiaxed, CET. The same occurs at = 46 min. Which as noted coincides with the zone in the curve where $\Delta T = \Delta T_{mln} = 1.195°C$. The similarity in behavior of $\Delta T(t)$ *and* $f_s(t)$, from the beginning to the end of solidification, indicating that it can express a law of growth based on the variation of ΔT or solid fraction with time:

$$v(t) = \frac{dR(t)}{dt} = \mu_1 . \frac{d[\Delta T(t)]}{dt} \quad \text{or} \quad v(t) = \frac{dR(t)}{dt} = \mu_2 . \frac{d[f_s(t)]}{dt} \tag{3}$$

where t is the time, R(t) is the radius of the grain to equiaxed growth, v(t) is the velocity of the grain to equiaxed growth, ΔT is the instantaneous actual supercooling, fs is the solid fraction, μ_1 *and* μ_2 are constants that are experimentally determined.

In order to calculate the velocity of equiaxed growth as a function of time, first we determine the variation of radius with time. If a linear variation of the radius is assumed with time:

$$R(t) = \mu[\Delta T(t)] + C \tag{4}$$

The initial conditions are: $R(0) = r^* \rightarrow \Delta T(0) = 0$

$R(t_f) = R_{Average} \rightarrow \Delta T(t_f) = \Delta T_{Max.}$

where r* is the critical radius of the nuclei formed, and $R_{Average}$ is the final average radius of equiaxed grains. Figure 4 shows the calculated values of equiaxed grain radius versus time during solidification for three different positions of the thermocouples located in the equiaxed zone of the samples.

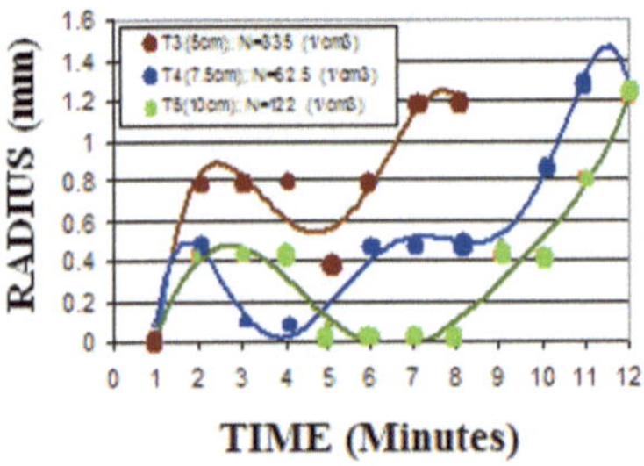

Figure 4. Equiaxed radius vs. time.

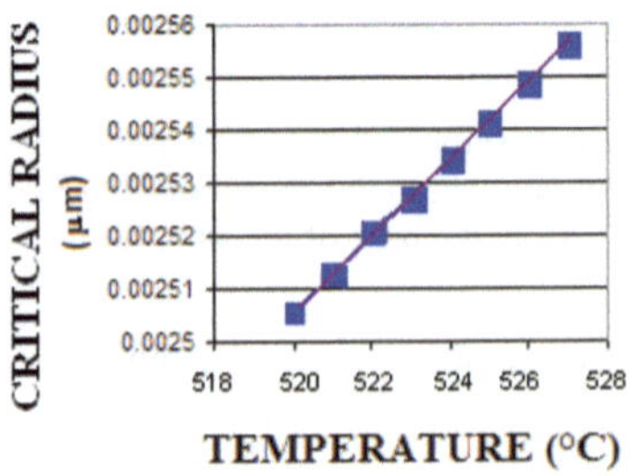

Figure 5. Critical radius vs. temperature.

To determine the μ constants in equations (3), the following conditions were used:

$R(t = 0) = 0$

$R(0) = r^* = \mu(0) + C$

Si $\mu(0) = 0 \rightarrow C = r^*$

$R(t_f) = R_{promedio} = \mu.\Delta T_{máximo} + C$

$$\mu = \frac{R_{promedio} - C}{\Delta T_{máximo}} = \frac{R_{promedio} - r^*}{\Delta T_{máximo}}$$

The critical radius according to the theories of homogeneous nucleation in the temperature range between the liquidus temperature of the matrix and the melt of pure Zn [20] was calculated. This assumption was made because no nucleating agent was added and in addition, the composites were prepared in glass molds. The calculation results are shown in Figure 4.

As can be seen in Figure 5, a value of $r^* = 2.5*10^{-7}$ cm for composites was obtained. This value of r * becomes negligible in the calculation of the μ constant in equation (10), which is:

$$\mu \approx \frac{R_{Average}}{\Delta T_{Max}} \quad \left(\frac{cm}{°C}\right) \tag{5}$$

Experimentally it was found that, μ is equal to $\mu \approx 0.35 cm/°C$ for ZA matrix with SiC composites and $\mu \approx 0.48 \ cm/°C$ for ZA matrix with Al_2O_3.

To calculate the average growth rate of equiaxed grain knowing the value of the μ constant, the variation of the radius and ΔT with time, the following expression was used: $v(t) = \frac{dR(t)}{dt} = \mu.\frac{d[\Delta T(t)]}{dt}$ = growing velocity of equiaxed grain.

In Figure 6 are showed the results of calculating of average velocity of growth of the equiaxed grains. Is possible to observe that there is a variation in the growth rate over time as solidification occurs. Furthermore, it can be seen in the same figure that the growth rate is higher for thermocouple position closer to where the CET occurs, and also its duration in time is shorter.

Growth rates in the two positions of the sample in the upper part, beginning with a lower velocity and greater time duration; and in these cases the growth rate is higher in sections where the radii of growth are higher. If is assumed valid the expression given by Rappaz [21]

$$\frac{df_s(t)}{dt} = n(t)4\pi\bar{R}(t)^2 v(t)\Psi(R) \tag{6}$$

where $\Psi(R)$ = is the effective normalized surface of the solid/liquid interface, calculated analytically by Rappaz and Zou Jie [22], and which for spherical grains ranges between $1 \geq \Psi(R) \geq 0$. If f_s (t) is cleared from the equation (2) and is derived with respect to time is obtained by the following expression:

$$\frac{df_s(t)}{dt} = \frac{d}{dt}\left(\frac{\Delta T(t)}{\Delta T_{máx}}\right) \tag{7}$$

Equating the expressions (6) and (7) yields:

$$\frac{1}{\Delta T_{máx}}\frac{d}{dt}\left(\frac{\Delta T(t)}{dt}\right) = n(t).4\pi\bar{R}(t)^2 v(t)\Psi(R) \tag{8}$$

As $v(t) = \mu.\frac{d[\Delta T(t)]}{dt}$, the expression (8) becomes:

$$\frac{1}{\Delta T_{máx}}\frac{d}{dt}\left(\frac{\Delta T(t)}{dt}\right) = n(t).4\pi\bar{R}(t)^2.\mu.\frac{d[\Delta T(t)]}{dt}.\Psi(R) \tag{9}$$

From this, it is follows that:

$$\frac{1}{\Delta T_{máx}} = n(t).4\pi\bar{R}(t)^2.\mu.\Psi(R) \tag{10}$$

Then, grain density versus time is obtained using the following equation:

$$n(t) = \frac{1}{4.\pi.\bar{R}(t)^2.\mu.\Delta T_{máx}.\Psi(R)}\left(\frac{grains}{volumen}\right) \tag{11}$$

From the analysis of the samples is known that $n(t_{final})$ and $R(t_{final})$ of grains in the positions of each thermocouple, also known $\Delta T_{Max.}$ at the end of solidification, and the value of μ for each experience. $\Psi(R)$ depends on geometrical factors. The value for each thermocouple position was calculated and the change was determined as a function of mean radius. Rearranging equation (11) the grain density versus time is obtained by:

$$n(t) = \frac{K}{\overline{R}(t)^2 . \Psi(R)} \left(\frac{grains}{volume}\right) \text{ ó } \left(\frac{1}{mm^3}\right) \tag{12}$$

where: $K = \frac{1}{4.\pi.\mu.\Delta T_{máx}}$. As it is assumed that nucleation occurs when $T_{Liquidus}$ is reached; in that period of time nuclei have the critical radius, $R = r^* = 2.5.10^{-8} mm \;\; y \;\; \Psi = \Psi(r^*) \rightarrow n(t = 0)$.

The following range of values were determined experimentally for $\Psi(R)$:

$$0.151 \leq \Psi(R)_{ZA27-SiC} \leq 0.431$$

$$0.154 \leq \Psi(R)_{ZA27-Al_2O_3} \leq 0.267$$

To validate the procedure, for each thermocouple position, n (t) versus time was plotted as shown in Figure 7. It can he seen from this figure that, at time t = 0 minutes when $T_{Liquidus}$ is reached, it has the density of nuclei with critical radius r *, as time passes, when the $T_{Solidus}$ is reached, there was obtained the final density of grains in the position of each thermocouple in the sample. It can also be seen from Figure 7 that, the density of nuclei with critical radius when to achieve $T_{Liquidus}$ is greater in the position of the thermocouple located in the zone where the CET occurs, (1×10^{10} nuclei/mm^3), and tapering to a value of 404.17 nuclei/mm^3 in the position of the next thermocouple (7.5cm) and 24.88 nuclei/mm^3 in the position of the last thermocouple (at 10 cm from the base of the sample). This is also observed in the other two positions of thermocouples, but in a less abrupt form. To validate the proposed model, in Table 1 is presented the comparison between the values of final density of grains for each position calculated with the law of growth, shows measured at each thermocouple position and the discrepancy between the two values. It can be seen in Table 1, the final density of grains calculated with this model is 14 % less than the experimental final density in the position of the thermocouple where the CET occurs. Likewise, in the other two positions the error is smaller. Similar results were obtained in other experiments.

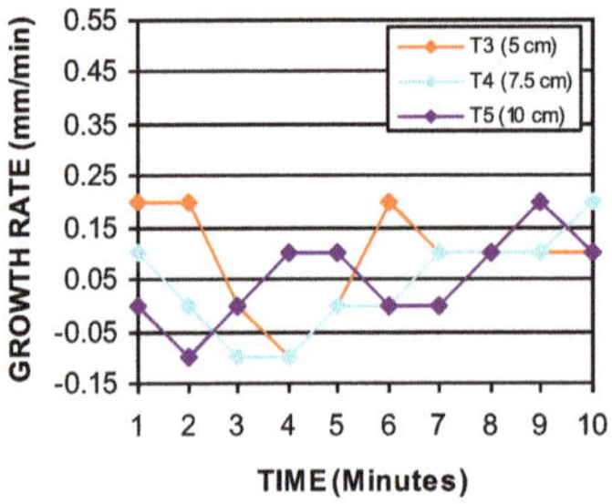

Figure 6. Changes in the growth rate over time.

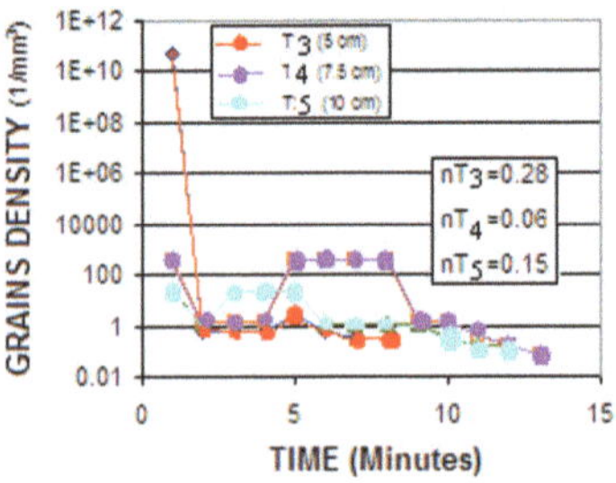

Figure 7. Nuclei density with time.

Table 1. Comparison between the values of final density of grains calculated with the growth law proposed and the values measured in the position of each thermocouple.

Position in the sample (cm)	N° of thermocouple	n(t) final experimental	n(t) final model	Discrepancy
5	T_3	0.33	0.28	0.05
7.5	T_4	0.06	0.06	0
10	T_5	0.12	0.15	0.03

Conclusions

In summary, the main results can be stated as follows:

1. The model predicts the final number of grains, the average diameter and the final density of grain on the basis of experimental data of temperature versus time during directional solidification of MMCs.
2. The assumption that $R(t) = \mu[\Delta T(t)] + C$, allows to determine that when the $\Delta T = \Delta T_{máximo}$, the average radius is maximum.
3. Due to the above assumption, also due to the assumption that the changes in solid content over time follows the equation $\frac{df_s(t)}{dt} = n(t)4\pi\overline{R}(t)^2 v(t)\Psi(R)$, the number of nuclei decreases with time.
4. The density of grains with time follows the law $n(t) = \frac{K}{\overline{\overline{R}}(t)^2.\Psi(R)}$, where $K = \frac{1}{4.\pi.\mu.\Delta T_{máx}}$.

Acknowledgements

This work was supported by PICT-2011-1378 of the National Agency for Promotion of Science and Technology. Thanks are due to the Argentinean Research Council (CONICET) for the financial support.

References

1. T.W. Clyne, "Numerical treatment of rapid solidification", *Metall. Trans.*,15B (1984), 369-381.

2. B. Giovanola, Ph.D Thesis, Ecole Polytechnique Lausanne, Switzerland, N° 647 (1986).

3. B. Giovanola and W. Kurz in B. H. Fredriksson (ed.): "State of the Art of Computer Simulation of Casting and Solidification Processes", *Les Editions de Physique,* Paris, (1986), 129-135.

4. C.G. Levi and R.Mehrabian, "Heat flow during rapid solidification of undercooled metals droplets", *Metall. Trans.,* 13 A (1982), 221-234.

5. S.C. Flood and J.D. Hunt, "Columnar and equiaxed growth I, A model of a columnar front with a temperature dependent velocity", *J.Cryst. Growth*, 82 (1987), 543-552.

6. T.W. Clyne, “The use of heat flow modeling to explore solidification phenomena”, *Metall. Trans.,* 13B (1982), 471-478.

7. M. Rappaz, B. Carrupt, M. Zimmermann and W. Kurz, “Numerical-simulation of eutectic solidification in the laser treatment of materials”, *Helv. Phys. Acta*, 60 (1987), 924-936.

8. W.Oldfield, “A quantitative approach to casting solidification: freezing of cast iron”, *Trans. ASM,* 59 (1966), 945-960.

9. M. Rappaz, Ph. Thevoz, J. Zou Jie, P. Gabathuler, and H. Lindscheid, in H. Fredriksson, (ed.): “State of the Art of Computer Simulation of Casting and Solidification Processes”, Paris, Les Editions de Physique, 277 (1986).

10. D.M. Stefanesku, and C. Kanetkar, in H. Fredriksson (ed.): “State of the art of computer simulation of casting and solidification processes”, Paris, Les Editions de Physique, 255 (1986).

11. Ph. Thevoz, J. Zou Jie, and M. Rappaz, in “Solidification Processing”, 168 (1988) London, The Institute of Metals (1987).

12. D.M. Stefanesku, and C. Kanetkar, in “Computer Simulation of Microstructural Evolution”, (ed. D.J. Srolovitz), Warrendale, PA, Met. Society of AIME (1986).

13. M. Rappaz, and D.M. Stefanescu, in “Solidification Processing of Eutectic Alloys”, (ed. Stefanescu, D.M. et al.), 133; Warrendale, PA, Metallurgical Society of AIME (1988).

14. H. Fredrisson and L. Svensson, in “The Physical Metallurgy of Cast Iron”, (ed. H. Fredriksson and H. Hillert), Mater. Res. Soc. Symp. Proc., 34, 273; New York, North Holland (1985).

15. M. Castro, J. Lacaze, and G. Lesoult, in “Erstarrung Metalisher Werkstoffe”, (ed. Sahm, P.R.), 225; Bad Nauheim, DGM Informationsgesellschaft Verlag (1988).

16. S.C.Flood and Hunt: Metals Handbook, Edition 9ª, Chapter 15 "Casting", (1988), 130.

17. I. Maxwell, and A. Hellawell, “A simple model for grain refinement during solidification”, *Acta Metall.,* 23 (1975) 229-237.

18. Ph. Thevoz, J. Desbiolles, and M. Rappaz, “Modeling of equiaxed microstructure formation in casting”, *Metallurgical Transactions*, 20 A (1989), 311-322.

19. C. Degand, D.M. Stefanescu, G. Laslaz, “Solidification Science and Processing”, Edited by I. Ohnaka and D.M. Stefanescu, The Minerals, Metals and Materials Society (1996).

20. C. Ford, “Nucleation theorems, the statistical mechanics of molecular clusters, and a revision of classical nucleation theory”, *Phys. Rev.,* 56 (1997) 5615-5629.

21. M. Rappaz, “Modeling of Microstructure Formation in Solidification Processes”, *Int. Mat.,* 34 (1989), 93-123.

22. J. Zou Jie: PhD Thesis, “Modelisation de la solidification equiaxe de la fonte grise”, Ecole Polytechnique Lausanne, Switzerland, N° 774 (1988).

23. A.E. Ares and C.E. Schvezov, Columnar-to-Equiaxed Transition in Metal Matrix Composites Reinforced with Silicon Carbide Particles, *Journal of Metallurgy*, (2013), Article ID 628495, 12 pages, http://dx.doi.org/10.1155/2013/628495, Hindawi Publishing Corporation.

24. A.E. Ares and C.E. Schvezov, "Metal Matrix Composites Directionally Solidified", in Proceedings of Advanced *Composites for Aerospace, Marine, and Land Applications*, 2014 TMS Annual Meeting, edited by T. Sano, M. Peretti, T. Srivatsan, p.115-124.

EPD Congress 2015
Edited by: James Yurko, Antoine Allanore, Laura Bartlett, Jonghyun Lee, Lifeng Zhang, Gabriella Tranell, Yulia Meteleva-Fischer, Shadia Ikhmayies, Arief Suriadi Budiman, Prabhat Tripathy, and Guy Fredrickson
TMS (The Minerals, Metals & Materials Society), 2015

COMBINED EFFECTS OF SILICON (Si) AND LOW TEMPERATURE ANNEALING ON THE TENSILE PROPERTIES OF CARTRIDGE (70/30) BRASS WITH NICKEL (Ni) AND IRON (Fe) CONTAMINANTS

Adegbenjo, Adewale Oladapo[1], Adegbola, Adekunle Amos[2], Raji, Fatai Olasunkanmi[2], Ladiipo, Olufemi Akinbiyi[2], Adeboje, Taiwo Bode[2], Ibitoye, Simeon Ademola[3]

[1]Department of Materials and Metallurgical Engineering, University Of Pretoria, +27, South Africa.
[2]Department of Mechanical Engineering, The Polytechnic, Ibadan, Oyo State, +234, Nigeria.
[3]Department of Materials Science and Engineering, Obafemi Awolowo University, Ile Ife, +234, Nigeria.

Keywords: Cartridge Brass, Contaminants, Alloying Element, Tensile Properties, Annealing

Abstract

The combined effects of the addition of Silicon (Si) as alloying element and low temperature annealing on the tensile properties of 70/30 brass with Nickel (Ni) and Iron (Fe) contaminants is investigated in this paper. Melts of cartridge brass were made to which 0, 0.5, 1, 2, 3 and 4 wt % of Si was added. These were sand cast into rods of 600 mm by 10 mm diameter and were thereafter machined into standard tensile test samples. Some of these samples were annealed in a muffle furnace at temperatures of 250, 300, 350, 400 and 450 °C respectively and the others were used as control. Subsequently, all the heat treated and non heat treated samples were subjected to tensile tests on a Hounsfield Extensometer and the load – extension plots retrieved were analyzed. Microstructural characterizations of the samples were carried out using Accu-Scope Optical Microscope. The results showed that cartridge brasses subjected to alloying and annealing treatments had improved tensile and yield strengths. However, the tensile properties were increased and maintained within acceptable limits at stress relieve annealing temperatures.

Introduction

Brasses are a class of indispensable engineering materials. They are versatile and cost-effective materials; ideal for designing complex and long lasting components [1]. Brass has been made for almost as many centuries as copper but has only in the last millennium been appreciated as an engineering alloy. They are used to manufacture many components because of their unique combinations of properties [2, 3]. Brasses combine good strength and ductility with excellent corrosion resistance and superb machinability.

The C26000 alloy, containing 70 % copper and 30 % zinc, is predominantly used for cartridge cases and has outstanding record of excellent formability coupled with high strength [4]. It has been adjudged the best known brass in the group of alpha brasses due to the ease with which the alloy can be deep drawn for the manufacture of cartridge cases [5, 6].In production processes, use of scraps have become a commonly acceptable practice, especially, when the maximum economy is required. However, this leads to an increase in the impurity content of various alloys. Recycled cartridge brass is such an example in which iron impurity drastically reduced the ductility and formability of the semi-finished products [4].This development had in many ways compromised the integrity and applicability of cartridge brass in various engineering service environments.

Different studies had been made to find a solution to this setback in order to enhance the properties of 70 / 30 brass reported that traditionally, modifier elements like Si and Mn were added as melt refiner to reduce the iron percentage of brass melt [6]. In investigating the effects of Si and Co on the recrystallization behavior of Cu-10%Zn it was shown that while the addition of silicon to Cu-10%Zn alloys increases the number of nucleation sites, the addition of minute amounts of cobalt to Cu-10%Zn-Si alloys inhibit grain growth [4].

In addition, an experimental study of cartridge brass L68 containing Pb and Fe indicates that alloying with Mn or Si leads to an increase in the volume fraction of intermetallic compounds of the type $MnSi_3$ and Fe_5Si_3 with a reduction in strength characteristics [8]. They observed that alloying elements and impurities may profoundly affect microstructures of Cu and its alloys during annealing by changing the recovery and recrystallization processes and concluded that thermomechanical treatment (TMT) can significantly improve the ductility of 70 / 30 brass containing 0.35 % iron impurity as a result of particle coarsening caused by TMT affecting the recrystallization of the brass sheets.

Therefore, from the fore-going, the purpose of this present study specifically however, is to investigate the combined effects of Si and low temperature annealing on the tensile and yield strengths of 70 / 30 brass containing Ni and Fe contaminants.

Experimental Procedure

The starting materials used in this study were melts of cartridge (70 /30) brass to which 0, 0.5, 1, 2, 3 and 4 wt % of Si were added respectively. These were sand cast into rods of 600 mm by 10 mm diameter. The elemental analysis of the prepared cartridge brass without any alloying addition was obtained by X-ray fluorescence (XRF). Standard tensile test specimens were machined from the cast brass rods using a lathe machine. Some of the machined tensile specimens were loaded in five successions into a digitally controlled muffle furnace and annealed in the temperature range of 250 – 450 °C. They were held at these temperatures for one hour and then cooled in air so as to retain the structures at these temperatures.

All the prepared tensile specimens (heat treated and non-heat treated) were tested for tensile and yield strengths on a Monsanto Hounsfield Extensometer at a load of 2000 kg and the load versus extension plots retrieved were analyzed accordingly. A 0.2 % offset was used to compute the yield strengths of the prepared brass samples, brass being a non-ferrous alloy and as such does not show a marked yield point.

Representative samples from the prepared specimens were subjected to gentle grinding on abrasive silicon carbide papers of successive finer grades of 240, 320, 400, and 600 grits lubricated with water. Polishing of the specimens was carried out on a 20.3 cm rotation disc of a BUEHLER Ecomet Polisher covered with velvet cloth and impregnated with 600 grit carborundum. This was followed by a final polishing using a 1 μm diamond paste colloidal suspension on the polishing cloth. The specimens were etched with an etchant containing 30 ml distilled water, 2 g ferric chloride, 10 ml hydrochloric acid and 60 ml ethanol. After about 3 – 5s, the etchant was washed off under running water. The specimens were microscopically observed using OLYMPUS CK 40M Accuscope Optical Microscope. The micrographs were taken using an attached Fametech USB PC digital camera with DCM 35 (350 K pixel USB 1.0) and Minisee software was used to capture and analyze the images.

Results and Discussion

Chemical Analysis

The result of the elemental analysis on the prepared 70 / 30 brasses from XRF is presented in Table I. The Cu and Zn compositions are 69.1205 and 29.5363 wt % respectively. Traces of Ni, Fe, and Pb were also found.

Table I. Chemical Composition (wt %) of The Prepared Virgin Cartridge Brass

Element	Cu	Ni	Fe	Pb	Zn
Wt %	69.1205	0.9715	0.3517	0.0100	29.5363
Error (%)	1.0771	0.0358	0.0586	0.0017	0.4962

Effect of Si on Tensile Properties of Cartridge Brass

The tensile test result of the non heat–treated cartridge brass alloyed with Si is presented in Figure 1. The addition of 0.5 wt % Si to the prepared cartridge brass resulted in an increase in the tensile properties of the brass.

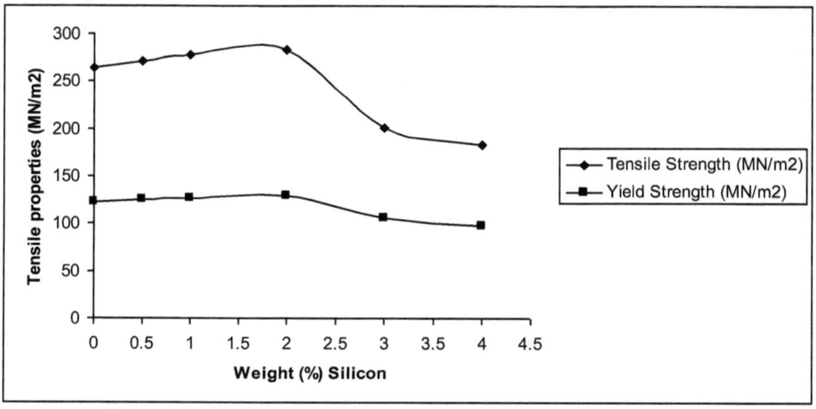

Figure 1. Tensile properties of non heat treated cartridge brass at varied weight percents of silicon

The increase was progressive as the Si content increased until peaks of 288.90 MN/m^2(1.9 wt % Si) and 129.40 MN/m^2 (2 wt % Si) were attained for the tensile and yield strengths respectively. There is however a rapid decrease in the properties thereafter up to the 3 wt % Si added to the brass, this decrease continued steadily after then till the 4 wt % Si added to the brass.

This could be due to an increase in the volume fraction of intermetallic compound of the type Ni_5Si_2 or Ni_3Si (as observed in Figure 2) with a reduction in strength characteristics as this precipitate phase is characteristically brittle with large additions of Si to cartridge brass, hence the observed loss of strength [6].

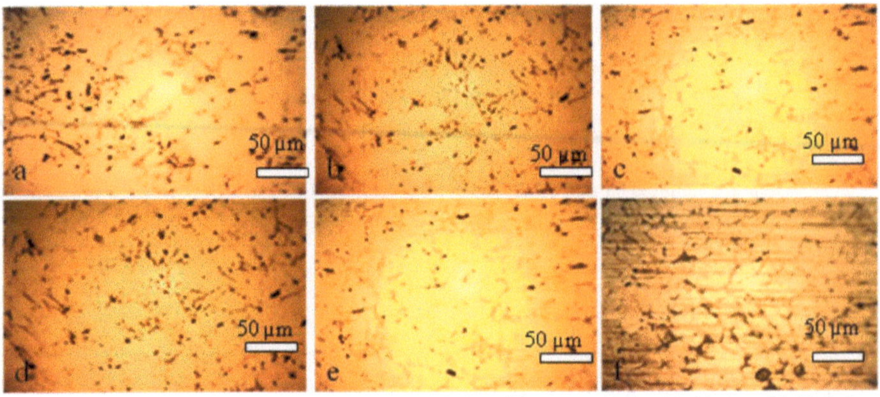

Figure 2. Microstructures of cartridge brass containing silicon at 400X with (a) 0 wt % (b) 0.5 wt % (c) 1wt % (d) 2 wt % (e) 3 wt % and (f) 4 wt %

Effect Of Annealing On Tensile Properties Of 70 / 30 Brass Alloyed With Si

Tensile test results on 70/30 brass alloyed with varied amounts of Si from 0 wt % to 4 wt % and annealed at different temperatures ranging from 250 °C to 450 °C presented in Figure 3 showed that the highest tensile and yield strengths of 312.50 and 133.20 MN/m^2 respectively at 2.7 and 2.1 wt % of Si added respectively to the brass matrix were obtained at 300 °C.

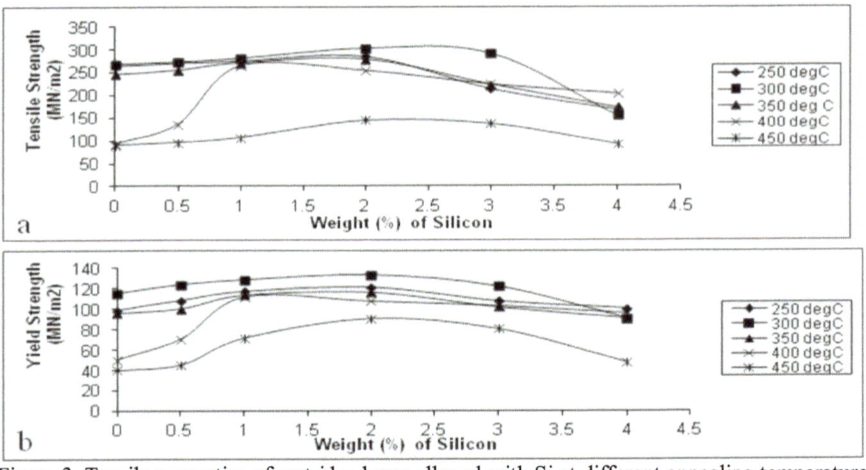

Figure 3. Tensile properties of cartridge brass alloyed with Si at different annealing temperatures (a) tensile strength (b) yield strength

A rapid drop in the measured properties was however noticed after the peaks with increased Si addition as the annealing temperature entered into the recrystallization region, that is, above 400

°C.The observed increase in the tensile properties of 70/30 brass was due to the presence of a second microstructural constituent formed with Ni impurity located on the α alpha grain boundaries as shown on the microstructures in Figures 4 and 5 [4].

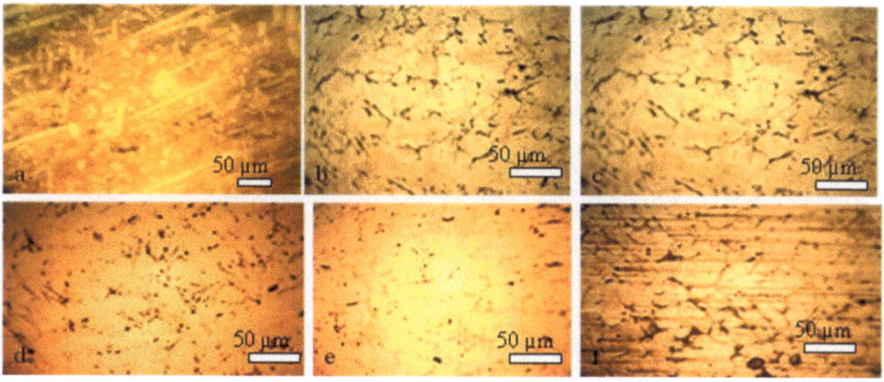

Figure 4. Microstructures of Cartridge Brass Containing (a) 0 wt % (b) 0.5 wt % (c) 1 wt % (d) 2 wt % (e) 3 wt % and (f) 4 wt % silicon at 250 °C (400X)

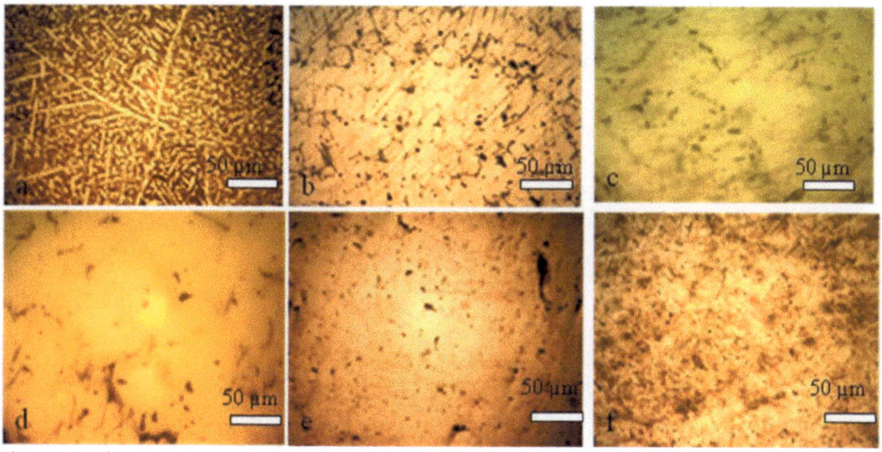

Figure 5. Microstructures of cartridge brass containing (a) 0 wt % (b) 0.5 wt % (c) 1 wt % (d) 2 wt % (e) 3 wt % and (f) 4 wt % silicon at 450 °C (400X)

Conclusion

The results obtained in this study show that the tensile properties of cartridge brass improved with the addition of Si as alloying element in the brass matrix. This is attributed to the presence of intermetallic compounds formed by Si with Ni contaminant in the brass matrix which built strong barrier to dislocation movement. Also, the combined effect of annealing and alloying with Si on the yield and tensile strengths of 70 / 30 brass has been shown to be beneficial at stress relieve annealing temperatures with optimal output at 300 ^{0}C.

References

1. Richard Flinn and Paul Trojan, *Engineering materials and their applications* (Boston: Houghton Mifflin Co., 1990), 45.

2. K. Ken, "Brass in focus," *Engineering Designer*, 30 (3) (2004), 6–9.

3. George T. Murray, *Introduction to engineering materials – behaviour, properties, and selection* (New York, NY: Marcel Dekker, Inc., 1993), 28.

4. N. Aghaie–Khafri and A. Mohebati–Jouibari, "Thermo mechanical treatment of 70/30 brass containing iron impurity," *Journal of Materials Science*, 41 (22) (2006), 7585 – 7589.

5. A.O. Adegbenjo, "Effects of selected processing variables on some mechanical properties of cartridge brass" (M.Sc. thesis, Obafemi Awolowo University, 2011), 2-5.

6. N.A. Titarev, L.I. Mitina and E.T. Mironenko, "Effect of manganese and silicon on the properties and structure of wear – resistant brass," *IZV. V.U.Z. Tsvetn Metall*, 10 (2) (1982), 105-110.

7. K. Oishi, I. Sasaki, and J. Otani, "Effect of silicon addition on grain refinement of copper alloys," *Matter. Lett.,* 49 (2003), 2280 –2286.

8. G. Edmonds, "Wrought brass alloy," *Transactions of the Metallurgical Society of AIME* 230, 3 (1985), 267.

EPD Congress 2015
Edited by: James Yurko, Antoine Allanore, Laura Bartlett, Jonghyun Lee, Lifeng Zhang, Gabriella Tranell, Yulia Meteleva-Fischer, Shadia Ikhmayies, Arief Suriadi Budiman, Prabhat Tripathy, and Guy Fredrickson
TMS (The Minerals, Metals & Materials Society), 2015

EFFECT OF LIME ON ALUMINA EXTRACTING PROPERTY OF CALCIUM ALUMINATE SLAG

Wang Bo[1], Zhang Yubing[1], Ma Lijuan[2], Sun Huilan[1]

[1]School of Materials Science and Engineering, Hebei University of Science and Technology; Shijiazhuang, 050018, China
[2] Zhengzhou Railway Vocational and Technical College, Zhengzhou 450042, China

Keywords: MgO; calcium aluminate clinker; leaching ratio; $C_{12}A_7$; Q-phase

Abstract

The effect of lime ratio on phase composition and alumina leaching property of MgO-contained calcium aluminate clinker were investigated by analytical reagent, the XRD analysis of its mechanism was discussed. The results showed that: when the sintering temperature was 1350 ℃, C/A was 1.4, A/S was 1.3 and the content of MgO was 4%, the increase of the C/A promoted the formation of $C_{12}A_7$ and restrained the formation of Q-phase. The alumina-leaching ratio of the clinker (A/S was 1.3) was increased from 74.33% to 91.34% when the content of CaO was 5.21%.

Introduction

With the development of steel and aluminum industry, the shortage of iron ore and bauxite is getting worse[1, 2]. The lime-sintering process can be used to deal with low-grade bauxite, fly ash and iron-bearing-bauxite, which has become the research focus[3]. The best phase composition of the clinker obtained by lime-sintering process contains both $12CaO\cdot 7Al_2O_3$ and γ-$2CaO\cdot SiO_2$, the former has a better leaching properties of alumina, and the latter could cause the self-disintegration of clinker[4]. Although there many advantages in lime sintering process, the lime ratio is comparatively higher. In order to solve this problem, Wei Xiao et al.[5] showed that the best stoichiometric for formation of $C_{12}A_7$ was 1.4 rather than 1.71. Zhiying Li and Nianbing Zhang et al[6] analyzed the effect of C/A (CaO/Al_2O_3, molar ratio, excluding CaO in $2CaO\cdot SiO_2$)on leaching rate based on intensified sintering mechanism, the results showed that peak of alumina-leaching ratio appeared when C/A was 1.4, which can reduce the vast requirement of lime on a certain extent .Wu Zhang [7]considered that the best C/A of clinker was 1.6 when A/S was 1, and the best C/A of clinker was about 1.4 when A/S was 3~5, the alumina-leaching ratio of clinkers was above 90% with the best C/A.

However, the author found that MgO, which contains in lime, has enormously negative effects on leaching property of the clinker, the leaching ratio of clinker decreased from 90% to about 70% when the content of MgO was 2%. Because $20CaO\cdot 13Al_2O_3\cdot 3MgO\cdot 3SiO_2$ (Q phase) will be formed by MgO, CaO, Al_2O_3 and SiO_2[8-12], whose leaching ratio was only about 70%. A great deal of research was carried out to increase the alumina-leaching ratio of MgO-contained calcium aluminate clinker. Bo Wang [13]considered that 3% addition of Na_2O could improve the leaching ratio of clinker from 71.90% to 81.03%. However, the application of this method was inhibited to some extent since the volatile issue of Na_2O.

There are a lot of alumina contained material phases, calcium aluminates-12CaO·7Al_2O_3 ($C_{12}A_7$) and sulphoaluminate-3CaO·3Al_2O_3·$CaSO_4$ (C_4A_3S, C-CaO, A-Al_2O_3, S-SO_3) are two kinds of phases with good alumina leaching performance. Huilan Sun et al[4] found that alumina-leaching ratio of $C_{12}A_7$ contained in calcium aluminate slag could be up to 92%. Kenneth p. Goodboy[14] provided that C_4A_3S formed when limestone was mixed and sintered with bauxite and calcium sulfate in the correct proportion, the alumina-leaching ratio of C_4A_3S in sodium carbonate solution could be up to 90% as well.

Therefore the pure chemical reagents were used as the raw material in this paper. And the phases, which contained alumina, were transformed from Q-phase to $C_{12}A_7$ or C_4A_3S with the addition of $CaSO_4$, to achieve the purpose of increasing the alumina-leaching ratio of clinker and eliminating the negative effect of MgO. The research emphasis on the effect of $CaSO_4$ on leaching performance, phase composition and self-disintegration of clinker, and the mechanism was analyzed by XRD.

Experiment

1 Experimental materials and equipment

The raw materials used are $CaCO_3$, Al_2O_3, MgO, SiO_2 and $CaSO_4·2H_2O$, and all these reagents are analytical grade in the experiment.
The experimental equipments are electronic balance, SFM-II planetary mixer, KSL-1700X box-type high temperature sintering furnace, Crusher, SFM-I planetary ball mill， Constant temperature water bath, Malvern laser particle size analyzer, XRD (MAX-2500/Rigaku Jap).

2 Preparation of calcium aluminate clinker

The raw materials weighed in a certain proportion and mixed for 2h by SFM-I planetary ball mill at a speed of 250 $r·min^{-1}$. The mixture was then melted at 1350 °C（10 $°C·min^{-1}$ / <1000 °C, 15 $°C·min^{-1}$ / >1000 °C） in the KSL-1700X box-type high temperature sintering furnace for 2 h. The clinker was taken out when the temperature was below 200 °C. Malvern laser particle size analyzer and standard sieve were used to analyze the self-disintegrating properties of clinker. The powder sample was analyzed by XRD (MAX-2500/Rigaku Jap), voltage: 40 kV, current: 150 mA, Cu $K\alpha_1$ (λ=0.154056 nm), scanning range: 10 ~70 °, scanning Speed: 2 $°·min^{-1}$.

3 Leaching of calcium aluminate clinker

The leaching experiments were carried out in the constant temperature water bath under the following leaching conditions: leaching temperature was 80°C , leaching time was 30 min, L/S ratio was 10, stirring rate was 300 $r·min^{-1}$ and the leaching solution was Na_2CO_3 (80 $g·L^{-1}$). 100 mL of leaching solution was accurately measured and was transferred into the flask-3-neck and preheated to 80 °C. Then 10 g of clinker was poured into the flask-3-neck, stirring and timing of leaching. Once the certain time was reached, a certain amount of leaching solution was taken out, drying-filtered and cooled. Chemical composition analysis method was used to analyze the concentration of Al_2O_3 in filtrate.

Results and discussions

The effect of $CaSO_4$ on calcium aluminate clinker system with different C/A

The literatures show that Al_2O_3 was mainly existed in the form of $C_{20}A_{13}M_3S_3$ in clinker[15], and the alumina-leaching ratio was decreased to about 70%. In order to eliminate the negative effect of MgO as much as possible, the content of MgO was set to 4% in this paper. In order to expend the application of lime-sintering process, the A/S was set to 1.3. The emphasis was laid on the effect of the addition of $CaSO_4$ on calcium aluminate clinker (C/A was 1.4 and 1.7) and alumina leaching performance under the sintering system, which was described in 2.2. The leaching ratio results and XRD pattern of partial clinker are shown in Fig. 1-3.

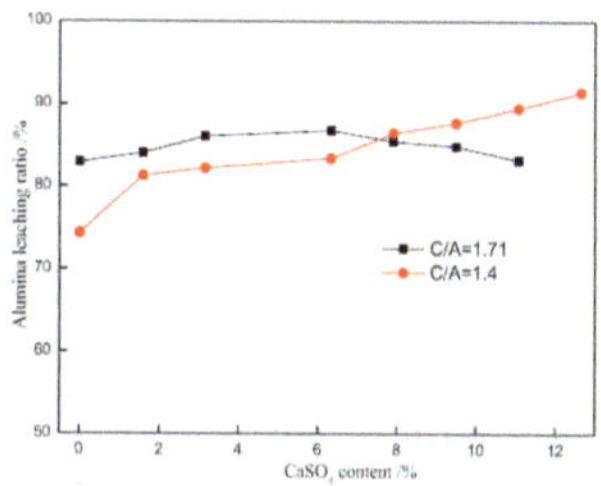

Fig. 1 Leaching ratio of clinker with different $CaSO_4$ addition (conditions: A/S 1.3, MgO 4%, temperature 1350 ℃)

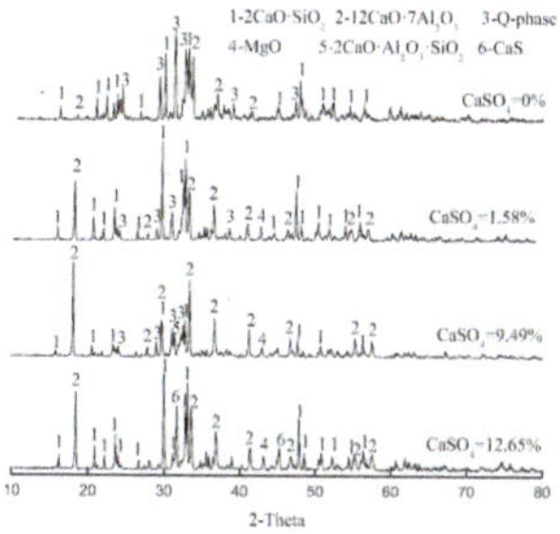

Fig. 2 The XRD pattern of clinker with C/A=1.4 (conditions: A/S 1.3, MgO 4%, temperature 1350 ℃)

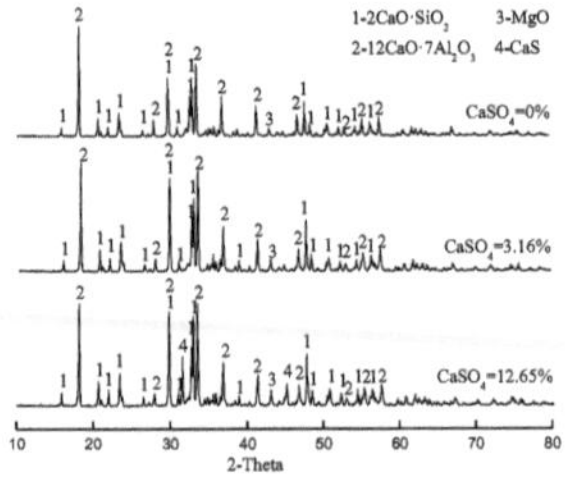

Fig. 3 The XRD pattern of clinker with C/A=1.7
(conditions: A/S 1.3, MgO 4%, temperature 1350 ℃)

1 The effect of $CaSO_4$ on calcium aluminate clinker system when C/A is 1.4

Fig. 1 shows that the leaching ratio of clinker increases with the increasing content of $CaSO_4$ when the C/A is 1.4. Alumina-leaching ratio of clinker increases obviously when the addition of $CaSO_4$ is 1.58%, the alumina-leaching ratio of clinker can be up to 91.34%, when the content of $CaSO_4$ is up to 12.65%. Fig. 2 shows that the main phases of the clinker are C_2S, $C_{12}A_7$ and Q-phase when there is no $CaSO_4$ in this case. The alumina-leaching ratio of clinker is only 74.33% since the high content of Q-phase. The peak intensity ratio between Q-phase and $C_{12}A_7$ decreases with the increasing addition of $CaSO_4$, because the transformation of Q-phase to $C_{12}A_7$ makes the content of $C_{12}A_7$ increase, the MgO separated from Q-phase exists in free state which increases the alumina-leaching ratio. The main diffraction peak variation of clinker (C/A=1.4) with different content of $CaSO_4$ is shown in Table 1.

Table 1 The main diffraction peaks variation of clinker with different content of $CaSO_4 \cdot 2H_2O$
(conditions: A/S 1.3, C/A 1.4, MgO 4%, temperature 1350 ℃)

The content of $CaSO_4$,%	Q-phase (d=2.8806Å)	$C_{12}A_7$ (d=4.8945Å)	Peak intensity ratio of Q-phase and $C_{12}A_7$
0	1250	72	17.36
1.58	470	990	0.48
9.49	314	1117	0.28
12.65	0	1191	0

In addition, the XRD results show that there is still no sulfate in the clinker even the content of $CaSO_4$ is 12.65%. So we suspected that the decomposition reaction of $CaSO_4$ during the sintering process caused the extra addition of lime to the clinker system. A contrast experiment was carried out, same materials, same sintering conditions, the difference is one experiment includes $CaSO_4$, and other includes CaO, to test the suspect. The amount of CaO in other experiment is equal to the content of CaO which is generated by decomposition of $CaSO_4$ at theoretical conditions, the results is shown in Fig. 4.

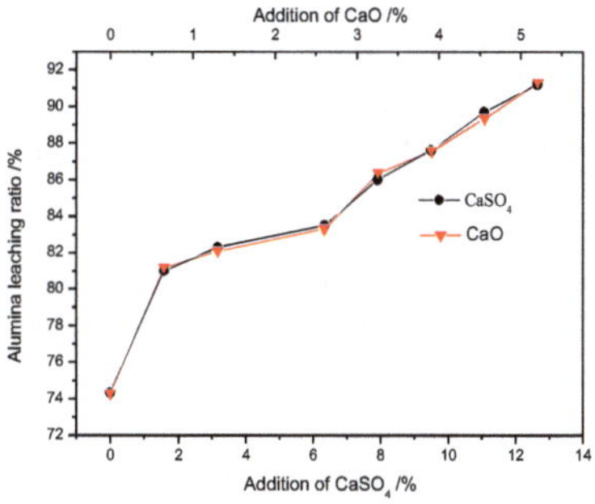

Fig. 4 The comparition of leaching ratio between CaO and $CaSO_4$
(conditions: A/S 1.3, C/A 1.4, MgO 4%, temperature 1350 ℃)

Fig. 4 shows that the alumina-leaching ratio with CaO is basically the same with the results of adding $CaSO_4$. When combined with the XRD results, it's obvious that $CaSO_4$ and CaO have the same effect mechanism on increasing the alumina-leaching ratio, because both of them increase the C/A of clinker, restrain the formation of Q-phase, and accelerate the formation of $C_{12}A_7$.

2 The effect of $CaSO_4$ on calcium aluminate clinker system when C/A is 1.71

It can be seen from Fig. 1 that the alumina-leaching ratio increases slowly, and then, decreases slowly with the increasing addition of $CaSO_4$ when the C/A is 1.71. The alumina-leaching ratio of clinker is almost the same (85%) under different conditions. The maximum alumina-leaching ratio of clinker (86.70%) can be obtained when the content of $CaSO_4$ is 6.33%. Semi-quantitative analysis based on the XRD results in Fig.3 shows that the main phases of the clinker are C_2S, $C_{12}A_7$ and MgO before $CaSO_4$ is added. Because of the high C/A of clinker, there is no Q-phase. When the content of $CaSO_4$ is continued to improve, the main phases and their contents are still the same. Restraint of the formation of Q-phase is the key to the high alumina-leaching ratio. A little of CaS is formed, the content of and $C_{12}A_7$ reduces and alumina-leaching ratio reduces slightly when the content of $CaSO_4$ is up to 12.65%. The reduction reaction of $CaSO_4$ and C, which contained in graphite crucible, causes the formation of CaS, the probable reaction equations are as follows[16]:

$$\frac{1}{2}CaSO_4+C\rightarrow\frac{1}{2}CaS+CO_2\uparrow \quad (1)$$

$$CaSO_4+\frac{5}{2}C\rightarrow CaS+\frac{3}{2}CO_2+CO\uparrow \quad (2)$$

The leaching ratio is unchanged when graphite crucible is replaced with corundum crucible, and there is no CaS any more. High C/A insures the full transformation of alumina to $C_{12}A_7$ when the C/A is 1.71. The research (N.I.Eremin)[17] of the effect of MgO on the alumina-leaching ratio showed that the negative effect of MgO is proportion to the content of MgO, and the lower the C/A is, the higher negative effect of MgO is, and the negative effect can be reduced greatly when the C/A is more than 1.8. The $CaSO_4$ has no effect on the alumina-leaching ratio of clinker when the C/A is 1.71.

The particle size analysis and calculation of these two groups of clinker show that self-disintegration of two groups are both 100%. This result proves that high content of Ca_2SiO_4 provides good self-disintegration when A/S is low.

Conclusions

The $CaSO_4$ can significantly improve the leaching properties of calcium aluminate clinker containing MgO. The addition of $CaSO_4$ plays a role in providing CaO and promoting the transformation of Q-phase to $C_{12}A_7$.

Acknowledgements

The authors greatly acknowledge the financial support of the National Nature Science Foundation of China (No: 51104053), and the National Nature Science Foundation of Hebei Province (No: E2012208047).

References

[1] P. Smith, "The processing of high silica bauxites-review of existing and potential processes," *Hydrometallurgy*, 98(1)(2009), 162-176.

[2] R. Paramguru, P. Rath, V. Misra, "Trends in red mud utilization–A review," *Mineral Processing & Extractive Metall Rev*, 26(1)(2004), 1-29.

[3] J. Grzymek, et al. The new way of alumina lixiviation from sinters containing 12CaO.7 Al_2O_3 in J. Grzymek's Method [M]. Light Metals 1988: 129-133.

[4] W. Bo, et al., "Effect of material ratio on leaching and self-disintegrating property of calcium aluminate slag," *Journal of Northeastern University: Natural Science*, 29(11)(2008), 1593-1596.

[5] Eva Selic, Jan-Dirk Herbell. "Utilization of Fly Ash from Coal-fired Power Plants in China," Journal of Zhejiang University(Science A), 9(5)(2008): 681-687.

[6] T. Zhi-fang, B. Shi-wen, Y. Hai-yan, "Leaching kinetics of non-constant temperature process of calcium aluminate slag under microwave radiation," *The Chinese Journal of Nonferrous Metals*, 16(2)(2006), 357-362.

[7] X. Wei, et al., "Study on the process of extracting alumina from the middling grade bauxite using lime sinter," *Journal of Guizhou University of Technology(Natural Science Edition)*, 05)(2008), 41-43.

[8] L. Zhi-ying, et al., "Research on the intensified sintering mechanism of middle-grade bauxite and its digestion performances," *Light Metals*, 12)(2009), 14-17.

[9] Z. Wu. Phase composition and leaching law of the sinter of CaO-Al_2O_3-SiO_2 system [D]. Shenyang; Northeastern University, 2011.

PREPARE FOR U_3O_8 FROM AMMONIUM URANYL CARBONATE USING MICROWAVE CALCINATION

Liu bing-guo[1, 2], Peng jin-hui[1, 2], Zhang li-bo[1, 2], Hu jin-ming[3], Guo sheng-hui[1,2], Kong dong-cheng[3]

1. Faculty of Metallurgy and Energy Engineering, Kunming University of Science and Technology, Kunming Yunnan, 650093, PR China
2. Key Laboratory of Unconventional Metallurgy, Ministry of Education, University of Science and Technology, Kunming Yunnan, 650093, PR China
3. No.272 Nuclear industry Factory, China National Nuclear Corporation, Hengyang, 421002, PR China

Keywords: Ammonium uranyl carbonate; Triuranium octaoxide; Calcination; Microwave.

Abstract

The novel technology of calcination to prepare triuranium octaoxide (U_3O_8) from ammonium uranyl carbonate in microwave fields was investigated on the basis of the temperature rising characteristics of ammonium uranyl carbonate, triuranium octaoxide (U_3O_8), and their mixture. The result of experiments show that ammonium uranyl carbonate had weak capability to absorb microwave energy, while triuranium octaoxide has the very strong capability to absorb microwave energy and the sample temperature increased rapidly with an increase mixture ratio of triuranium octaoxide. The optimal calcination conditions were as follows: microwave power 700 W, calcination time 10 min and 60 g in this experiments range, respectively. Under these conditions the value of total uranium and U^{4+} of triuranium octaoxide was 84.28% and 31.02%, respectively. It is feasible to prepare triuranium octaoxide by calcination from ammonium uranyl carbonate, which mixed with small amounts of triuranium octaoxide under microwave fields.

Keywords

Ammonium uranyl carbonate; Triuranium octaoxide; Calcination; Microwave.

Introduction

Uranium is an important in nuclear fuel and as the complexity of the atomic structure and the abundance of energy levels [1]· Uranium dioxide, due to important uranium-oxygen systems is widely used as nuclear fuel in light water reactors. They are manufactured by calcinaing (thermal decomposition) the ammonium uranyl carbonate to triuranium octaoxide, followed by reduction to uranium dioxide using cracked ammonia at about 700℃[2]. Triuranium octaoxide is currently Being manufactured by calcination from ammonium uranyl carbonate or ammonium durante etc. Calcination process is the key process in preparation of triuranium octaoxide, as it determines the quality of the product and hence the economy of the industry. Currently ammonium uranyl carbonate is calcined in static tunnel kilns or semi-dynamic electric kilns, which has a number of disadvantages, such as high energy consumption, poor product quality due to the non-uniform temperature distribution in the same section of the reactor, and longer duration of the process. Additionally, human contact with uranium and its compounds are highly toxic which can cause progressive or irreversible renal injury and in acute cases may lead to kidney failure/death, which necessitates faster and safer processing of these compounds.

Microwaves are a form of electromagnetic radiation with frequencies in rang of 0.3-300GHz. Microwave processing is an alternative heating method which has the potential to overcome conduction problems normally faced in the conventional heating process in addition to its advantage of selective heating. Microwave heating has been accepted to be a promising method for rapid volumetric heating, higher reaction rates and reduction in reaction times compared with conventional heating methods [3]. In addition microwave has the potential to heat the materials directly and at higher heating rates than conventional heating [4]. The heating of industrial materials by microwaves has been under investigation for many years and has recently been applied in the field of extractive metallurgy [5-7]. However possibilities of the preparation of triuranium octaoxide from ammonium uranyl carbonate using microwave induced calcinations has not been reported in the literature.

The purpose of the present work is to study temperature rising behavior of the ammonium uranyl carbonate and relative materials in microwave fields. The effects of microwave irradiation time, power, and weight of sample on the value of total uranium and U^{4+} of triuranium octaoxide were investigated systematically, to indentify the optimal calcination conditions.

Experimental

Materials and method

Nuclear purity ammonium uranyl carbonate used in the present study was obtained from No.272 Nuclear industry Factory, China National Nuclear Corporation, which hand particle size less than100μm. Triuranium octaoxide powders were prepared by means of thermal decomposition of ammonium uranyl carbonate, at temperature of 873 K at atmospheres pressure. The X-ray diffraction pattern of the calcined powder indicated the composition of U_3O_8, with the corresponding disappearance of the ammonium uranyl carbonate ammonium uranyl carbonate-U_3O_8 mixture of composition 0%, 5%, 10%, 15%, 20%, 25% and 30% is prepared for experimentation by mixing calculated quantities of ammonium uranyl carbonate-U_3O_8 and grinding the mixture in a stainless steel ball mill for an hour.

Equipment for microwave heating

Microwave reactor employed in the present study is manufactured by Key Laboratory of Unconventional Metallurgy, Ministry of Education, Kunming University of Science and Technology, which has the ability to alter the power intensity from 0-3000W and to operate at a frequency of 2.40GHz. The system is equipped with a water-cooled condenser. Temperature is controlled with the help of the temperature controller which alters the microwave power to match the set temperature. The crucible that hold the sample material is made of ceramic, while the temperature is measured using a type K thermocouple. A schematic of the microwave-heating system is shown in Figure.1.

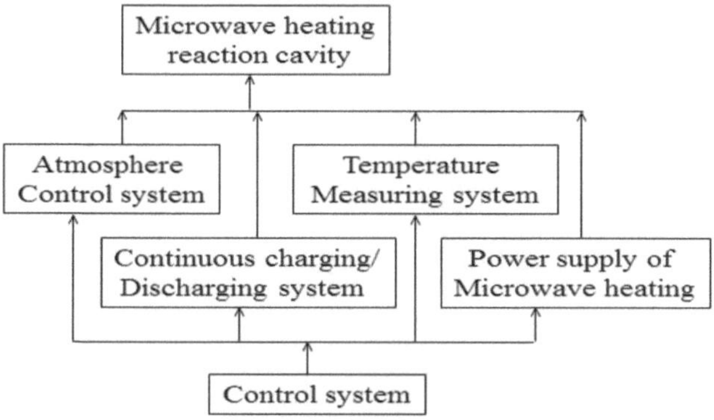

Figure.1 Schematic diagram of microwave heating equipment

Microwave heating behavior for the ammonium uranyl carbonate and related materials

It is well known that materials can be categorized into three principal groups [8] according to the interaction with microwave: transparent which are low loss materials, where microwave pass through without any losses; conductors where microwaves are reflected and cannot penetrate; and absorbing which are high loss materials, where microwaves are absorbed depending on the value of the dielectric loss factor [9]. Thus, it is essential to study the capability of ammonium uranyl carbonate to absorb microwave energy for assessing its suitability to microwave heating for performing thermal decomposition reaction.

A sample of 10g was placed in the crucible, which is positioned at the center of the microwave chamber. The tip of the thermocouple is placed at the center of the sample for every test. The effect of microwave irradiation time on the temperature rise of sample at an applied microwave power of 700 W, corresponding to different mixture ratios of ammonium uranyl carbonate-U_3O_8 is presented in Figure.2. It can be observed from the figure that sample temperature increases rapidly with an increase in the proportion of U_3O_8, in the mixture, depicting its ability to strongly absorb microwave energy in presence of a weak absorber, ammonium uranyl carbonate. A similar observation has been reported be Wei Ming Bao et al., [10]. It is well established that, the addition of a material which has good microwave absorbing capacity with poor microwave absorbing capacity, can have significant effects on the microwave absorption characteristics of the poor absorber [11]. It can be seen that a 20% addition of ammonium octaoxide with ammonium uranyl carbonate, resulted in an excellent increase in the sample temperature of 1100 K in 220 seconds. This well indicates the process feasibility to prepare triuranium octaoxide by calcination of ammonium uranyl carbonate, under microwave fields.

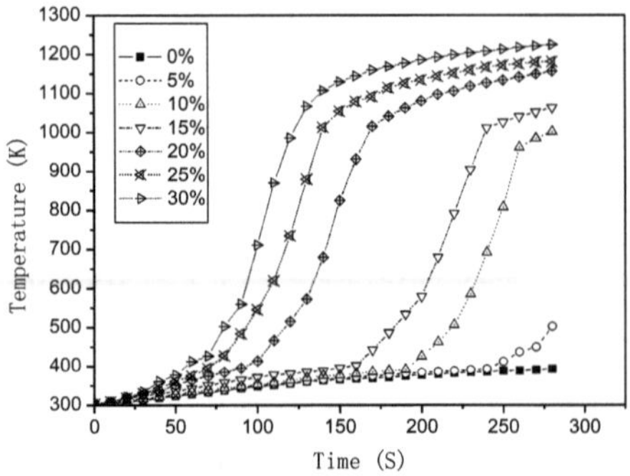

Figure.2. Curve of increasing temperature of ammonium uranyl carbonate and relative materials in microwave field (P=700 W)

1-ammonium uranyl carbonate; 2,3,4,5,6,7- ammonium uranyl carbonate-U_3O_8 mixture of composition 5%, 10%, 15%, 20%, 25% and 30% , respectively

Results and discussion

With the preliminary experiments showing a significant increase in the temperature rise of the sample, with addition of 20% U_3O_8, in comparison with addition of 15% with ammonium uranyl carbonate, a 20% addition of U_3O_8 is chosen to assess the effect of other parameters.

The effect of microwave irradiation time in the total uranium and U^{4+}

Experiments are carried out to assess the effect of microwave irradiation time on the value of total uranium and U^{4+} of triuranium octaoxide. The calcination is performed with U_3O_8-ammonium uranyl carbonate mixture of composition 20%, at 700W for six different microwave irradiation times of 4, 6, 8, 10, 12, 14 minutes, respectively. Figure.3 shows the effect of microwave irradiation time on the value of total uranium and U^{4+} of triuranium octaoxide. The results indicated that the value of total uranium and U^{4+} of triuranium octaoxide increase sharply with increase in microwave irradiation time until 6 minutes, and reaches an asymptote with further increase in the microwave irradiation, indicating complete conversion of ammonium uranyl carbonate within 10 minutes. Hence 10 minutes of processing time are chosen as the optimal time for calcination.

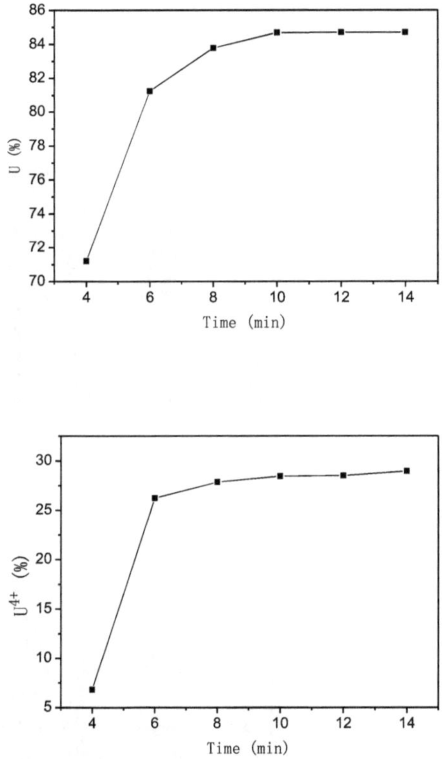

Figure.3 Relationship between microwave irradiation time on the total uranium and U^{4+} of triuranium octaoxide

The effect of microwave power on the total uranium and U^{4+}

The calcinations process is a temperature dependent process and the rate of calcinations increase with increase in the calcinations temperature. In a microwave induced process the temperature of the calcinations process can be altered with the variation in the applied microwave power. The effect of microwave power on the value of total uranium and U^{4+} of triuranium octaoxide is represented in Figure. 4. The Figure. 4 shows that the value of total uranium and U^{4+} of triuranium octaoxide significantly increases with increase in the applied microwave power, with the rate of increase significantly higher at lower microwave power than at higher microwave power, with the rate of conversion reaching a plateau at higher microwave applied power. A significant rise in the value of total uranium and U^{4+} is evident from the figure with the value of total uranium and U^{4+} of triuranium octaoxide reaching 84.10% and 27.43%, respectively for microwave irradiation time 10 minutes, at a microwave power of 460 W, as compared to the rise beyond. The increased rate of conversion with increase in the applied microwave power is due to the proportional increase in temperature of the sample, which reduces the activation energy for the thermal degradation which promotes the rate of reaction. As the conversion reaches an asymptote at a microwave applied power of 700 W, an optimal microwave applied power is chosen as 700 W in the present study.

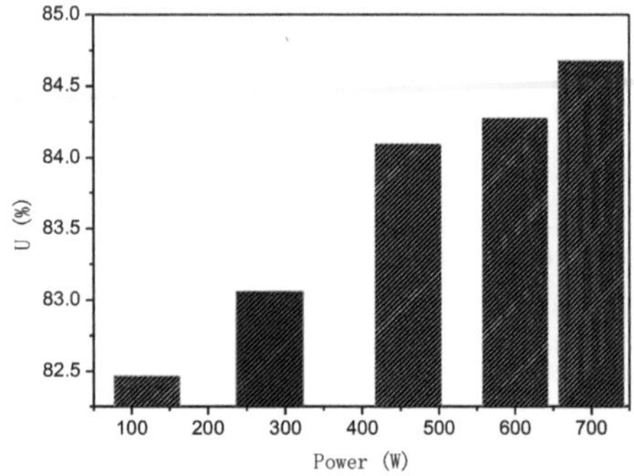

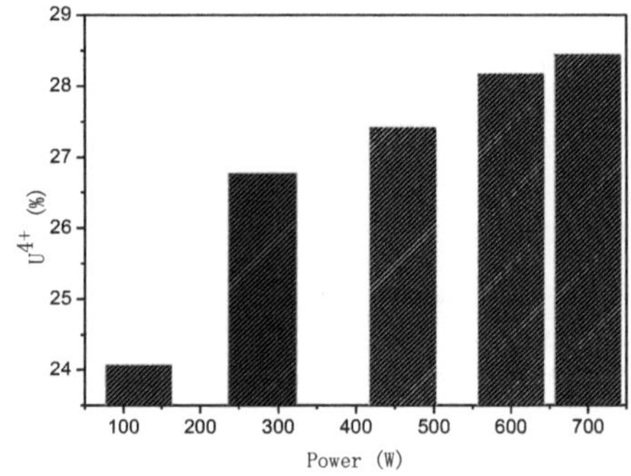

Figure.4 Relationship between microwave power on the total uranium and U^{4+} of triuranium octaoxide

The effect of weight of the sample on the total uranium and U^{4+}

Figure.5 shows that the variation in value of total uranium and U^{4+} of triuranium octaoxide corresponding to the change in the weight of sample at an applied microwave power 700 W and an irradiation time of 10 minutes. It can be observed from the figure that the weight of sample of U_3O_8-ammonium uranyl carbonate mixture of composition 20%, do not show any significant effect on the value of total uranium and U^{4+} of triuranium octaoxide in the range of test covered in the present study (30-60g). Although microwave heating is a volumetric heating, for opaque material, most the microwave will be absorbed on the surface of the material which gets converted as heat. The difference in temperature at the surface to the inside temperature could cause the heat dissipation through the mode of conduction heat transfer. As the heat transfer barrier increases with increase in the sample size, a marginal variation in the conversion could be expected. As the variation in the sample weight covered in the present study is not significant the effect of weight of the sample on the conversion rate is insignificant. In the work, the weight of sample 60 g was chosen.

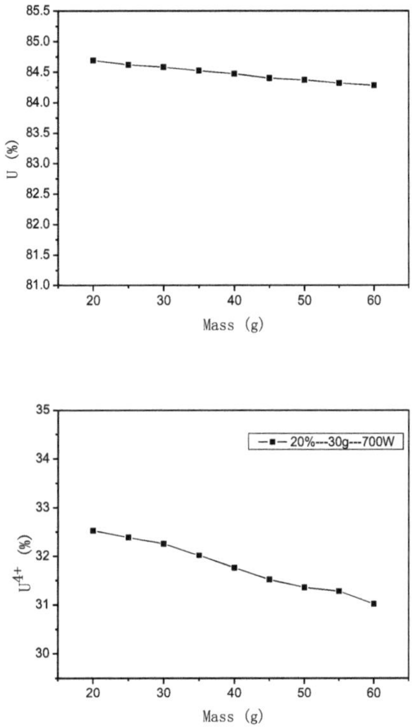

Figure.5 Relationship between weight of sample on the total uranium and U^{4+} of triuranium octaoxide

Conclusions

Ammonium uranyl carbonate has weak capability to absorb microwave energy, while triuranium octaoxide has the very strong capability to absorb microwave energy and the sample temperature increased rapidly with an increase mixture ratio of triuranium octaoxide.The value of total uranium and U^{4+} of triuranium octaoxide increases with increase in microwave power and irradiation time, while the increases in the weight of the sample do not show any significant effect. The optimal condition is identified to be an applied microwave power 700W, with 10 minutes microwave irradiation time at a maximum sample of 60 g respectively. Under these conditions the value of total uranium and U^{4+} of triuranium octaoxide is 84.28% and 31.02%, which meet next procedure, indicating the feasibility to prepare triuranium octaoxide by calcination from ammonium uranyl carbonate under microwave fields.

Acknowledgements

Financial support for this work from the National Natural Science Foundation Council of China (50734007), the National Technology Research and Development Program of China (863Program) and Project (2011FZ038) Supported by the Applied Basic Research of Yunnan Province are gratefully acknowledged.

Reference

1. Yi xiang Duan et al., "Uranium emission spectra with power microwave plasma source,"*Ananytica Chimica Acta*, 532 (1) (2005), 47 – 54.

2. Kun Woo Song et al., "Grain size control of UO_2 pellets by adding heat-treated U_3O_8 particles to UO_2 powder," *Journal of Nuclear Materials*, 317 (2,3) (2003), 204 – 211.

3. D.A. Jones et al., "Microwave heating applications in environmental engineering," *Resource, Conservation and. Recycling*, (34) (2002), 75 – 99.

4. A.S. Mujundar (Ed.), 2nd ed., Handbook of Industrial Drying, vol. 1/2, (Marcel Dekker Inc., New York, 1995).

5. Standish et al., "Microwave application in the reduction of the metal oxides," *Journal of Microwave Power and Electromagnetic Energy*, 25 (3) (1996), 177 – 180.

6. Haque, K.E. "Microwave irradiation treatment of a refractory gold concentrate," (Proceedings International Symposium on Gold Metallurgy, Winnipeg, Canada, 1987), 327 – 339.

7. Kelly, R. M and Rowson, N .A, "Microwave reduction of oxidized ilmenite concertrates," *Minerals Engineering*, 8(11) (1995), 1427 – 1438.

8. D.E Clark, D. C. Folz, and J. K.West. "Processing materials with microwave energy," *Materials Science and Engineering A*, 287(2) (2000), 153 – 158.

9. T.J. Appleton et al., "Microwave technology for energy-efficient processing of waste," *Applied Energy*, 81 (1) (205), 85 – 113.

10. Wei-ming BAO, Bao-xiang CHANG, and Ze-hong GUO. "The research for applying microwave denigration on the conversion of high-enriched uranium," *Atomic Energy Scinece and Technology*, 29(3) (2005), 268– 274.

11. Jing-jing YANG et al., "Microwave absorbing properties and electric field distribution of conductor-dielectric compound," (The Proceedings of ISAPE 2008, Kunming, Yunnan, China), 673 – 676.

2015 EPD Congress

SOLAR CELL SILICON

2015 EPD Congress

SYMPOSIUM: SOLAR CELL SILICON

Silicon Production and Refining

Gabriella Tranell,
Norwegian University of Science and Technology

EPD Congress 2015
Edited by: James Yurko, Antoine Allanore, Laura Bartlett, Jonghyun Lee, Lifeng Zhang, Gabriella Tranell, Yulia Meteleva-Fischer, Shadia Ikhmayies, Arief Suriadi Budiman, Prabhat Tripathy, and Guy Fredrickson
TMS (The Minerals, Metals & Materials Society), 2015

DIRECTIONAL GROWTH OF BULK SILICON FROM SILICON-ALUMINUM-TIN MELTS

Yaqiong Li[1, 2, 3], Yi Tan[1, 2], Kazuki Morita[3]

[1] School of Materials Science and Engineering, Dalian University of Technology, Dalian 116024, China
[2] Key Laboratory for Solar Energy Photovoltaic System of Liaoning Province, Dalian University of Technology, Dalian 116024, China
[3] Department of Materials Engineering, Graduate School of Engineering, The University of Tokyo, Tokyo 1538505, Japan

Keywords: Bulk Si, Si-Al-Sn alloy, directional solidification process

Abstract

Molten Si-Al alloy, with the addition of Sn, was used for metallurgical grade Si purification by a low-temperature solidification method, with the aim of improving the recovery rate of Si. In this ternary melt, bulk Si was grown with a planar front by a directional solidification process under a well-controlled thermal gradient and growth rate conditions, thereby reducing the contamination of solvent metals. Effects of cooling rate, amount of Sn addition, and temperature gradient on the quality of bulk Si were determined. The criterion for constitutional supercooling for single-phase growth from multicomponent melts was employed to evaluate the growth process.

Introduction

Continued development of the photovoltaic (PV) industry has led to a high demand for solar-grade silicon (SoG-Si), and keeping its production cost low is essential for the widespread use of Si solar cells [1]. Until now, a number of efforts have been devoted to purify metallurgical grade Si (MG-Si) by metallurgical methods. Si purification by alloy solidification offers several advantages over other methods, such as a refining temperature lower than the Si melting temperature (i.e., 1687 K) and an enhanced solid/liquid segregation tendency of impurities; hence, this method has the potential to further reduce energy consumption and environmental impact.

When considering the alloy solidification process, one focus is to evaluate the refining efficiency amongst various solvent metals. In this regard, Cu[2], Ni[3], Sn[4-6], Fe[7,8], and Al[9-12] have been extensively investigated. Another focus is the method for separating Si from the melt, which is of great significance ensuring Si purity and reducing metal loss. However, very few studies on these aspects have been reported to date. Recently, the Morita group proposed a bulk Si growth method, which was originally used to produce thin layers for semiconductor manufacturing, known as Liquid-Phase Epitaxy (LPE). This approach can afford high-quality bulk Si and prevent its contamination by the melt. Nishi et al.[13] obtained bulk Si crystals in 44.7mol%Si-Al melt in the temperature range of 1273-1173 K, which decreased the Al impurity to the level of its solid solubility in Si (about 200 ppmw at 1173-1273 K). Ohshima et al. and Ma et al. also applied the same method to successfully produce bulk Si from Si-Cu[14] and Si-Sn[6], respectively. From these results, Si-Sn alloy, with a moderated liquidus slope, revealed a larger growth rate, indicating that Sn favored to bulk Si growth. Therefore, the Si-Al alloy with the addition of Sn was proposed as a ternary alloy system aimed at improving bulk Si growth by the directional solidification process.

A basic understanding of Si directional growth from the ternary alloy Si-Al-Sn was investigated in this work. The criterion for constitutional supercooling, which was a morphological investigation of the solid/liquid interface, was studied in consideration of *G/R* (*G* was the temperature gradient, and *R* was the growth rate of bulk Si) and alloy compositions.

Experimental

Bulk Si growth from Si-Al-Sn alloy

High purity Si (99.99999 wt% pure), Al shots (99.9 wt% pure), and Sn particles (99.9 wt% pure) were used as raw materials. The compositions of the alloys are Si-61.5mol%Al-5mol%Sn, Si-59.2mol%Al-10mol%Sn, Si-56.1mol%Al-15mol%Sn, Si-53.5mol%Al-20mol%Sn, Si-50.5 mol%Al-25mol%Sn, and Si-47.8 mol%Al-30mol%Sn, the liquidus temperature of which were all 1173 K. A total mass of 0.3 g of metals were first weighed and pre-melted by an induction furnace.

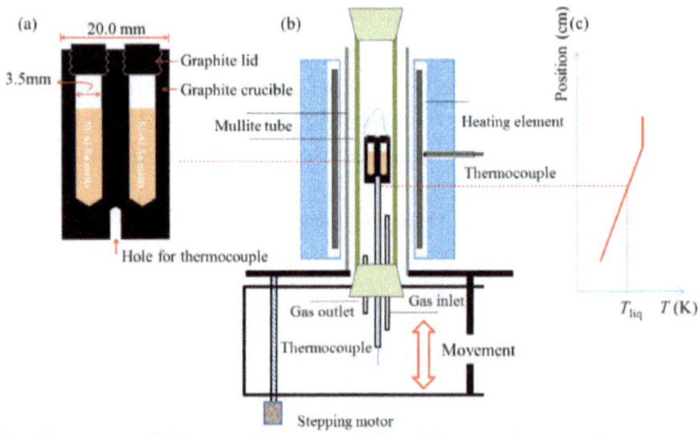

Figure 1. Schematic diagram of (a) graphite crucible, (b) experimental apparatus for a directional solidification process, and (c) a temperature profile for the center of a hot-zone.

A SiC electric furnace with a stepping motor and Proportional-Integral-Differential control system was used for the directional growth of Si crystals as shown schematically in Figure 1. Firstly, the pre-melted alloy in a dense graphite crucible with a lid (99.999 wt% pure, inner diameter 3.5 mm, height 20 mm) was placed into a mullite tube, and then melted at 1223 K for 30 min. Secondly, the sample was moved downwards and cooled to 1073 K at a constant cooling rate, which was controlled by the lowering rate of the mullite tube and the temperature gradient of the furnace. The lowering rate was varied in the range of 0.02-0.08 mm/min under the temperature gradient of 3.13-4.69 K/mm. Finally, the sample was removed from the top of the furnace and quickly quenched in water in order to evaluate the state of solidification. A Pt30%Rh-Pt6%Rh thermocouple was used and positioned in the center of the crucible to measure the temperature changes during the solidification process. All experiments were carried out in an Ar atmosphere (99.999 wt% pure) at a flow rate of 200 mL/min, in which the moisture, CO_2, and O_2 were eliminated by passing successively through silica gel, soda lime, magnesium perchlorate, and magnesium turnings heated at 823 K.

After the directional solidification process, the above mentioned samples were cut parallel to the temperature gradient in two equal parts, followed by a polishing treatment, and observed by Optical Microscopy (OM, Olympus BX60, Japan).

Results and Discussion

Principle of bulk Si growth from Si-Al-Sn alloy melts by directional solidification method

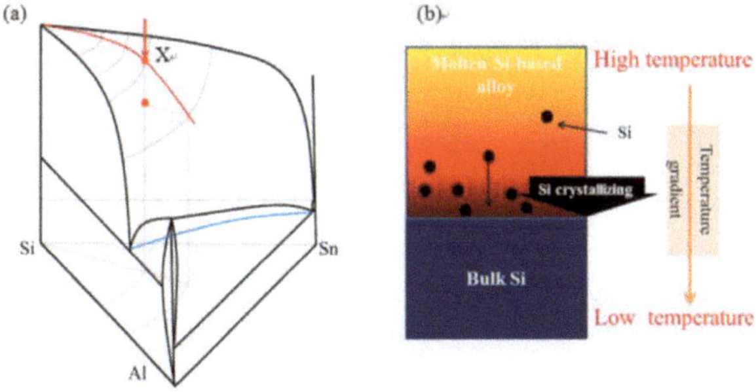

Figure 2. (a) Three-dimensional phase diagram of Si-Al-Sn, and (b) schematic image of bulk Si growth from alloy melt by a directional solidification method.

Figure 2 (a) illustrates a three-dimensional phase diagram of Si-Al-Sn in which Si with a composition of X could precipitate as a first solid phase from the ternary alloy melt as the temperature decreases. As shown in Figure 2 (b), when the ternary melt was placed in a temperature gradient region and moved downward at a low constant rate, the low temperature melt region at first became oversaturated Si and then overcame any kinetic barriers for Si crystals nucleation and growth [15]. With continuous fall in temperature, the amount of bulk Si was further increased, and grew directionally. During this process, the driving force for Si growth was considered to be the slow cooling of the saturated melt.

For the binary alloy system, the kinetics of Si growth has been discussed by assuming that the Si growth was mainly controlled by the diffusion transport of Si in the melts, and so the growth rate of bulk Si has been successfully expressed by the steady-state diffusion equation for Si in binary alloy system, as shown in Eq.(1),

$$V = D_{\text{Si in melts}} \cdot \frac{\partial X_{\text{Si in melts}}}{\partial T} \cdot \frac{\partial T}{\partial x} \quad (1)$$

Here, the D_{Si} in melt denotes the diffusion coefficient of Si in the alloy melts, the $\frac{\partial X_{\text{Si in melts}}}{\partial T}$ and $\frac{\partial T}{\partial x}$ are the liquidus slope and temperature gradient, respectively. Therefore, the growth rate of Si was determined by (1) diffusion coefficient of Si in melts, (2) liquidus slope, and (3) temperature gradient. The first two terms denote the physical properties of the alloy, and the last is the performance of the heating equipment. As compared with binary alloy, the ternary alloy system became complicated because the liquidus was a surface and solute diffusive interactions may have occurred. The Si growth from Si-Al-Sn melt was then investigated.

Table I. Experimental conditions and results.

No.	Composition (mol%)			Lowering rate (mm/min)	Cooling rate (K/mm)	G (K/mm)	Growth rate $R\times10^3$ (mm/min)	G/R (10^{-3} K·min/mm^2)	Bulk Si length (mm)	Interface morphology
	Si	Al	Sn							
J8-1	28.9	56.1	15	0.08	0.2504	3.13	1.90	1647.77	0.75	plane
J8-2	26.5	53.5	20	0.08	0.2504	3.13	15.12	207.01	5.97	coarse
J8-3	24.5	50.5	25	0.08	0.2504	3.13	3.32	942.88	1.31	in-between
J8-4	22.2	47.8	30	0.08	0.2504	3.13	1.70	1848.66	0.67	in-between
J4-1	28.9	56.1	15	0.04	0.1252	3.13	1.98	1582.84	1.62	in-between
J4-2	28.9	56.1	15	0.04	0.1252	3.13	1.98	1930.95	1.33	in-between
J2-1	30.8	59.2	10	0.02	0.0626	3.13	1.05	2983.84	1.81	plane
J2-2	28.9	56.1	15	0.02	0.0626	3.13	0.67	4669.67	1.16	plane
J2-3	26.5	53.5	20	0.02	0.0626	3.13	0.63	4932.88	1.09	plane
J2-4	22.2	47.8	30	0.02	0.0626	3.13	0.88	3558.49	1.52	plane
J4-3	26.5	53.5	20	0.04	0.1516	3.79	9.30	407.31	6.21	coarse
J4-4	33.5	61.5	5	0.04	0.1516	3.79	2.10	1802.51	1.40	plane
J4-5	33.5	61.5	5	0.04	0.1516	3.79	2.10	1801.87	1.40	plane
J2-5	26.5	53.5	20	0.02	0.0758	3.79	0.93	4094.91	1.21	plane
J2-6	24.5	50.5	25	0.02	0.0758	3.79	1.32	2866.16	1.73	in-between
J2-7	22.2	47.8	30	0.02	0.0758	3.79	1.01	3747.48	1.33	in-between
J2-8	30.8	59.2	10	0.02	0.0938	4.69	1.21	3877.40	1.35	plane
J2-9	26.5	53.5	20	0.02	0.0938	4.69	3.87	1210.59	4.33	coarse

The solid/liquid interface morphology was markedly changed with varying cooling rates, and classified as three types as shown in Figure 3. The cooling rate was determined by the product of lowering rate and temperature gradient.

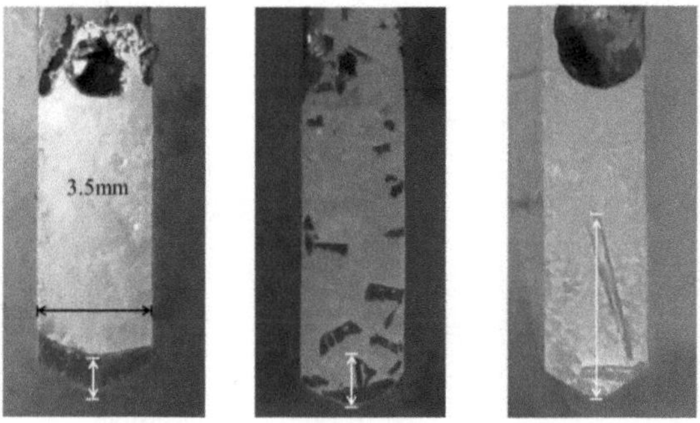

Lowering rate:	(a) 0.02mm/min	(b) 0.04mm/min	(c) 0.08mm/min
Cooling rate:	0.063 K/min	0.126 K/min	0.252 K/min
Sample No.:	J2-3	J4-2	J8-2

Figure 3. Photos of Si-53.5mol% Al-20mol%Sn samples with varying lowering rate (Temperature gradient of 3.13 K/mm; Temperature range of 1173-1073 K)

In Figure 3 (a) bulk Si with a planar interface was formed under steady growth solidification at a cooling rate of 0.063 K/min, and the length of the bulk Si was about 1.09 mm. As the cooling rate increased, the in-between interface was developed with dendritic crystals as shown in Figure 3 (b), while the rough interface was found in Figure 3 (c) obtained at a cooling rate of 0.252 K/min. Analysis by OM of the magnified views of the interface are shown in Figure 4. No inclusions were found in the bulk Si of sample J2-3, while some metal inclusions were left in the bulk Si of sample J4-2 and J8-2, indicating that the occurrence of interfacial breakdown at a high cooling rate was accompanied by trapping of solvent melts. This is because Si atoms near the interface have enough time to diffuse and solidify in a stable state at a slow cooling rate, forming a planar interface on a macroscopic scale. A high cooling rate could cause constitutional supercooling of the melt ahead of the solid Si front, which is able to trigger nucleation on available nucleating particles, resulting in the instability of the interface and the reduction of bulk Si. According to these three types of interface morphologies, the criterion for constitutional supercooling was applied to investigate the effects of G/R and alloy compositions on the bulk Si growth. Here, R was the growth rate of bulk Si, which was determined by the thickness of the bulk Si during the duration of directional solidification experiments. This was listed in Table I. The value of G/R was plotted against the content of Si in Si-Al-Sn alloy as shown in Figure 5 together with the reported results in the Si-Al system.

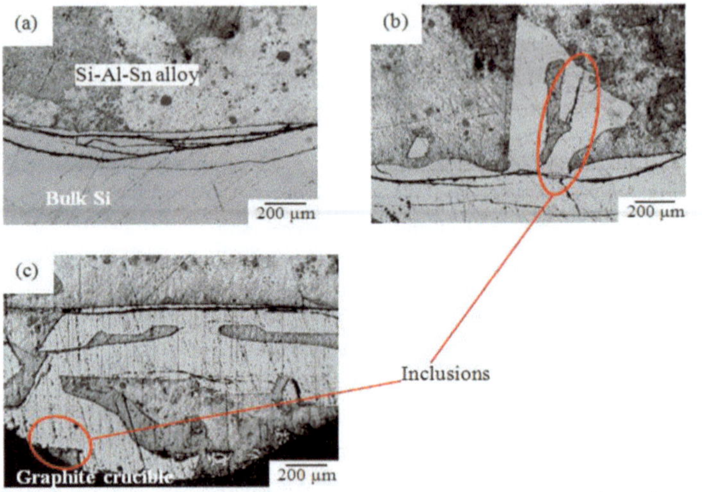

Figure 4. OM images of Si-53.5mol%Al-20mol%Sn with varying lowering rate. (a) 0.02mm/min, (b) 0.04 mm/min, and (c) 0.08 mm/min.

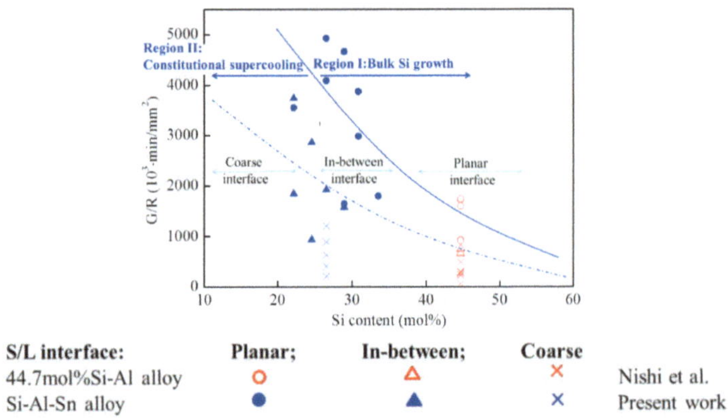

Figure 5. Relationship between the solid/liquid interface and growth conditions.

As shown in Figure 5, there are two separate regions according to whether or not constitutional supercooling occurred. In Region I, stability conditions were provided for bulk Si with a planar interface growing from melts, and the instability of the interface could be eliminated by the application of a sufficiently large temperature gradient and sufficiently low growth rate, namely, a larger value of *G/R*. With the increasing content of Si, the requirement of *G/R* was obviously decreased. In Region II, constitutional supercooling occurred which was produced by rejection of solvent in front of the growing front, and the interface became increasingly unstable, possibly forming a coarse interface. Therefore, the constitutional supercooling became more pronounced in Si-Al-Sn melts with decreasing content of Si, and the unstable growth morphologies occurred more easily.

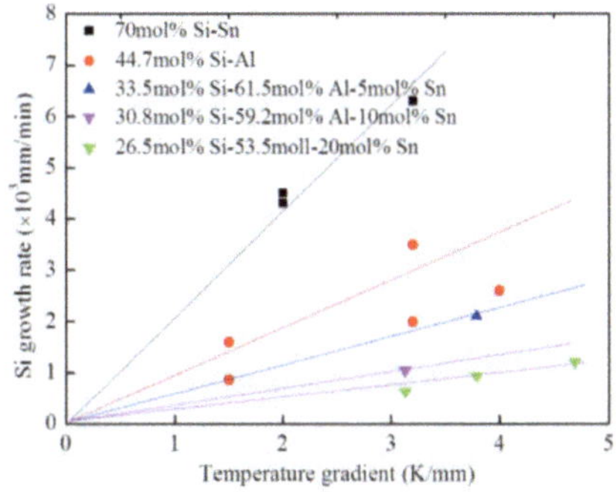

Figure 6. Relationship between Si growth rate and temperature gradient in various alloy systems.

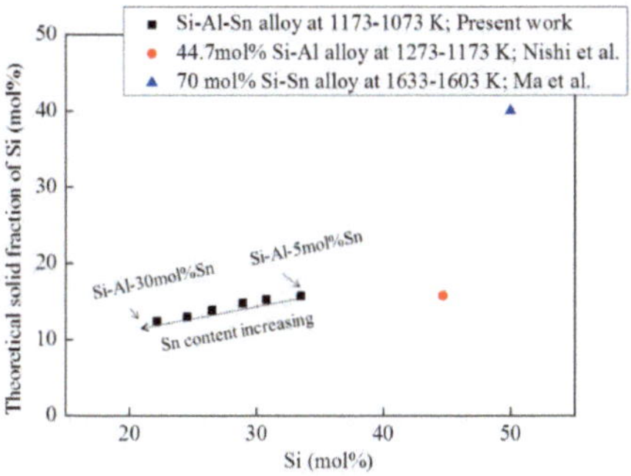

Figure 7. Theoretical solid fraction of Si in Si-Al-Sn, 70mol%Si-Sn and 44.7mol%Si-Al alloy systems.

Based on the above results, the growth rate of Si in Si-Al-Sn melts under stable growth conditions was plotted against the temperature gradient in Figure 6, together with the reported results in 44.7mol%Si-Al and 70mol%Si-Sn alloy systems. The lowering rate in these experiments was fixed at 0.02 mm/min. For the Si-Al-Sn alloy system, the growth rate of Si has a linear relationship with the temperature gradient, and it increased with the increasing temperature gradient. Additionally, with increasing Sn content, Si-Al-Sn alloy with reducing content of Si showed a lower growth rate, which revealed that the Si growth rate was also limited by the content of Si. As compared with 70mol%Si-Sn and 44.7mol%Si-Al alloy systems, the growth rate of Si in Si-Al-Sn alloy tends to be a little lower, but with a similar order of magnitude. It can be explained using the theoretical solid fraction of Si as shown in Figure 7. The 70mol%Si-Sn alloy precipitated the highest amount of solid Si at a temperature range of 1633-1603 K. More Si could be precipitated from the melts per time unit, namely, high Si growth rate. While Si-Al-Sn alloy with liquidus temperature of 1173 K revealed a low solid fraction of Si at temperature range of 1173-1073 K, therefore resulting in a low Si growth rate.

Conclusions

Bulk Si was obtained from the Si-Al-Sn alloy system by the directional solidification method, which could be applied to separate Si from the solvent. Based on the effect of *G/R* and alloy compositions on the solid/liquid interface, Si-Al-Sn alloy with high content of Si and/or high *G/R* ratio were found to be beneficial for bulk Si growth with a planar interface; otherwise, constitutional supercooling occurred, resulting in an unstable interface. The Si growth rate increased monotonously with the Si content and the temperature gradient.

Acknowledgements

This program is financially supported by National Natural Science Funds of China (Grant No.51104028).

References

[1] S. Pizzini, "Towards solar grade silicon: Challenges and benefits for low cost photovoltaics", *Solar Energy Materials and Solar Cells*, 94 (2010), 1528-1533.

[2] A. Mitrašinović and T. Utigard, "Refining Silicon for Solar Cell Application by Copper Alloying", *Silicon*, 1 (2009), 239-248.

[3] Z. Yin et al., "Solvent refining of silicon using nickel as impurity getter", *Canadian Metallurgical Quarterly*, 50 (2011), 166-172.

[4] X. D. Ma et al., "Phase relations and thermodynamic property of boron in the silicon-tin melt at 1673K", *Journal of Alloys and Compounds*, (529) 2012, 12-16.

[5] X. D. Ma et al., "Removal of Boron from Silicon-Tin Solvent by Slag Treatment", *Metallurgical and Materials Transactions B*, (2013), 1-6.

[6] X. D. Ma et al., "Si growth by directional solidification of Si-Sn alloys to produce solar-grade Si", *Journal of Crystal Growth*, 377 (2013), 192-196.

[7] S. Esfahani and M. Barati, "Purification of metallurgical silicon using iron as an impurity getter part I: Growth and separation of Si", Metals and Materials International, 17 (2011), 823-829.

[8] S. Esfahani and M. Barati, "Purification of metallurgical silicon using iron as impurity getter, part II: Extent of silicon purification", *Metals and Materials International,* 17 (2011), 1009-1015.

[9] T. Yoshikawa et al., "Boron removal by titanium addition in solidification refining of silicon with Si-Al melt", *Metallurgical and Materials Transactions B*, 36 (2005), 837-842.

[10] T. Yoshikawa and K. Morita, "Removal of phosphorus by the solidification refining with Si–Al melts", *Science and Technology of Advanced Materials*, 4 (2003), 531.

[11] T. Yoshikawa and K. Morita, "Refining of Si by the Solidification of Si-Al Melt with Electromagnetic Force", *ISIJ International*, 45 (2005), 967-971.

[12] T. Yoshikawa and K. Morita, "Continuous solidification of Si from Si-Al melt under the induction heating", *ISIJ International*, 47 (2007), 582-584.

[13] Y. Nishi et al., "Control of Si Crystal Growth during Solidification of Si-Al Melt", *Materials Transactions*, 51 (2010), 1227-1230.

[14] Y. Ohshima et al., "Effect of Solidification Conditions on Si Growth from Si-Cu Melts". *Supplemental Proceedings: Materials Processing and Energy Materials,* 1 (The Minerals, Metals & Materials Society, Hoboken, New Jersey, 2011), 677-684.

[15] C. Lesher and D. Walker, "Cumulate maturation and melt migration in a temperature gradient", *Journal of Geophysical Research: Solid Earth*, 93(1988), 10295-10311.

EPD Congress 2015
Edited by: James Yurko, Antoine Allanore, Laura Bartlett, Jonghyun Lee, Lifeng Zhang, Gabriella Tranell, Yulia Meteleva-Fischer, Shadia Ikhmayies, Arief Suriadi Budiman, Prabhat Tripathy, and Guy Fredrickson
TMS (The Minerals, Metals & Materials Society), 2015

PREPARATION OF SOLAR GRADE SILICON PRECURSOR BY SILICON DIOXIDE ELECTROLYSIS IN MOLTEN SALTS

Liangxing Li[1, 2], Jinzhao Guan[1], Aimin Liu[1], Zhongning Shi[1], Michal Korenko[3], Junli Xu[4], Bingliang Gao[1], Zhaowen Wang[1]

[1]School of Materials and Metallurgy, Northeastern University, Mail Box 117, Shenyang, China
[2]School of Construct and Resources Engineering, Jiangxi University of Science and Technology, Nanchang, China
[3]Institute of Inorganic Chemistry, Slovak Academy of Sciences, Dubravska cesta 9, 84536 Bratislava, Slovakia
[4]School of Science, Northeastern University, Shenyang 110004, China

Keywords: Silica, molten salt electrolysis, aluminum-silicon alloy, solar grade silicon

Abstract

Al-Si alloy, a precursor of solar grade silicon, was prepared by direct electrolysis in cryolite molten salt at 950 °C using high purity silica as material, liquid aluminum as the cathode and high purity graphite as the anode. The electrochemical behavior of Si(IV) ion was investigated using cyclic voltammetry method. The electrolysis products were characterized by XRD, SEM/EDS and ICP. The results indicate that the reduction process of Si(IV) on tungsten electrode is a two-step process and there is about 0.6 V gap between the two steps. The contents of boron and phosphorus in the aluminum-silicon alloy are 3 ppmw and 8 ppmw, which will make the directional solidification purification effectively and reduce the cost of preparation solar grade silicon from metallurgical grade silicon.

Introduction

Solar energy as a renewable energy, with inexhaustible potential, and without pollution, attracts more and more attentions over the world. Using the photoelectric converter in order to convert solar energy into electrical energy is so far the most common use of the solar cell.[1, 2] At present, more than 95% of the solar cells are using silicon as substrate. In the past few decades, demands for photo-voltaic (PV) cells have significantly increased.[3, 4] In photo-voltaic industry, Si materials have been widely employed for the fabrication of commercial solar cells for decades. Most of Si raw materials used for solar cells are high-cost solar grade (SOG), which act as barriers for the cost reduction of cell fabrication. Therefore, up to date, various methods of the production of solar grade silicon have been proposed.[5]

The production technologies of solar grade silicon mostly include Simens process, silane process and metallurgical process.[6] Directional solidification is a usual metallurgical route to remove impurities from metallurgical grade (MG) Si,[7] this approach is very effective to remove metallic impurities such as Cu, Al and Fe due to their small segregation coefficients (ratio of the impurity content in solid Si to that in liquid Si), but, this method is useless for the impurity with large segregation coefficients, such as phosphorus and boron (k_P=0.35, k_B=0.8).[8, 9]

In recent years, the development of an electrochemical reduction technique using molten salt has been proposed, particularly for the production of metal from its oxide.[10] Molten salt electrolysis as new production processes for SOG-Si, which allows reduction of the economic cost and energy consumption, and probably has a sufficient productivity.[11] In these investigations, SiO_2[12-14] or K_2SiF_6[15] dissolved in molten salt electrolyte, like LiF-NaF-KF, LiF-KF, or NaF-AlF_3[16], was used in order to obtained elemental Si. One of the most serious problems of these attempts, however, was the fact that Si was electrodeposited in the solid form[10]. This makes it difficult to separate the Si deposit from the molten salt, since the conductive properties of silicon is poor.

The objective of this paper is to present a molten salt electrolysis for the production of Al-Si alloy as a precursor of solar grade silicon with the low boron and phosphorus contamination.

Experimental

Silica (SiO_2≥99.99%, mass fraction) was used as raw material. NaF salt was dried before experiments under vacuum at 473℃ for 24 hours to remove moisture. Anhydrous AlF_3 was prepared by AlF_3•$3H_2O$ using dehydration techniques under the conditions of vacuum at 1050 °C. NaF-AlF_3-Al_2O_3 salt mixture was used as the electrolyte. Each experiment about 200 g of salt were brought into high purity graphite crucibles inside the medium-frequency induction furnace under argon atmosphere and at a constant temperature of 950 °C. The anode was prepared by connecting a 15 cm height and 20 cm diameter high purity cylindrical graphite to a stainless steel rod. This assembly was covered with quartz to prevent short-circuiting of electrodes and undesired interactions with the gases formed inside the cell vessel. Pure liquid aluminum was used as the cathode. A schematic diagram of the experimental apparatus was shown in Figure 1.

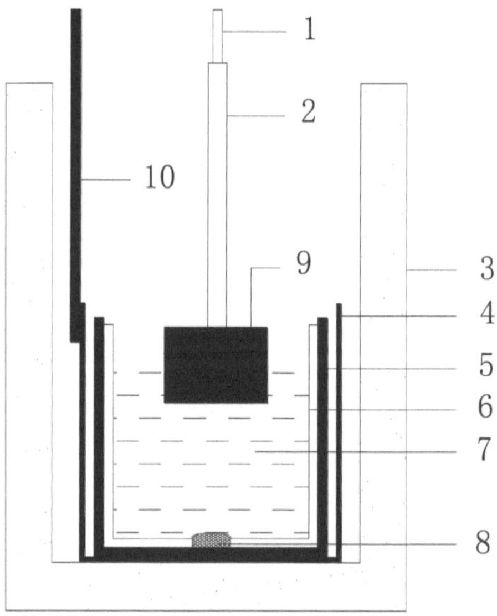

Figure 1. A schematic diagram of the experimental apparatus
1-anode rod; 2-corundum tube; 3-electric furnace; 4-iron crucible; 5-graphitecrucible; 6-corundum crucible; 7-electrolyte; 8-aluminum liquid; 9-graphite anode; 10-cathode rod.

In the cyclic voltammetry experiment, the electrochemical measurement equipment is AUTOLAB PGSTAT30 electrochemical workstation by a computer using the research software

GPES. The schematic view of the measurement cell was shown in Figure 2. Electrochemical measurements were performed were performed under Ar atmosphere. Tungsten wire (diameter 1 mm) was chosen as working electrode, the counter electrode was the graphite crucible (diameter, 100 mm; height, 105 mm), and the reference electrode was a platinum wire (diameter 0.5 mm).

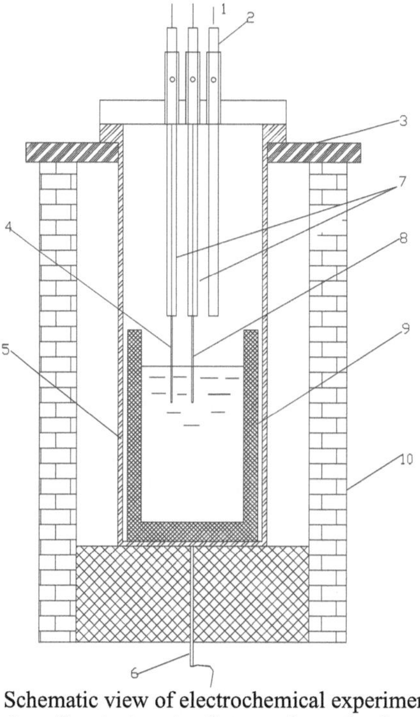

Figure 2. Schematic view of electrochemical experiment set-up
1-Ar gas; 2-gas conduit; 3-cooling-jacket; 4-reference electrode; 5-stainless steel crucible; 6-thermocouple; 7-Corundum tube; 8-work electrode; 9-graphite crucible; 10-Silicon carbide heating element furnace.

Scanning Electron Microscope (SEM, SUPERSCAN SSX-550, acceleration voltage was 15 kV) was used to observe the morphology of the products, and the composition of the sample was determined by an energy-dispersive X-ray spectrometer (EDS). The phase of the reduction products was identified by X-ray diffraction (XRD, X'pert Pro, PANalytical Co.) using Cu Kα radiation in the range of 10° to 90° (2θ) with a step of 5°/min. The boron and phosphorous content in the products were analyzed by inductive coupled plasma-atomic emission spectrometry (ICP-AES).

Results and Discussion

Cyclic Voltammetry

Cyclic voltammograms was carried out on tungsten electrode in the NaF-AlF_3 (the molar ratio are NaF/AlF_3=2.2 : 1)-Al_2O_3 (1.5 wt.%)-SiO_2 (3.0 wt.%) melts at 950 °C. Before adding SiO_2, a blank measurement was performed, as shown in Figure 3. The scan potential starts from 0 V to −2 V with a scan rate 10 mV/s, and the second inflexion potential is −2 V. There are two oxide current peaks (P_{a1} and P_{a2}) and two reduction current peak (P_{c1} and P_{c2}) on the whole cyclic voltammogram. The cathodic current arising at about −1.4 V is due to Al formation， the current peak of P_{a2} and P_{c2} corresponds to oxidation of the deposited aluminum and the reduction of Al^{3+} ions, respectively. P_{a1} and P_{c1} are the redox current peak of Na.

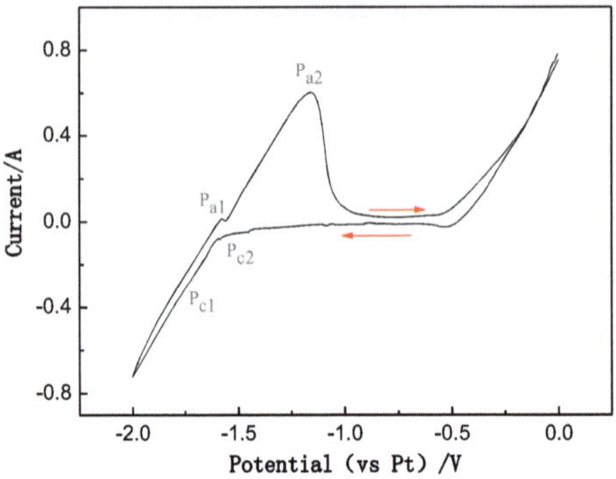

Figure 3. Cyclic voltammogram at the tungsten electrode in the NaF-AlF_3-Al_2O_3 (1.5 wt.%) molten salt at 950 °C, electrode area 0.316 cm^2, scan rate: 10 mVs^{-1}.

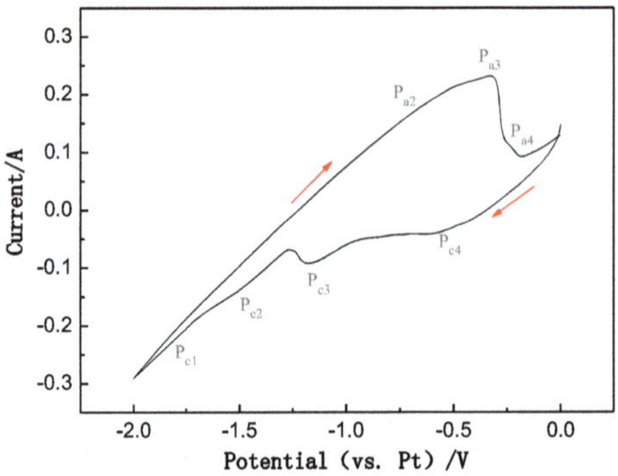

Figure 4. Cyclic voltammogram at the tungsten electrode in the NaF-AlF_3-Al_2O_3 (1.5 wt.%)-SiO_2 (3.0 wt.%) molten salt at 950 °C, electrode area 0.316 cm^2, scan rate: 10 mVs^{-1}.

Figure 4 is a typical cyclic voltammogram at tungsten electrode in NaF-AlF_3-Al_2O_3 (1.5 wt.%)-SiO_2 (3.0 wt.%) molten salt. The scan potential section is from 0 V to 2 V, and the open circuit potential is −0.3 V before scanning. The deposition of silicon takes place at potential more positive than aluminum deposition. Figure 3 shows that the reduction potential of Al is begin at −1.4 V, therefore, the reduction current peak of P_{c3} corresponds to the reduction of silicon ions. It is noted that before the reduction current peak of P_{c3}, there is a small reduction peak P_{c4} at the potential of −0.55 V, and the oxidation peak P_{a4} is corresponding to the reduction peak P_{c4} in the cyclic voltammogram. The reduction peak P_{c4} can be attributed to reduction of silicon step reaction: Si(IV) + 2e$^-$ = Si(II), the product is soluble Si(II) ions. Therefore, the reduction peak P_{c3} at the potential of −1.15 V corresponding to reaction is: Si(II) + 2e$^-$ = Si. JIA Ming and LAI Yan-Qing[17] also pointed out that the reduction process of Si(IV) is a two-step.

Potentiostatic electrolysis

Al-Si alloys were prepared by potentiostatic electrolysis from the NaF and AlF_3 (molar ratio=2.2 : 1) containing 1.5 wt% of Al_2O_3 and 3 wt% of SiO_2 melts at 950 °C under Ar atmosphere. Al_2O_3 was added to stabilize Si(IV) species and to increase the solubility of SiO_2. The anode was high purity graphite. A liquid Al of 99.9 wt% purity was used as the starting material for the cathode, which could absorb electrodeposited Si up to 45 wt% at the experimental temperature. Electrolysis was carried out by potentiostatic electrolysis at −1.2 V for 6 hours, which corresponds to a current density of around 0.5 A/cm^2. After electrolysis, Al-Si alloys were held for 30 min under the melts at the working temperature for sufficient interdiffusion and shaping up. Then, the mixture was gradually cooled down to the room temperature, spherical bulk alloys can be obtained.

Characterization of the Al-Si alloy

Figure 5 shows the complicated XRD patterns of Al-Si alloys obtained by potentiostatic electrolysis. EDX analysis confirmed that atomic composition was Al : Si =87.23 : 12.77 in the alloy layer. The XRD pattern of the sample confirms the formation of $Al_{3.21}Si_{0.47}$ phase. From the analyses of the electrolyzed samples, the formation reactions of the Al-Si alloys and the corresponding equilibrium potentials will be summarized.

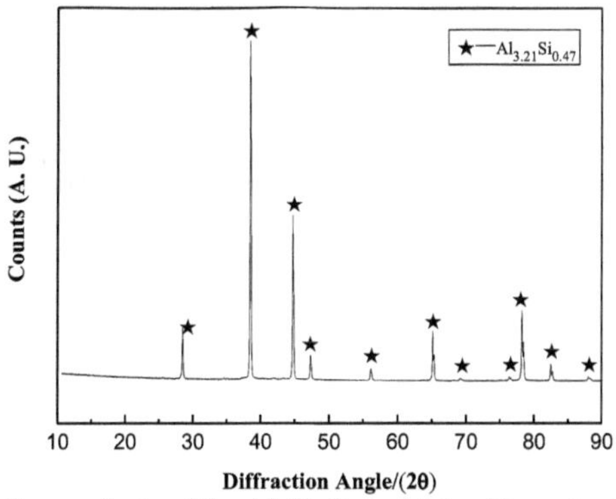

Figure 5. X-ray diffraction analysis of the Al-Si alloy obtained by potentiostatic electrolysis at −1.2 V in the Na_3AlF_6 (CR=2.2)-Al_2O_3 melts for 6 h at 950 °C.

The electrolysis products obtained were analyzed by SEM and EDX as shown in Figure 6. Figure 6(a) shows an SEM image of the cross-section of the Al-Si alloy which was obtained by potentiostatic electrolysis at −1.2 V. A red dashed circle region in Figure 6(a) was enlarged and shown in Figure 6(b). It can be seen that the cross-section existed three distinct layers. The EDX analysis performed simultaneously with SEM observation demonstrates that the white area consists of Al, Si, and Fe. Iron might be derived from the stainless steel anode rod. The black area in the middle is composed of Si crystals. It is well known that during the cooling down of Al-Si melt, the supersaturated Si is inclined to segregate.

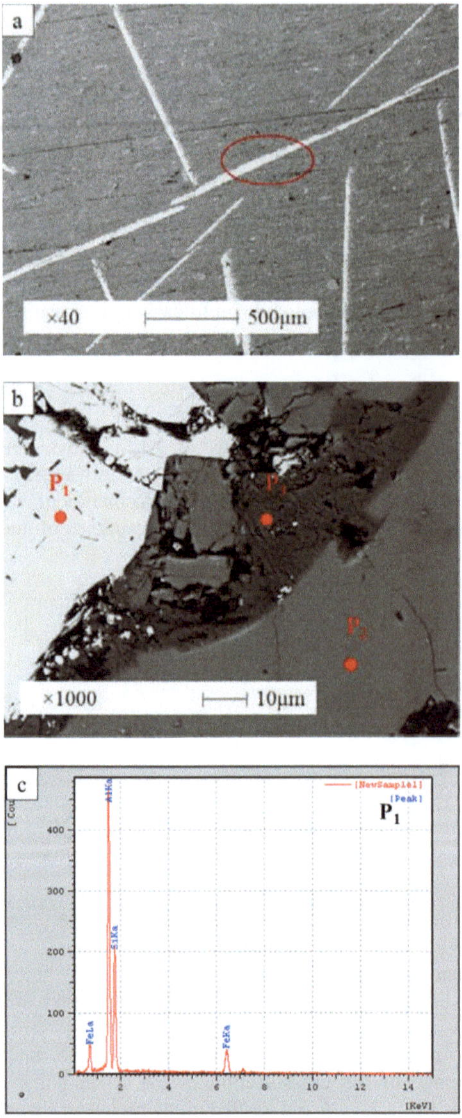

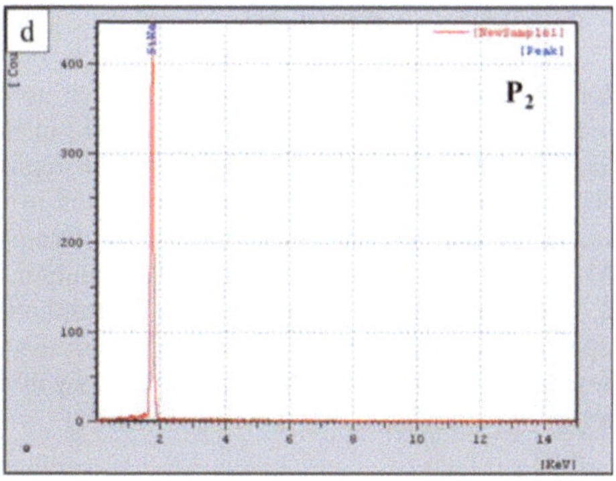

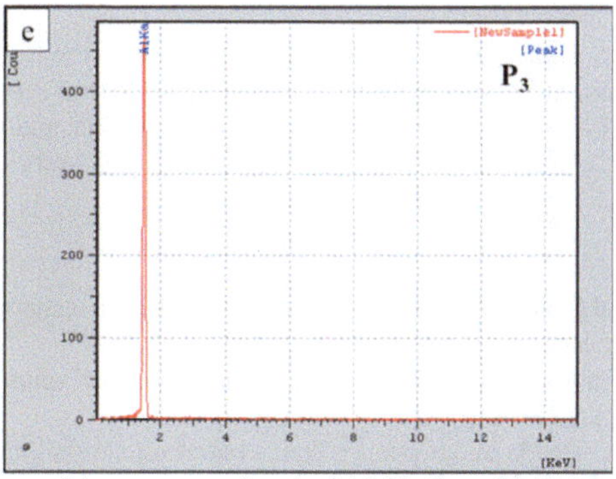

Figure 6. (a) SEM images of the cross-sectional of Al-Si alloy, (b) Enlarged SEM images of a Si product embedded in the Al-Si alloy in (a), a red dashed circle, (c) EDS energy spectrum of point P_1, (d) EDS energy spectrum of point P_2, (e) EDS energy spectrum of point P_3.

The mixture of Al-Si alloy and precipitated Si was dipped in a diluted HCl solution, Al and Fe can be easily removed by diluted HCl solution leaching. After acid leaching and deionized water cleaning, the flake-like Si crystals were obtained. The impurity contents of final Si products were determined by an ICP-mass, as shown in Table 1. It can be seen that the B and P impurities is very low in the Si products. The level of iron and nickel was found to increase slightly during the electrolysis due to brought by the oxidization of anode rod.

Table 1 Impurity contents of the Si products (ppmw)

Impurities	Si products
B	3.0
P	8.0
Al	1800
Fe	420
Ni	123
C	120

Conclusion

In this article, preparation of Al-Si alloy using high purity silica as material by electrolysis reduction was investigated. According to the cyclic voltammogram of molten $NaF-AlF_3-Al_2O_3-SiO_2$, the reduction process of Si(IV) on tungsten electrode is two-step mechanism involving an intermediate product Si(II). The first reduction peak at the potential of −0.55 V corresponds to the formation of Si(II), and the second reduction peak at −1.15 V corresponds to the formation of Si(IV). The product of Al-Si alloy was leached by diluted HCl solution, which make the silicon purity reaches 99.8 at.%. The content of impurity boron and phosphorus in the silicon are 3.0 ppmw and 8.0 ppmw, respectively, which has already reached the quality requirement of SOG-Si. It is easy to remove the impurities of Al, Fe, and Ni in Al-Si alloy by directional solidification to produce solar grade silicon.

Acknowledgements

The authors would like to acknowledge the financial support from the National Natural Science Foundation of China (Nos. 51322406, 51104042, and 51074046), the Program for New Century Excellent Talents (NCET-13-0107), Ministry of Education of China, and the Fundamental Research Funds for the Central Universities (No.N120402009 and No.N120405006).

References

1. Parida B, Iniyan S, and Goic R, "A review of solar photovoltaic technologies," *Renewable and Sustainable Energy Reviews*, 15(3), (2011), 1625-1636.
2. Dennis Elwell and Robert S. Feigelson, "Electrodeposition of solar silicon," *Solar Energy Materials*, 6(2), (1982), 123-145.
3. T.M. Bruton, "General trends about photovoltaics based on crystalline silicon," *Solar Energy Materials & Solar Cells*, (2002), 3-10.
4. Müller Aet al., "Silicon for photovoltaic applications," *Materials Science and Engineering: B*, 134(2-3),(2006), 257-262.
5. Gu X, Yu X, and Yang D,"Low-cost solar grade silicon purification process with Al-Si system using a powder metallurgy technique," *Separation and Purification Technology*, 77(1), (2011), 33-39.
6. Braga A F Bet al., "New processes for the production of solar-grade polycrystalline silicon: A review," *Solar Energy Materials and Solar Cells*, 92(4), (2008), 418-424.
7. Su Het al., "Preparation, microstructure and dislocation of solar-grade multicrystalline silicon by directional solidification from metallurgical-grade silicon," *Transactions of Nonferrous Metals Society of China*, 22(10), (2012), 2548-2553.
8. Ding Zet al., "Boron removal from metallurgical-grade silicon using lithium containing slag," *Journal of Non-Crystalline Solids*, 358(18-19), (2012), 2708-2712.
9. Morita K and Miki T., "Thermodynamics of solar-grade-silicon refining," *Intermetallics*, 11(11-12), (2003), 1111-1117.
10. Lee S, Hur J, and Seo C., "Silicon powder production by electrochemical reduction of SiO_2 in molten $LiCl-Li_2O$," *Journal of Industrial and Engineering Chemistry*, 14(5), (2008), 651-654.
11. Oishi Tet al., "Process for solar grade silicon production by molten salt electrolysis using aluminum-silicon liquid alloy," *Journal of the Electrochemical Society*, 158(9), (2011), E93-E99.

12. Yasuda Ket al., "Direct electrolytic reduction of solid SiO_2 in molten $CaCl_2$ for the production of solar grade silicon," *Electrochimica Acta*, 53(1), (2007), 106-110.
13. Cho S K, Fan F F, and Bard A J., "Formation of a silicon layer by electroreduction of SiO_2 nanoparticles in $CaCl_2$ molten salt," *Electrochimica Acta*, 65(2012), 57-63.
14. Xiao Wet al., "Rationalisation and optimisation of solid state electro-reduction of SiO_2 to Si in molten $CaCl_2$ in accordance with dynamic three-phase interlines based voltammetry," *Journal of Electroanalytical Chemistry*, 639(1-2), (2010), 130-140.
15. Cai Jet al., "Purification of metallurgical grade silicon by electrorefining in molten salts," *Transactions of Nonferrous Metals Society of China*, 22(12), (2012), 3103-3107.
16. Haarberg G Met al., "Electrodeposition of silicon from fluoride melts," *Electrochimica Acta*, 100(2013), 226-228.
17. JIA Minget al., "Electrodeposition Behavior of Silicon From Na_3AlF_6-LiF Melts," *Acta Physico-Chimica Sinica*, 27(5), (2011), 1108-1115.

EPD Congress 2015
Edited by: James Yurko, Antoine Allanore, Laura Bartlett, Jonghyun Lee, Lifeng Zhang, Gabriella Tranell, Yulia Meteleva-Fischer, Shadia Ikhmayies, Arief Suriadi Budiman, Prabhat Tripathy, and Guy Fredrickson
TMS (The Minerals, Metals & Materials Society), 2015

EFFECT OF TEMPERATURE IN EXTRACTION OF HIGH PURITY AMORPHOUS SILICA FROM RICE HUSK FOR SILICON PRODUCTION.

C. E. Ilochonwu[1], C. O. Nwajagu[1], I. U.Onyenanu[2], E. I. Nwonye[1]

[1]Scientific Equipment Development Institute Enugu, Nigeria.
[2]Department of Mechanical Engineering, Anambra State University, Nigeria.

Keywords: Silica, Rice Husk Ash, Burning

Abstract

This work shows how pure silica was derived directly from rice husks ash (RHA) in different temperatures for reduction to Silicon (Si). The processes to form Si materials are usually complex, costly and energy-intensive. The silica in the ash undergoes structural transformations depending on the conditions of combustion. A high purity silica was produced at 650°C after burning the rice husk (RH) for up to 7 hours. The purity of the silica was determined by dissolving the RHA with alkali solution to form sodium silicate solution of pH 11.0. This was added to hydrochloric acid of pH 1.5 which lowered the pH of the solution to 4.0 to form silica gels. The silica gels produced were washed and dried to powder and then characterized using XRD and EDX techniques. This study was done to reduce energy waste and environmental pollution in Nigeria as silica is used to develop many materials.

1. Introduction

Globally, approximately 550 million tonnes of rice paddy is produced each year. On average 20% of the paddy is husk, giving an annual total production of 120 million tonnes [1]. Rice husk biomass waste is very much abundant in Ebonyi State, Nigeria and other parts of northern Nigeria. This waste material can be found elsewhere and oftentimes we can see piles of rice husks at the back of the rice mill where they are stacked for disposal or some are thrown and burned on road sides to reduce its volume. This waste can be converted into fuel for domestic and wealth creation for other applications [2].

Rice husks are one of the largest readily available but most under-utilized biomass resources, being an ideal fuel for electricity generation. Rice husk is unusually high in ash compared to other biomass fuels close to 20%. The ash is 92 to 99.99% silica, highly porous and lightweight with a very high external surface area. Its absorbent and insulating properties are useful to many industrial applications [3].

The silica in the ash undergoes structural transformations depending on the conditions (time, temperature etc.) of combustion. At 550°C to 700°C amorphous ash is formed and at temperatures greater than this crystalline ash is formed. These types of silica have different properties and it is important to produce ash of the correct specification for the particular end use. The technology of converting rice husk ash to amorphous silica is to dissolve it in acid/base solutions to form precipitate or gels. This can be washed and dried to yield the intended product [1].

2. Materials and Methods

2.1 Preparation and Burning of Rice Husk

This study was developed by using the facilities and chemical laboratory at the Civil and Environmental Engineering Department of Worcester Polytechnic Institute Massachusetts, USA. The rice husk (RH) used in this investigation was supplied by BSG HandCraft USA which they are using as needed to improve filter bed performance of wheat beer or other high protein mashes as normal usage up to 5%. The rice husk is not treated with anything before burning it to ashes. First the rice husk was sieved in a sieving system that collected the retained rice husks in the USA Standard Test Sieve No.8 of 2.38mm. 30 pieces of rice husk were measured with the average length of 8.64mm, width of 2.09mm and thickness of 0.17mm.

Based on literature data, amorphous silica is produced when burning rice husks in the presence of oxygen for a long time at temperature below 700°C [4]. So, an incineration temperature of 550°C, 600°C and 650°C was chosen. The husks were burned inside a muffler furnace. For each burning 1000g of rice husks was put inside a steel basket and placed inside the furnace under static air for 7 hours at heating rate of 7.01°C/min. After each burning the rice husk ashes in the furnace were allowed to cool down overnight and then collected for chemical treatment.

2.2 Silica Extraction

Silica was extracted from rice husk ash (RHA) after each burning at different temperature, 250ml portions of 1 N NaOH were added to 50g of RHA samples and boiled in a covered 500ml beaker for 1 hour with constant stirring using magnetic stirrer to dissolve the silica and produce a Sodium Silicate solution [5]. The solutions were filtered through Whatman No.41 ashless filter paper after the solution was allowed to cool to room temperature.

Deionized water was added to residue silicate solution to get the pH up to 11. Acidic solution of hydrochloric was prepared by adding 6M hydrochloric acid to 100ml of deionized water until pH 1.5 was reached. The acidic solution was added with constant stirring to silicate solution of pH 11 until pH 4 was reached and gel formation was promoted by incubating it at room temperature for overnight [6].

The silica gels produced were crushed inside the beakers, dispersed in deionized water of 100ml and the slurry was centrifuged at 5000rpm for 10min to remove soluble salt, the supernatant were discarded and the gels were subjected to washing and then dried at 80°C for 24 hours to obtain pure powder silica.

2.3 Analyses

X-ray diffraction (XRD) patterns of ground silica powder samples were obtained by using a Bruker Advanced X-ray Solution D8 Advance at Life Sciences and Bioengineering Center of Worcester Polytechnic Institute (WPI) MA, USA. Using an acceleration voltage of 40KV and current of 40mA. 2-Theta was scanned from 5° to 55°, at a scanning speed of 2sec/step in the increment of 0.05.

After XRD measurement the silica powder samples were coated with carbon of 15nm in thickness to aid conductivity for the energy dispersive X-ray (EDX) and scanning electron microscope (SEM) measurement to characterize the particle morphologies of the sample.

3. Results and Discussion

3.1 Burning of Rice Husk

The table I below summarizes the percentage loss in weight after burning RH at different temperatures.

Table I. Different burning temperatures of rice husk (RH).

Rice husk specimen.	Temperature (°C).	Weight of RH before burning (g).	Weight after burning (g).	% loss.	Ash color after burning.
A	550	1000	181.7	81.83	Dull White
B	600	1000	180.3	81.97	White
C	650	1000	192.6	80.74	White

3.2 Characterization of rice husk ash

3.2.1 XRD Analysis

The reactivity and health issue of silica is based on its crystal structure. Crystalline silica is less reactive to chemical treatment and may serve as carcinogenic while the amorphous silica is rather very reactive and with no harmful effect to human [7]. The comparison of XRD spectra of the untreated and treated RHA at 550°C, 600°C and 650°C are shown in figure 1, 2, and 3 respectively. While in figure 4 and figure 5 we have XRD spectra comparison for the all untreated rice husk ash and treated rice ash at temperature of 550°C – 650°C. From the figure 1, 2, 3, 4 and 5 a broad peak spanning 2-theta angle range of 22.75 to 24 degree confirms the amorphous nature of silica from rice husk. In figure 3 it indicates that RHA at 650°C shows broad peak of 2-theta of 23.15 degree both on the untreated and treated with a line count difference of 29.

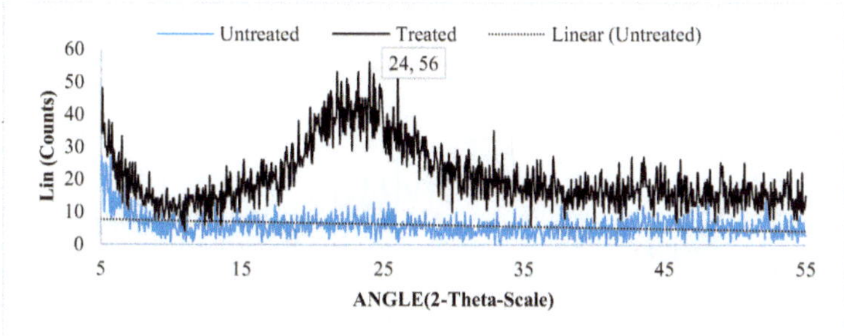

Figure 1. X-ray diffraction spectra of rice husk ash at 550°C.

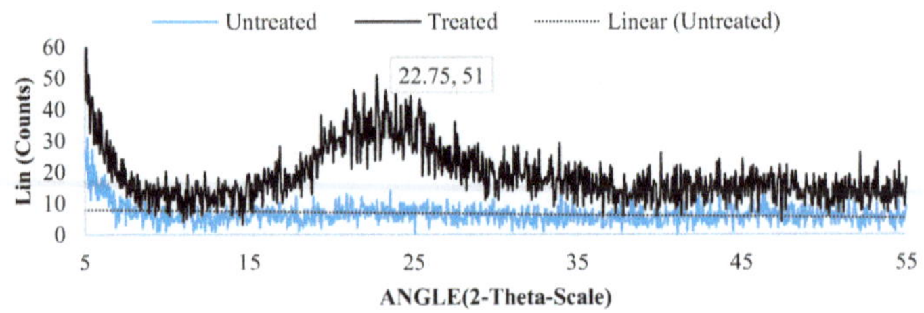

Figure 2. X-ray diffraction spectra of rice husk ash at 600°C.

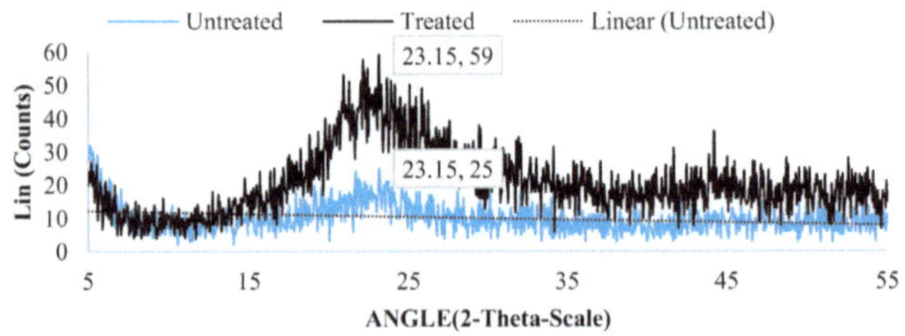

Figure 3. X-ray diffraction spectra of rice husk ash at 650°C.

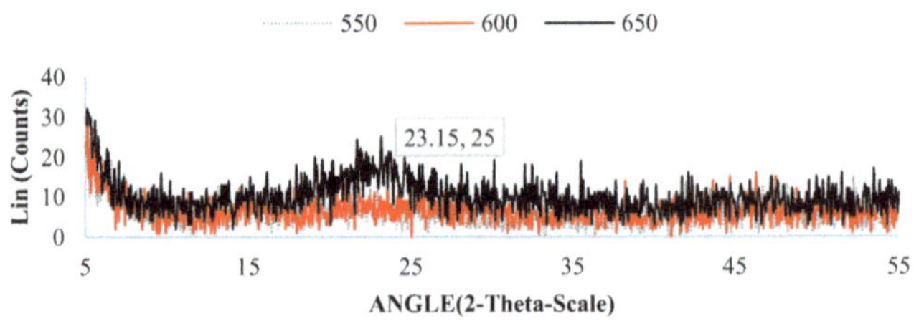

Figure 4. X-ray diffraction spectra of untreated rice husk ash at 550°C – 650°C

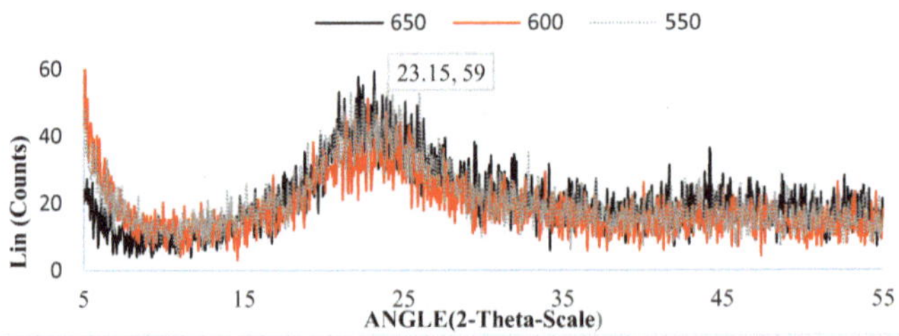

Figure 5. X-ray diffraction spectra of treated rice husk ash at 550°C – 650°C.

3.2.2 SEM Analysis

The untreated RH silica saves the unique 3D porous structure that comes from the outer epidermis of natural rice husks. SEM image of RH in figure 6 reviews RH silica from outer surface of a rice husk at low magnification. The image confirms that the nanoporous structure is well organized. The morphological features of the untreated and treated RHA at 600°C and 650°C are shown in figure 7 and figure 8 respectively. From these figures the RHA shows a porous and multifaceted particle shape and size. Knowing that the main components of rice husk includes silica, cellulose, hemi-cellulose, and lignin [3]. The porous and honeycomb morphology seen in the figure 7 and 8 can be attributed to the burning out of the organic components in the rice husk during combustion. The solid hydrated silica subsequently polymerizes to form a skeletal silica network which may explain the flaky and honey comb-like structure in the SEM picture of figure 7 and 8.

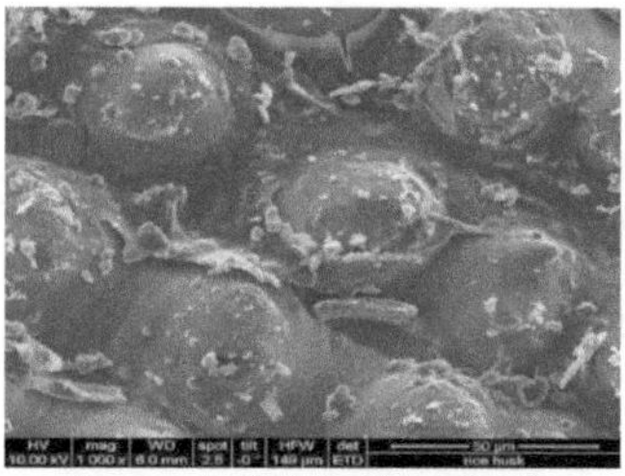

Figure 6. SEM of rice husk.

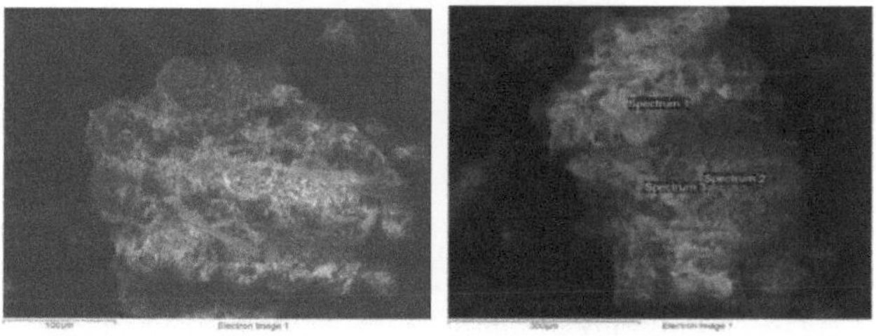

a b

Figure 7. SEM of (a). Untreated RHA and (b). Treated RHA both at 600°C with 3 spectra.

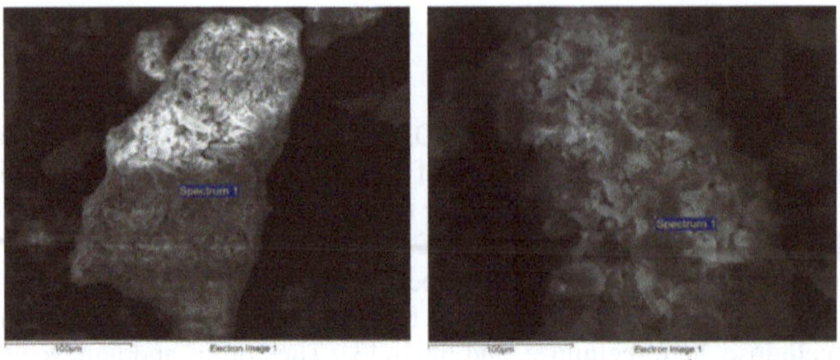

a b

Figure 8. SEM of (a). Untreated RHA and (b). Treated RHA both at 650°C.

3.2.3 EDX Analysis

Burning of RH to RHA at 650°C was more effective for elements (carbon, calcium, potassium, and phosphorus) removal than at 550°C and 600°C as shown in table I and table II for ash colour change and compound percentage respectively. The corresponding EDX spectra of points identified in figure 7 and 8 for RHA gotten at 600°C (untreated and treated) and 650°C (untreated and treated) respectively are shown in figure 9, 10, 11 and 12. The analysis revealed a high silica rich solid with porous and sponger or honeycomb morphology. See table II for the percentages of silica extracted after burning followed by chemical treatment was much higher at 650°C. Before chemical treatment the ash gotten at 650°C has shown that you can have purity silica as observed in XRD and EDX and when treated with chemical a much more high-purity amorphous silica is obtained.

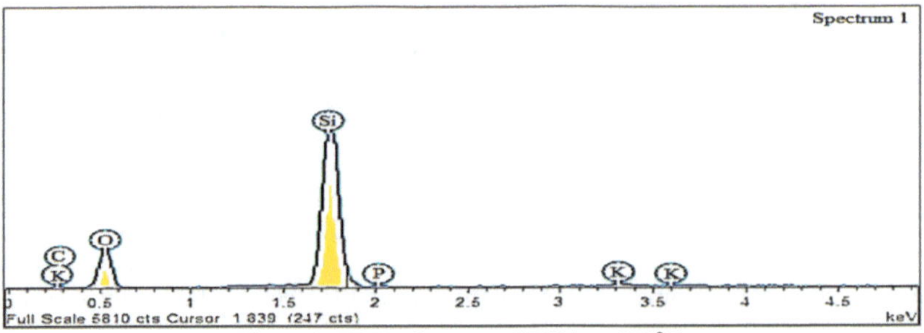

Figure 9. EDX of untreated RHA at 600°C.

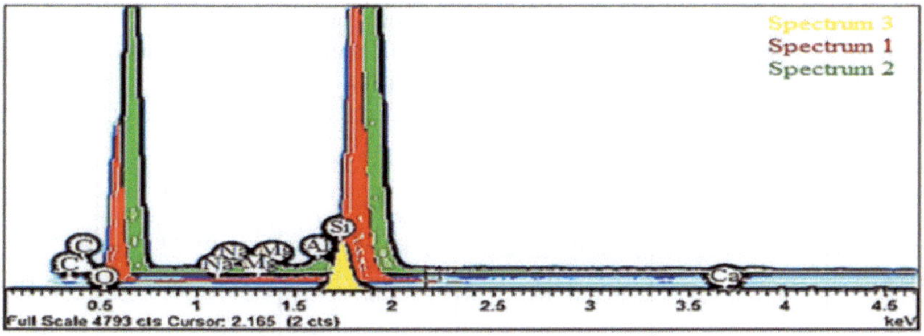

Figure 10. EDX of treated RHA at 600°C.

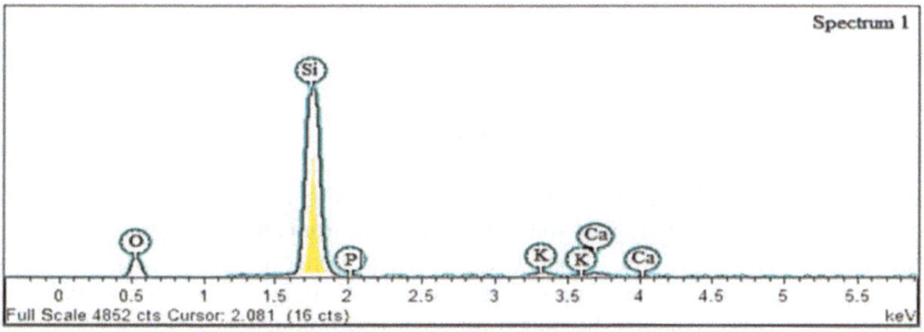

Figure 11. EDX of untreated RHA at 650°C.

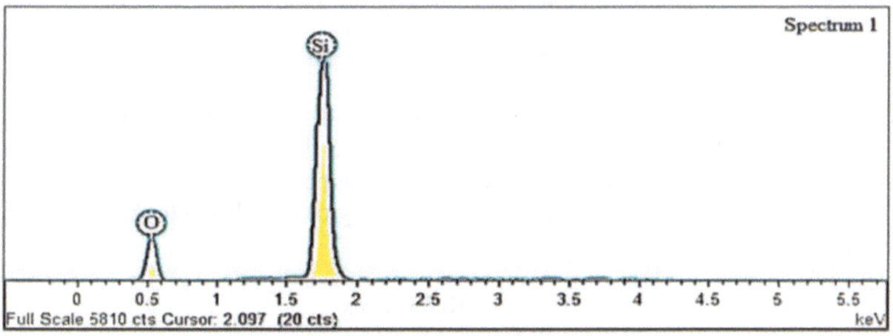

Figure 12. EDX of treated RHA at 650°C.

Table II. Percentage of compound in the RHA.

Element.	650 Treated RHA.	650 Untreated RHA.	600 Treated RHA	600 Untreated RHA	Formula
O	-	-	-	-	-
C	-	-	-	10.71	CO_2
Si	100	95.54	97.89	87.34	SiO_2
P	-	1.35	-	1.08	P_2O_5
S	-	-	-	-	SO_3
K	-	1.89	-	0.88	K_2O
Ca	-	1.22	2.11	-	CaO

4. Conclusions

The authors investigated the effect of temperature in extraction of silica from rice husk and were able to extract high-purity amorphous silica at 650°C. And from analysis it agreed with literature that rice husks have lots of silica when burnt in the presence of air. Extensive work is still required in order to gain sufficient knowledge and understanding in silica extraction by varying the pH levels and silica reduction to silicon for solar cell production. Also, extensive work on burning RH at 650°C and varying the burning time should be done knowing fully well that it is more advantageous to work with amorphous silica than crystalline silica from quartz.

5. Acknowledgments

The authors would like to thank National Agency for Science and Engineering Infrastructure (NASENI)-WORLD BANK STEP B program and the Africa University of Science and Technology (AUST) Abuja Nigeria for their financial support. The efforts of technical staff of Civil and Environmental Engineering, Worcester Polytechnic Institute MA, USA is also acknowledged.

References

1. Marvin Dodson, "Chemically reactive concrete stains" US Patent No: 6500249B1.
2. Alexis T. Belonio, *Rice Husk Gas Stove Handbook* (Central Phillippine University, Iloilo City, Phillippines 2005), 15-17.
3. D. S. Jung et al.., *Recycling rice husks for high-capacity lithium battery anodes.* (www.pnas.org/cgi/doi/10.1073/pnas.1305025110).
4. N. Yalcin and V. Sevinc, "Studies on silica obtained from rice husk," *Ceramics International. 27* (2001), 219-224.
5. S. R. Kamath and A. Proctor, "Silica gel from rice hull ash: preparation and characterization," *Cereal Chem. 75,* (1998), 484-487.

6. U. Kalapathy, A. Proctor and J. Shultz, "An improved method for production of silica from rice hull ash," *Bioresource Technology 85,* (2002), 285-289.
7. K. K. Larbi, "Synthesis of High Purity Silicon from Rice Husks" (M.Eng. thesis, University of Toronto, 2010), 50-65.

2015 EPD Congress

SYMPOSIUM: SOLAR CELL SILICON

Crystallization and Mechanical Properties

Arief Budiman,
Singapore University of Technology and Design

EPD Congress 2015
Edited by: James Yurko, Antoine Allanore, Laura Bartlett, Jonghyun Lee, Lifeng Zhang, Gabriella Tranell, Yulia Meteleva-Fischer, Shadia Ikhmayies, Arief Suriadi Budiman, Prabhat Tripathy, and Guy Fredrickson
TMS (The Minerals, Metals & Materials Society), 2015

NUMERICAL MODELING OF STRESS DISTRIBUTION IN A BI-GRAIN SMALL SCALE SILICON INGOT INCLUDING CRUCIBLE DEFORMATION

Sylvain Gouttebroze[1], Mohammed M'Hamdi[1]

[1] SINTEF MK, Forskningsveien 1, Oslo, 0340, Norway

Keywords: Silicon, Solidification, Stress, Finite Element, Dislocation, Crystal plasticity

Abstract

In a previous work, a small scale Bridgman furnace has been used to study silicon bi-grain crystallization at different cooling rate. This work expands the analysis by studying the mechanical interaction between the crucible and ingot during the solidification and cooling. The thermal model is based on a 2D-axisymetric heat-transfer model. The flux histories are then transferred to the ingot-crucible 3D-model. Anisotropic Elastic and Crystal Plasticity model are used to model the silicon deformation. Four different assumptions are applied to model the mechanical contact crucible-ingot and three grain misorientations are considered. The results show the strong impact of the alumina crucible contraction on the stresses and deformations in the silicon ingot.

Introduction

Commercial module efficiencies range from 14% to 21% for monocrystalline Si modules and from 12% to 18% for multi-crystalline Si modules [1]. The reason for this is the higher defect content in multi-crystalline wafers produced by directional solidification compared to Czochralski grown Si ingot. Recently, a novel approach [2-3] has been tested to bridge the efficiency gap to mono-crystalline Si by using seeded-growth technique in directional solidification of mc-Si. Mono-like crystalline silicon ingots are then fabricated using a Cz mono-seed layer with carefully controlled solidification conditions to maintain the mono-like growth. Unfortunately this technique has resulted in a degradation of the performance for wafers cut from the top part of the ingot compared to standard crystallization procedures. This new process has a tendency to generate extra-large dislocation clusters at grain boundaries that expand and propagate vertically. As a result, the top part of the ingot has a very high dislocation density leading to lower efficiency. It is, therefore, crucial to understand the stress build-up during crystallization.

As the interactions between structural defects and impurities during seeded mono-like and multi-crystalline directional growth processes are still not well understood, a small-scale experiment has been developed at NTNU [4-5]. In that work, multi-crystalline and bi-crystal silicon ingots were pulled at different speeds. The analysis of this experiment was complemented by a Finite Element analysis of the directional solidification furnace [6] including the heat transfer and mechanical deformation of the ingot. The aim of this work is analyze the impact of constrains imposed by the crucible during solidification and their effects on the silicon ingot deformation. The work presented here extends the previous numerical model by including the deformation of the alumina crucible.

Model description

The simulations have been carried out using Sisim, a heat and mass transfer model coupled with thermo-elasto-viscoplastic model [7, 8].

Thermal model

In order to accurately model the crystallization of the ingot (40-mm height and 40-mm diameter solidified in 7 hours), a two-step approach has been used (see Reference 6 for more details). In the first step, an axisymmetric mesh is built where all the furnace parts are included. This global furnace model is used to predict the heat flux history on the external surfaces of the crucible-ingot reduced model. Then, the computed heat fluxes are applied as boundary conditions in a 3D thermo-mechanical model for the crucible and ingot system.

Mechanical model

The present work investigates the effect of the crucible deformation on the stresses and deformation in the ingot. The ingot is modeled either as elastic or plastic material. In both cases, the constitutive model is based on Crystal Plasticity (see Reference 6 for more details). In the CP model, the elastic case is retrieved by simply setting the initial dislocation density to zero (instead of 10^4 m^{-2}) in the multiplication law, therefore preventing plastic deformation.

Except for the thermal expansion coefficient (see Figure 1), all mechanical properties (e.g. anisotropic stiffness matrix for silicon) are assumed constant. But the deformation of the ingot will lead to dislocation multiplication/propagation. Density, Young's modulus and Poisson's ratio for the alumina are set to 3800 kg/m^3, 330 GPa and 0.22, respectively. Please refer to Ref. 6, 9 for silicon mechanical properties.

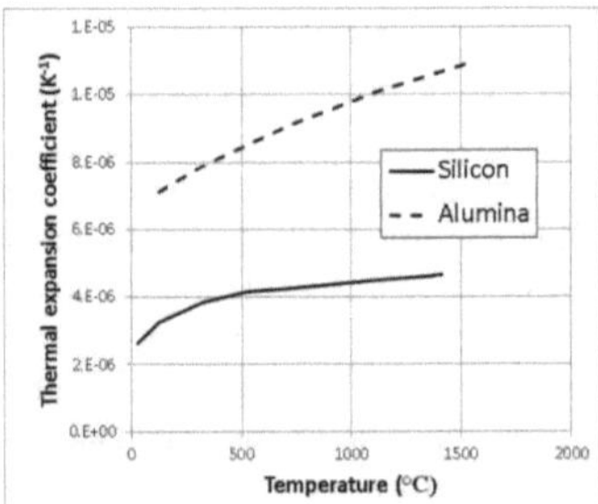

Figure 1. Thermal expansion coefficient of Silicon [10] and Alumina [11] as a function of temperature.

Mechanical contact between ingot and crucible

The mechanical contact between the ingot and the mold is modeled by a spring with specified stiffness that can either be constant or vary with gap/penetration distance. A force will therefore be applied between the two domains which is equal to the normal gap distance (or penetration depth) multiplied by the spring stiffness (constant or as given in Figure 2). Four contact configurations have been investigated:

- crucible treated as a rigid body: weak contact between the bottom of the ingot and the crucible or sticking contact on all the contact surface crucible/ingot.
- crucible treated as an elastic material: weak or strong contact (distance dependent, see Figure 2) on all the contact surface crucible/ingot.

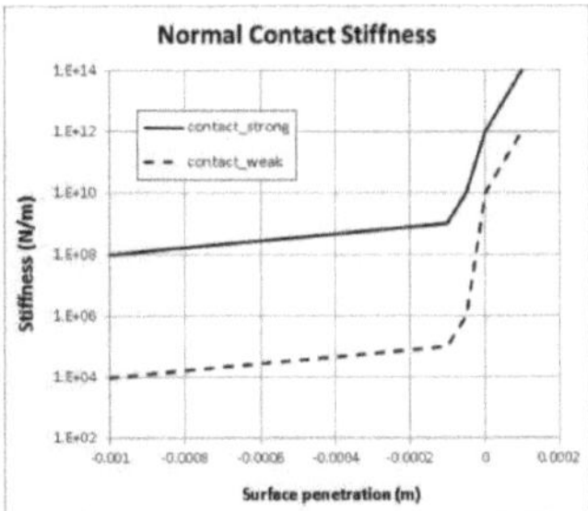

Figure 2. Normal contact stiffness between the crucible and ingot as a function of the penetration (or distance for negative values)

Results and discussion

The analysis of the results will focus on the stresses and deformation in the vertical central cross-section during solidification (at 21000 s corresponding to a total solidified fraction of 0.5) and at the end of the cooling stage (T=25°C). The study is divided into three parts: effect of contact properties with elastic deformation, with plastic deformation and grain misorientation effect.

<u>Effect of mechanical contact properties</u>

The first part of the study is focused on the elastic deformation of the ingot. Four cases are modeled: (a) with a rigid crucible where the contact is either weak with the bottom of the ingot or (b) sticky on all contact surfaces between the crucible and the ingot, (c) with crucible treated as an elastic body with either weak or (d) strong gradual contact as presented in previous section. The stress distributions in the central vertical cross-section are presented in Figure 3. One can notice the large differences in the maximum effective stress depending on the kind of applied contact. The imposition of a sticking contact in case (b) leads to a strong constraint of the material that uniformly deformed in the axial direction. For cases (a) and (c), the effect is limited to the periphery of the ingot. Interestingly, when the contact is weak, the crucible deformation does not significantly modify the stress state during the solidification. Case (d) is a combination of a strong lateral constraint (as in case b) with large deformation on the lower ingot periphery in contact with the crucible.

The impact of the type of contact applied is also visible when considering the residual stresses in Figure 4. The local deformation at the corners is clearly visible due to the amplified deformation of the ingot. The similarities between cases (a) and (c) are not extended during the cooling phase as the crucible will compress the ingot and generate high stresses due to the higher thermal expansion coefficient. This phenomenon has been experimentally observed and leads to fracture in the ingots. In case (b), the ingot keeps the same shape due to the sticking contact with the crucible and the effective stress is almost uniform in the domain. Therefore, based on this thermo-elastic study, cases (b) and (c) will be analyzed in the rest of the study.

When the plastic deformation is included in the model, one has to consider the initialization of the stress/strain state at the crystallization front. In the present work, the plastic strain is initialized to zero while the stress is inherited from the previous layer in order to represent a material that solidifies perfectly into the lattice structure of the already solid material. This hypothesis seems reasonable when considering a mono-crystal but might not be adequate for multi-crystalline material.

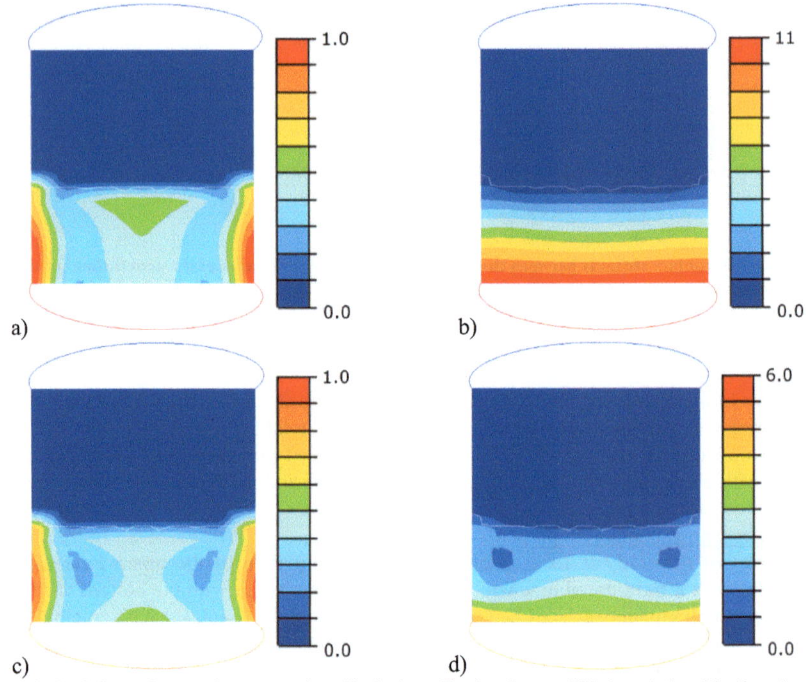

Figure 3. Undeformed central cross-section displaying effective Stress (MPa) and the thin line shows the position of the crystallization front at 21000s for elastic ingot and cases: a) rigid crucible, weak contact, b) rigid crucible, sticking, c) elastic crucible, gradual weak contact, d) elastic crucible, gradual strong contact

During solidification, as illustrated in Figure 5, the front curvature leads to locally high stresses for both cases. The coarse grid will adversely affect this local stress concentration. This local effect is not visible when the residual stress is considered (Figure 6). For the free surface case (a), a typical stress pattern is observed with higher stresses in the corners and on the upper surface that solidified last. The stress distribution when the crucible deformation is included is quite different with high stresses on the external surfaces (with a peak at the bottom corners) decreasing toward the center. The final dislocation density and deformed shape (Figure 7) provides additional information on the crucible deformation effect. Due to the contraction the crucible, the liquid silicon is compressed and pushed upward leading to a slightly elongated ingot shape (amplified 30 times in Figure 7), as well as the high dislocation densities for case (b). For case (a), the computed dislocation density is more uniform in the ingot with higher values in the area slightly above the constrained ingot bottom surface.

As the mesh is not extremely fine (due to cpu considerations), the discretization of the cylindrical surfaces of both ingot and crucible implies numerical approximations that will be amplified by the gradual contact strength. Therefore this work should completed by a more thorough study on the mesh effect in coupled crucible-ingot deformation. Nevertheless the present results illustrate the drastic changes in the deformation of the silicon ingot when the deformation of the alumina crucible is taken into account. Not only the level of stress and dislocation density are increased but their distributions in the ingot are significantly altered. It is therefore essential to include that aspect when analyzing experimental results.

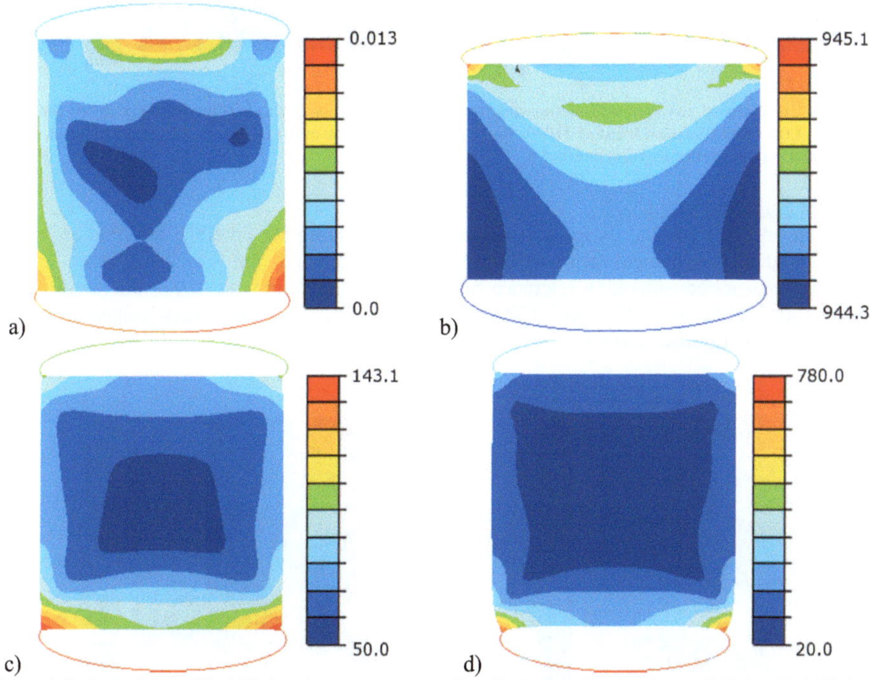

Figure 4. Deformed (amplified 30 times) central cross-section displaying effective Stress (MPa) at T=25°C for elastic ingot and cases: a) rigid crucible, weak contact, b) rigid crucible, sticking, c) elastic crucible, gradual weak contact, d) elastic crucible, gradual strong contact

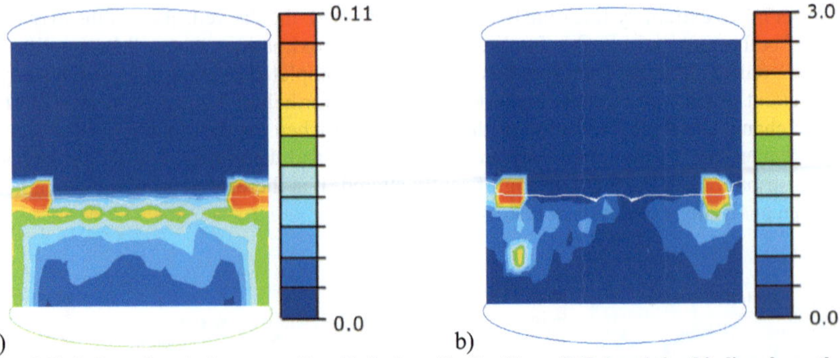

a) b)

Figure 5. Undeformed central cross-section displaying effective Stress (MPa) and the thin line shows the position of the crystallization front at 21000s for plastic ingot and cases: a) rigid crucible, weak contact, b) elastic crucible, gradual strong contact

<u>Effect of grain orientation</u>

Based on previous results, the model was run only including the crucible deformation with a strong gradual mechanical contact between the ingot and the crucible. Three crystal configurations are considered: a reference case with a single grain with <100> orientation, a bi-grain <100>/<111>, and a bi-grain <100>/<110>. The <100> grain is always located on the left hand side of the figures.

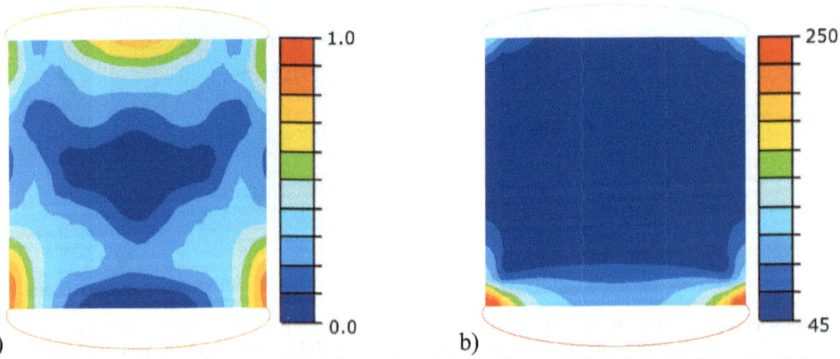

a) b)

Figure 6. Undeformed central cross-section displaying effective Stress (MPa) at T=25°C for plastic ingot and cases: a) rigid crucible, weak contact, b) elastic crucible, gradual strong contact

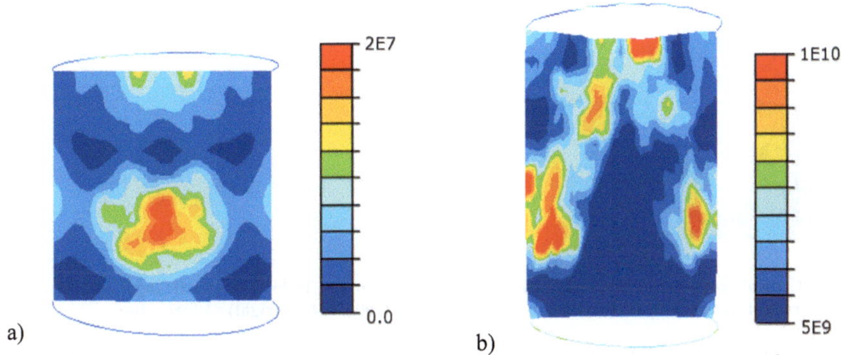

Figure 7. Deformed (amplified 30 times) central cross-section displaying Dislocation Density (m^{-2}) at T=25°C for plastic ingot and cases: a) rigid crucible, weak contact, b) elastic crucible, gradual strong contact

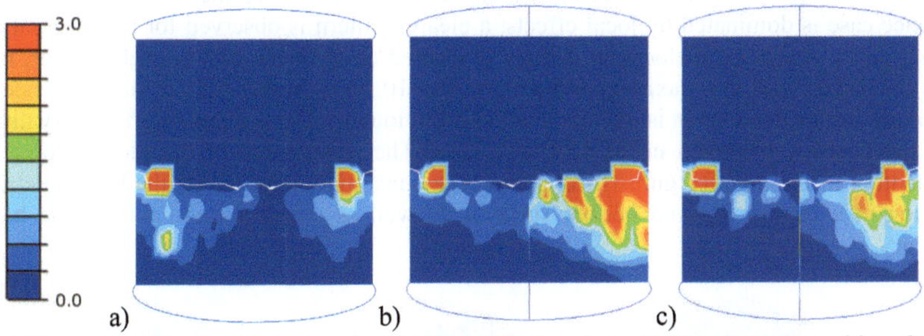

Figure 8. Central cross-section displaying effective Stress (MPa) and the thin line shows the position of the crystallization front at 21000s for plastic ingot, elastic crucible and gradual strong contact: a) mono <100>, b) <100>/<111>, c) <100>/<110>

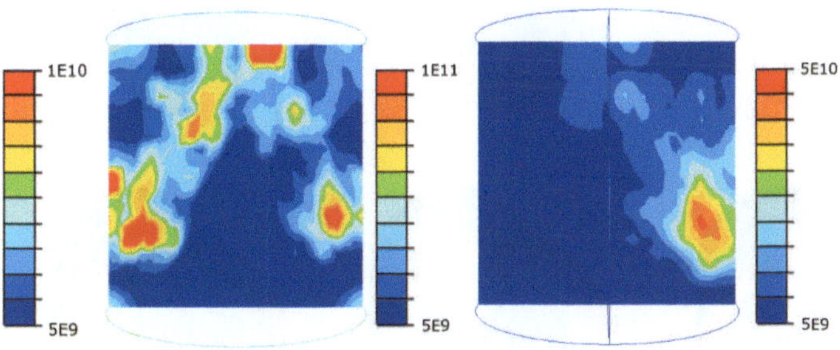

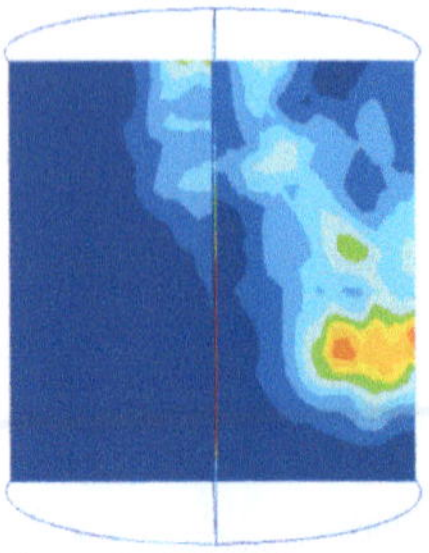

Figure 9. Central cross-section displaying Dislocation Density (m^{-2}) at T=25°C for plastic ingot and elastic crucible, gradual strong contact: (left) mono <100>, (center) <100>/<111>, (right) <100>/<110>

As illustrated on Figure 8, the local stress concentration due to the front curvature is amplified by a grain misalignment with the growth direction. The bi-grain structure on the right side reaches higher stresses even during the solidification phase. But it does not seem to affect the stress distribution on the other side of the bi-grain during solidification. The grain misorientation leads, however, to a very different dislocation density distribution as presented in Figure9. While the reference case is dominated by local effects, a clearer pattern is observed for the bi-grain cases. In both cases, a peak of dislocation density is located close to the bottom right corner, not far from the surface and the maximum values are 5 to 10 times higher than in the reference case. The origin of this defect area is still unclear but it obviously reflects a macro-effect very different from the A-shape observed in the reference case. The observation of the global stresses and deformation of the ingot (Figure 10) illustrates the constraints imposed by the bi-grain structure resulting in a curved top and bottom surfaces as well as stress discontinuity at the bi-grain boundary.

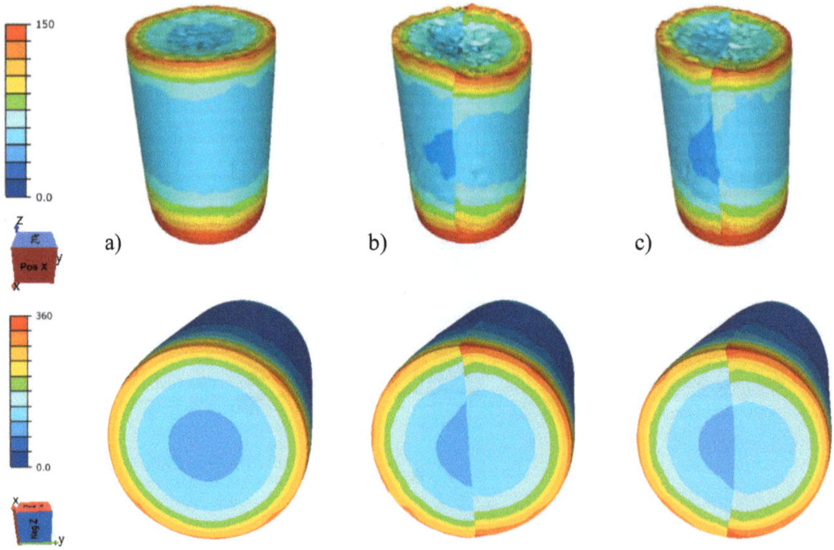

Figure 10. Deformed ingot (amplified 30 times), effective Stress (MPa) at T=25°C for plastic ingot and elastic crucible, gradual strong contact: a) mono <100>, b) <100>/<111>, c) <100>/<110>

Conclusion

A 3D mechanical model of the crucible-ingot system has been established including different mechanical contact properties, elastic and plastic deformations of the ingot. Moreover, three grain configurations have been considered. The effective stresses and dislocation density have been analyzed during solidification and at room temperature. The main conclusion is that when using alumina crucible with high thermal expansion coefficient, it is crucial to include the crucible deformation as it significantly alter ingot deformation due the compression imposed by crucible contraction. Introducing bi-grain structures results in a stress discontinuity at the interface are captured and an altered distribution of computed dislocation densities and ingot deformation.

Acknowledgments

This work has been performed within "The Norwegian Research Centre for Solar Cell Technology" (project number 193829), a Centre for Environment-friendly Energy Research co-sponsored by the Norwegian Research Council and research and industry partners in Norway.

References

1. A. Jäger-Waldau, "PV status report 2013", Report EUR 26118 EN
2. N. Stoddard et al., "Casting Single Crystal Silicon: Novel Defect Profiles from BP Solar's Mono2 TM Wafers", *Solid State Phenomena*, 131, (2008), 1-8.
3. A. Jouini et al., "Improved multicrystalline silicon ingot crystal quality through seed growth for high efficiency solar cells", Progress in Photovoltaics: research and applications, 20, (2012).
4. I. Brynjulfsen, L. Arnberg, and A. Autruffe, "Nucleation in small scale multicrystalline silicon ingots", *Journal of Crystal Growth*, 361 (2012), 206-211
5. A. Autruffe et al., "Impact of growth rate on impurities segregation at grain boundaries in silicon during Bridgman growth," *Journal of Crystal Growth*, 372 (2013), 180-188.
6. S. Gouttebroze et al., "Effect of Grain Orientation and Cooling Rate on Stress Distribution in a Small-scale Silicon Ingot", *Met. Mat. Trans. E*, 1(2), (2014), 180-186.
7. M. M'Hamdi, S. Gouttebroze, and H.G. Fjær, "Thermo-mechanical analysis of the ingot-crucible contact during multi-crystalline silicon ingot casting", *Journal of Crystal Growth*, 318 (2011), 269-274
8. M. M'Hamdi, S. Gouttebroze, and H.G. Fjær, "3D modelling of stresses and deformations during crystallisation of silicon accounting for ingot-crucible interactions", *Journal of Crystal Growth*, 362 (2013), 83-87
9. J. Cochard et al., "A multicrystal model for the viscoplastic deformation of Silicon at high temperature", Proceedings of EUPV Conference, Valencia (2008).
10. R. Hull, "Properties of Crystalline Silicon", INSPEC, London, 1999.
11. Y. S. Touloukian et al., "Thermophysical properties of matter", 1975, Vol. 13, p.176.

EPD Congress 2015
Edited by: James Yurko, Antoine Allanore, Laura Bartlett, Jonghyun Lee, Lifeng Zhang, Gabriella Tranell, Yulia Meteleva-Fischer, Shadia Ikhmayies, Arief Suriadi Budiman, Prabhat Tripathy, and Guy Fredrickson
TMS (The Minerals, Metals & Materials Society), 2015

MECHANICAL STRENGTH OF SILICON SOLAR WAFERS CHARACTERIZED BY RING-ON-RING TEST IN COMBINATION WITH DIGITAL IMAGE CORRELATION

V.A. Popovich[1,2], W. Geerstma[2], M. Janssen[2], I.J. Bennett[3], I.M. Richardson[2]

[1]Materials innovation institute M2i, Delft, The Netherlands, v.popovich@m2i.nl
[2]Delft University of Technology, Department of Materials Science and Engineering, Delft, The Netherlands
[3]Energy Research Centre of the Netherlands, Solar Energy, PV Module Technology, Petten, The Netherlands

Keywords: Silicon solar wafer, mechanical strength, fracture

Abstract

Avoiding wafer breakage is a big challenge in the photovoltaic silicon industry, limiting production yield and further price reduction. Special fracture strength tests suitable for thin silicon solar wafers and solar cells, to be used in combination with Weibull statistics, finite-element (FE) modelling and digital image correlation have been developed in order to study the mechanical stability of solar wafers. The results show that removal of the saw damage significantly increases the strength of crystalline silicon wafers. Furthermore, it was found that silicon crystallinity and the location where the wafer is extracted from the cast Si ingot have a significant effect on the strength, namely samples taken from the bottom of the ingot are 30% stronger than those taken from the top. The study also showed that there is a decrease in fracture strength when an anti-reflective SiN_x coating is applied, which is caused by high thermal stresses.

1. Introduction

Increase in silicon wafer size in combination with continuous wafer thickness reduction without strengthening the wafer leads to a high breakage rate during subsequent handling and processing steps and results in high costs [1, 2]. It is well known, that silicon is a brittle material that is easy to break during the in-line production process in which several loads are applied on the wafer surface and edges [3]. Microflaws generated from wafer sawing, impurity precipitations, structural defects, and residual stresses are the leading cause of degradation of mechanical strength in silicon wafers. Therefore, it is not only important to investigate the electrical properties of silicon solar wafers and cells, but also the mechanical properties, especially the strength.

The most common method for testing fracture strength of a silicon wafer is by a bending (uniaxial) test or a ring-on-ring (biaxial) test. However, standard mechanical testing methods are not capable of stressing a large area of the cell specimen uniformly. In our previous work [4] the 4-point bending test results of both single crystalline (cz) Si and multicrystalline (mc) Si samples were discussed in detail, however, an important restriction of the 4-point bending test is the quality of the edge (possible cracks, defects) which might induce premature fracture of the entire solar cell. The use of a biaxial (ring-on-ring) test is required in order to exclude the influence of edge defects and thus to more accurately investigate the effects of different processing conditions, such as surface texturing, and of impurities on the bulk fracture strength of silicon solar cells. Furthermore, biaxial stress is more consistent with the actual stress load in the silicon solar cells in the PV module, either due to snow or wind.

In this work the fracture strength of silicon wafers is measured by ring-on-ring tests in combination with finite element modelling validated by digital image correlation. Results are statistically evaluated by a Weibull analysis, which provides information on the flaw distribution in the sample. In this paper several aspects are described that affect mechanical strength, *i.e.* silicon wafer crystal structure, saw damage, and surface roughness.

2. Experimental conditions

2.1. Material preparation

The samples used for the ring-on-ring test were laser cut from 156 × 156 mm^2 single and multicrystalline silicon wafers. Laser cutting was selected in order to avoid unnecessary mechanical loading on the wafer. The standard samples were cut from the wafers into round shapes with a diameter of 21.44 mm (Figure 1 a). Samples with specific types of crystallinity were prepared in order to investigate the effect of crystallinity on the mechanical strength of the silicon wafer. In order to statistically evaluate the results, 10 neighbouring specimens (thus with the same crystallinity features) were prepared with three different grain morphology types, namely: one big grain, several large grains and many small grains (Figure 1 b).

a)

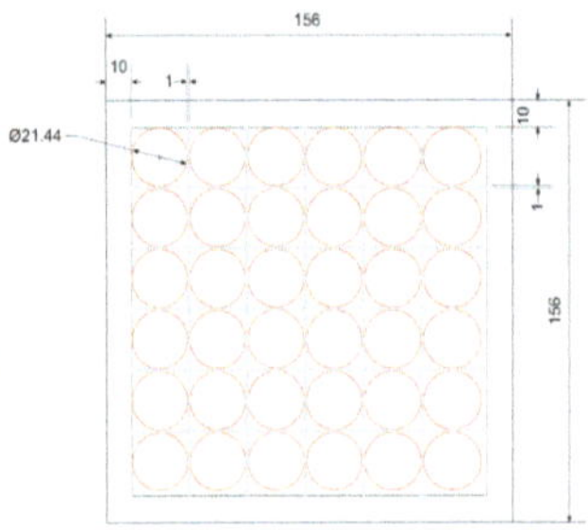

b)

Figure 1. a) Standard Si wafer cut-out pattern; b) examples of tested grain morphologies; left: one large grain, middle: several large grains, right: many small grains. (Red lines show the edge of the laser cut sample and black dotted lines indicate the loading ring position).

To analyze the effect of surface roughness three types of specimens were prepared also from neighbouring wafers. The surface condition of these specimens included:

- the as-cut state, thus including the saw-damage layer.
- a surface textured by etching for 30 s in an HF (10%) + HNO_3 (30%) + CH_3COOH (60%) solution. This serves two main purposes: to remove the damaged layer and to create a highly textured silicon surface in order to trap the light.
- a chemically polished surface (15 μm removal from both wafer sides in an HF+HNO_3 bath for 1 min.).

2.2. Ring-on-ring test configuration in combination with digital image correlation

A test configuration has been designed in accordance with ASTM standard C 1499-09 and it was also made compatible with a digital image correlation (DIC) system [5]. The equipment is shown schematically in Figure 2 and consists of a supporting ring and a loading ring.

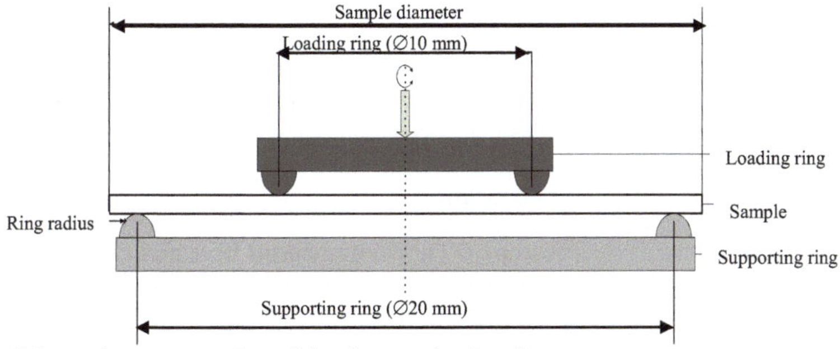

Figure 2. Schematic representation of the ring-on-ring bending setup.

The ASTM standard allows the use of different sample sizes [5] and the dimensions are determined using appropriate formula. However, deflections needed to fracture the silicon wafer specimens are expected to be significantly higher than the ASTM standard limitation of 25 % of the thickness, which causes a stress concentration at the loading ring. In order to solve this problem FE modelling using ANSYS software has been used to calculate the peak tensile stress at fracture (at the tensile side of the sample under the loading ring) from the applied load at fracture.

The DIC system was combined with the ring-on-ring test to measure the deformed shape as a function of the applied load (Figure 3 b), in order to validate the FE model. Digital image correlation is a technique that can be used to monitor the deformation of a sample by comparing a reference image with an image in the deformed state [6]. In this research a 2 camera setup was employed, Tests were performed using a 100 kN Instron 5500R tensile machine equipped with a 10 N load cell. The complete test setup with the ring-on-ring fixture and the two cameras of the DIC system is shown in Figure 3 a.

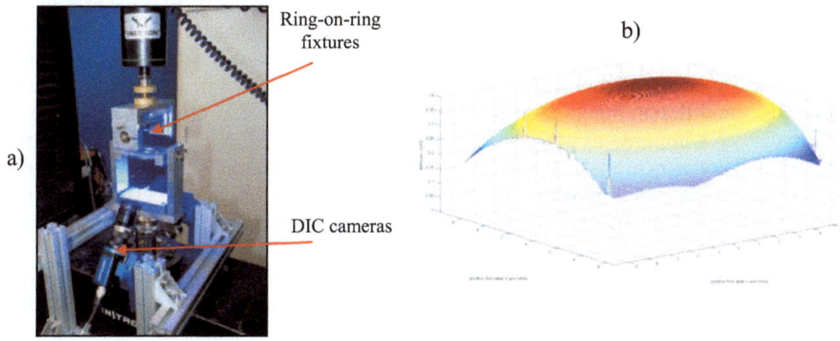

Figure 3. a) Ring-on-ring test setup with two cameras for the DIC system b) An example of the fitted DIC profile used to measure the deflection and the deformed shape of the samples.

The geometry of the supporting ring ensures a good visibility for the two DIC cameras during sample loading. The area visible to the cameras is limited to a square section of 12 × 12 mm^2 due to the positioning of the cameras on the tensile machine. The measurement system used in this research (Limess 3D Digital Image Correlation System with two 5 MPixel cameras) has a calibration procedure incorporated in the measurement and analysis software. The calibration was performed using a Limess A12 10×10 mm^2 calibration grid and a 3D residuum of 0.25 or lower was achieved for every calibration. The camera signal and analog signals were collected using a Limess DAQhw data acquisition box. The load and displacement of the tensile machine have been acquired on a 10 V signal, which corresponds to the load of 250 N and the displacement of 5 mm. The data acquisition is synchronized to that of the images.

2.3. Fourier Transform Infrared Spectroscopy

Fourier Transform Infrared Spectroscopy (FTIR) measurements were performed in a BRUKER Vertex 70 spectrometer in the transmission mode in order to measure the dissolved carbon and oxygen in the silicon wafers. Infrared light is focused onto a photodiode of a liquid nitrogen-cooled, wide band DLaTGS W/KBr detector with a spectral response of 250 to 12000 cm^{-1}. Measurements were taken from a 10×10 mm^2 area of the sample. Before measuring the samples, an FTIR spectrum for the background was collected in a temperature-stabilized chamber; this spectrum represents absorption from the atmosphere in the chamber. Then, one silicon sample with oxygen and carbon concentration below the detection limit was used as a reference sample. The output was recorded by OPUS software. Each spectrum is the average from 64 scans. The spectrum resolution is 0.5 cm^{-1}. During analysis, the background spectrum was subtracted from spectra of all the samples and the reference.
In order to exclude possible effects of the thickness, the samples and the reference were chosen with identical thicknesses of 182 μm.

3. Results and Discussion

3.1. Effect of wafer position in the cast mc-Si ingot and of silicon crystallinity on the mechanical strength

The mechanical properties of mc-Si samples have been tested to examine the influence of silicon crystallinity (grain morphology) and impurity concentration (qualitatively assessed as a function of the position within the cast mc-Si ingot). The ring-on-ring test results are shown in Table I.
The concentrations of oxygen and carbon were qualitatively assessed with FTIR for different positions in the ingot, see Figure 4. As can be seen, only two distinct peaks were found for Si wafers taken from different ingot positions, namely 1107 cm^{-1} and 605 cm^{-1}. The concentration of interstitial oxygen impurities corresponds to the peak at 1107 cm^{-1}. There is an increase in this peak height as the wafers are taken from higher positions from the mc-Si ingot, which points to an increasing oxygen concentration. The concentration of substitutional carbon impurities corresponds to the peak at 605 cm^{-1}, indicating that the concentration of carbon is higher near the bottom of the ingot.
Both the grain morphology and the impurity concentration are found to have a significant effect on the characteristic peak stress at fracture. The average characteristic peak stress shows an increase from the top towards the bottom of the ingot.
Carbon is one of the most important light element impurities in PV silicon and its concentration in mc-Si is usually rather high, due to the lower-quality silicon feedstock and the direct contact with crucible walls, and it is likely that carbon may affect the fracture strength of mc-Si wafers.
The results suggest that in samples with many small grains the characteristic peak stress increases by up to 30% as a result of the increasing carbon concentration from the middle to the bottom ingot position (Table I). It should be pointed out that the thermal history of the silicon at the different ingot positions might also play a role., due to residual stresses that develop as a result of thermal gradients during solidification.

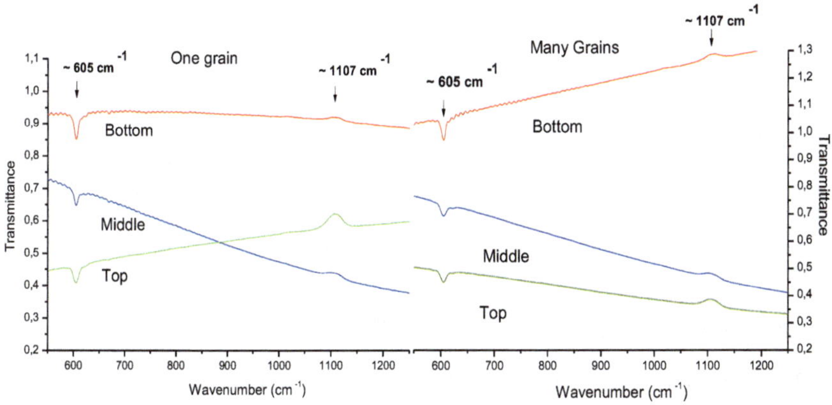

Figure 4. FTIR measurement showing the qualitative concentration of substitutional carbon (605 cm^{-1}) and interstitial oxygen (1107 cm^{-1}) as a function of mc-Si ingot position; left: samples with one large grain, right: many small grains.

Table I. Effect of wafer position in the mc-Si cast ingot and of grain morphology on the mechanical strength (average specimen thickness 182 μm).

Position in mc-Si ingot	Grain morphology	Characteristic peak stress at fracture (MPa)	Weibull modulus (-)
Top	many small grains	295	5.7
	several large grains	323	3.8
	one big grain	298	5.7
Middle	many small grains	282	3.7
	several large grains	320	3.8
	one big grain	351	4.3
Bottom	many small grains	399	6.3
	several large grains	375	7.9
	one big grain	361	5.1

Carbon is introduced from crucible graphite parts in the furnace and occupies substitutional lattice sites in silicon. The dissolved carbon atoms and their associated residual stresses and defects can affect the fracture strength in two ways: (1) through the formation of cracks during wire-saw cutting, which can be considered as an indentation process [7], and (2) by changing the propagation of cracks when a tensile stress is applied. In addition, carbon can enhance the nucleation of oxygen clusters at structural defects in mc-Si [8]. The presence of high levels of dissolved carbon atoms and the enhanced oxygen clusters may increase the surface energy of new surfaces, making cracking more difficult and thus improving the fracture strength.

The top of the ingot shows a decrease in the strength of the tested samples (Table I), which may be related to an increase of the interstitial oxygen concentration at the top of the ingot. Point defects are involved in the nucleation process of oxygen precipitates, represented by the SiO_2 phase. The higher oxygen concentration at the top of the ingot is likely to enhance the formation of SiO_2, which would facilitate the nucleation of dislocations at solidification and result in the lower strength found. A volume mismatch occurs as the precipitates grow in size and a variety of defects, including stacking faults, are associated with precipitate formation. Furthermore,

these defects attract fast-diffusing species and thus the concentrations of other (non) metallic impurities/inclusions are also expected to be higher in the top of the cast, which is expected to also reduce the strength.
The concentration of impurities is expected to be the lowest in the middle part of the cast ingot. Thus this middle part is the most representative to study the effect of grain morphology on mechanical strength. As can be seen from Table I, there is a clear trend in the results, namely the samples with more grain boundaries tend to have lower strength than samples with fewer grain boundaries. Hence, grain boundaries seem to represent weak areas where cracks can easily initiate and propagate. This result is in good agreement with the 4-point bending strength data from [3, 6].

3.2. Effect of damage layer removal of single crystalline silicon wafers

The effect of the saw-damage-removal depth on mechanical strength was investigated by means of single crystalline silicon samples with an initial thickness of 195 μm. The ring-on-ring results for three etching depths are shown in Table II.
It is clear that the strength of the wafer increases when the etch depth is increased. This large increase in strength can be attributed to removal of more small cracks due to the chemical polishing process. The results are in good agreement with previous 4-point bending results [6]. The modulus shows a decrease with increasing etching depth, which indicates more spread in the peak stress at fracture. The wafer with an etch depth of 55 μm shows the largest standard deviation for the sample thickness, which is due to non-uniform etching of certain areas and the creation of over-etched islands on the silicon wafer surface. Thus it is concluded that damage etch removal up to 35 μm is preferable in view of wafer strength improvement.

Table II. Effect of damage-layer removal on fracture strength.

Etch depth (μm)	Characteristic peak stress at fracture (MPa)	Weibull modulus (-)	Average specimen thickness (μm)	Standard deviation thickness (μm)
15	404	7.2	185	1.6
35	455	6.0	159	6.4
55	491	4.7	138	7.3

3.3. Effect of surface finish of single and multicrystalline silicon wafers

The effect of silicon wafer surface finish (roughness) was investigated for as-cut, textured and chemically polished conditions. The tests were performed on single crystalline silicon and the results are shown in Table III. From these results it can be seen that the strength increases when the surface quality/roughness is improved (thus for smoother surfaces). The lowest strength was found for specimens with an as-cut surface finish, which is in good agreement with the 4-point bending results from [7].
The fracture strength increases for specimens with textured surfaces, which can be explained by the removal of the damaged surface layer. Textured surfaces yield a lower Weibull modulus compared to as-cut surfaces, which points to a larger spread in the peak stress at fracture. The highest characteristic peak stress was found for specimens with polished surfaces, probably due to the smooth surface and further removal of the damaged layer. The Weibull modulus for polished samples has decreased compared to samples with textured surface.

Table III. Effect of surface finish of single crystalline wafers on fracture strength.

Surface quality	Characteristic peak stress at fracture (MPa)	Weibull modulus (-)	Average specimen thickness (μm)
as-cut	122	8.1	195
textured	347	9.1	176
polished	487	3.5	159

The effect of surface quality has also been investigated for mc-Si samples, the results of which are shown in Table IV. The same relationship that was found for single crystalline silicon samples can be seen for the stress at fracture. The magnitude of the difference between the peak fracture stresses for as-cut and textured samples is however much smaller. This can be attributed to preferential etching of the grain boundaries, which probably decreases the peak fracture stress of specimens. As can be seen from Table IV, the stress for polished mc-Si samples is the highest. The strength of both the textured and the polished mc-Si samples is somewhat lower than the equivalent single crystalline samples (Table III). The largest difference is found for textured samples, which is probably due to the damage etching itself. However the as-cut fracture strength for both types of silicon crystallinity cannot be directly compared due to the differences in cutting processes for cz-Si and mc-Si wafers. Furthermore, there is a different level of impurities and residual stresses resulting from solidification of single and multicrystalline silicon. Nevertheless, it can be concluded that the cutting process itself is the most detrimental (compared to other surface treatment procedures) for the fracture strength of both silicon crystallinity types.

Table IV. Effect of surface finish of multicrystalline wafers on fracture strength.

Surface quality	Characteristic peak stress at fracture (MPa)	Weibull modulus (-)	Average specimen thickness (μm)
as-cut	206	9.8	195
textured	223	9.4	182
polished	349	7.0	173

3.4. Effect of antireflection SiN_x layer on mechanical strength

The effect of an antireflection SiN:H coating on the mechanical strength of silicon wafers is shown in Table V. As can be seen, there is a significant decrease in the fracture strength after the antireflection coating (ARC) is applied. The Weibull modulus also shows a significant increase, suggesting a more uniform distribution of flaw sizes. This could be attributed to thermal stresses in the SiN_x coating. It should be noted that the SiN_x layer is approximately 100 nm thick. It was found by Tien *et.al.* [9], that for a temperature difference of 110 °C a residual tensile stress of 120 MPa is created in the silicon nitride layer. However, the high application temperature of the SiN_x coating used in this study (375 °C), will create even larger thermal stresses in the SiN_x layer, which contribute to the decrease in the stress at fracture. High residual stresses could cause fracture in the SiN_x layer, which consequently could result in a failure of the complete wafer. Furthermore, it was shown in [9] that during SiN:H deposition itself, hydrogen diffuses into the silicon substrate, which will distort the Si lattice and introduce further stresses into the silicon wafer.

Table V. Effect of antireflection SiN layer on fracture strength.

Surface quality	Characteristic peak stress at fracture (MPa)	Weibull modulus (-)	Average specimen thickness (μm)
textured with ARC	264	14.5	173
textured	347	9.1	176

4. Conclusions

The mechanical strength of multicrystalline (mc) and single crystalline (cz) silicon solar wafers was investigated using a ring-on-ring test in combination with digital image correlation. The study of the fracture strength of silicon solar wafers showed that:

- The ring-on-ring test combined with finite-element (FE) modelling developed within this research provides a new biaxial fracture strength test method for thin solar cell samples. The FE model was validated by digital image correlation. The use of this ring-on-ring test can be recommended for those applications where the surface properties of solar cells are of interest, such as the effects of crystallinity and the impurity concentrations on fracture strength.
- Damage-layer removal by etching significantly increases the strength of both multicrystalline (mc) and single crystalline silicon wafers.
- It was found that mc-silicon wafer crystallinity has a significant effect on the mechanical strength, *i.e.* the more grain boundaries, the weaker the silicon wafer.
- Samples taken from the bottom of the multicrystalline silicon ingot are up to 30% stronger than those taken from the top. This effect was most significant for samples with many grain boundaries. The decrease in strength for the top ingot location could be related to a high concentration of oxygen and other non-metallic impurities.
- There is a significant decrease in fracture strength when an anti-reflective coating is applied. The high application temperature of this SiN_x coating, 375 °C, induces high thermal stresses in the SiN_x layer, which are thought to be the cause of the decrease in the stress at fracture.

References

1. X.F. Brun, S.N. Melkote, Analysis of stresses and breakage of crystalline silicon wafers during handling and transport, Solar Energy Mater. & Solar Cells 93, 2009, pp. 1238–1247.
2. Budiman et al., Solar Energy Mater. & Solar Cells 130, 2014, p.303.
3. B.R. Lawn, Fracture of Brittle Solids, Cambridge University Press, 1993.
4. V.A. Popovich, A. Yunus, M. Janssen, I.M. Richardson, I.J. Bennett, Effect of silicon solar cell processing parameters and crystallinity on mechanical strength, Solar Energy Mater. & Solar Cells s 95, Issue 1, January 2011, pp. 97-100.
5. ASTM Standard C 1499-09, Standard test method for monotonic equibiaxial flexural strength of advanced ceramics at ambient temperature, American Society for Testing and Materials, West Conshohocken, PA, 2013.
6. Limess website, http://www.limess.com/index.php/en/digital-image-correlation
7. V.A. Popovich, A. Yunus, A.C. Riemslag, M. Janssen, I.J. Bennett, I.M. Richardson, Characterization of Multicrystalline Silicon Solar Wafers Fracture Strength and Influencing Factors, International Journal of Material Science, 3 (1), 2013.
8. P. Rupnowski, B. Sopori, Strength of silicon wafers: fracture mechanics approach, Int. J Fracture, 155, pp. 67-74, 2009.
9. C.-L. Tien and T.-W. Lin, Thermal expansion coefficient and thermomechanical properties of SiN_x thin films prepared by plasma enhanced chemical vapour deposition, Applied Optics, 51 (30), pp. 7229-7235, 2012.

EPD Congress 2015
Edited by: James Yurko, Antoine Allanore, Laura Bartlett, Jonghyun Lee, Lifeng Zhang, Gabriella Tranell, Yulia Meteleva-Fischer, Shadia Ikhmayies, Arief Suriadi Budiman, Prabhat Tripathy, and Guy Fredrickson
TMS (The Minerals, Metals & Materials Society), 2015

BEHAVIOR FOR NITROGEN AND IRON IN THE BOTTOM OF CASTING MULTICRYSTALLINE SILICON INGOT

Cong Zhang[1,2], KuixianWei[1,3*], WenhuiMa[1,2,3*], Jiao Li[3], YongnianDai[1,2]
1. Faculty of Metallurgical and Energy Engineering, Kunming University of Science and Technology, Kunming 650093, P. R. China;
2. The National Engineering Laboratory for Vacuum Metallurgy and State Key Laboratory of Complex Nonferrous Metal Resources Clean Utilization, Kunming 650093, P. R. China;
3. Key Laboratory for Nonferrous Vacuum Metallurgy ofYunnan Province and Engineering Research Center for Silicon Metallurgy and Silicon Materials of Yunnan Provincial Universities, Kunming 650093, P. R. China.

Keywords: multicrystalline silicon ingot; impurities behavior; diffusion; nitrogen; iron

Abstract

The paper studied the diffusion process of nitrogen and characteristic distribution of iron impurities in the bottom of cast crystalline silicon ingot along the direction of crystal growth. According to the dynamics analysis of the decomposition of Si_3N_4 coating, the diffusion process of nitrogen into multicrystalline silicon ingot during casting process should be divided into three steps. The diffusion length of nitrogen in multicrystalline silicon ingot is 700 μm. The results of EDS analysis indicate that the distribution characteristic of iron in multicrystalline silicon ingot is concerned with the diffusion process of nitrogen in this length. Based on the above research results, the distribution model of iron in casting multicrystalline silicon ingot was established and verified by the experimental results.

Introduction

Recently, using the cast multicrystalline silicon ingot for production of solar cells has been received more and more attention due to its low cost and simple technology [1]. As we known, the multicrystalline silicon with the minority carrier lifetime larger than 2 μs can meet the requirements of solar cell preparation. However, the low minority carrier lifetime area with the length of about 2~3 cm is present at the bottom of the casting multicrystalline silicon ingot, which greatly reduces the material utilization ratio and therefore increases the cost of solar cells.

There are many research on affect factors about the low minority carrier lifetime width, and the results indicates that the larger amount of impurities present in the bottom of casting multicrystalline ingot leads to the formation of low minority carrier lifetime area [2]. Naerland et al. [3] found that the low minority carrier lifetime areas of the multicrystalline silicon ingot had the higher level of iron impurity amount, which is the most important reason for reducing the life spans of minority carrier. The researchers speculated that the high iron area was caused by the iron diffusion form crucible and nitride coating to the crystal silicon. Gao et al. [4] studied the process of casting quasi-monocrystalline silicon ingot using the simulation method and the results indicate that the iron concentrations in the silicon ingot had two peaks values, one is cause by the iron diffusion from the crucible and nitride coating iron to the silicon crystal, the other one is caused by the iron diffusion from the melt to the silicon seed crystal. Yu[5] optimized the two peaks of the iron content in the bottom of quasi-single-crystalline silicon ingot, but the diffusion distance of the first iron peak was less than expected.

In order to investigate the formation reason of the low minority carrier lifetime area in the bottom of the casting ingot, the decomposition of Si_3N_4 coating and the diffusion process of nitrogen into multicrystalline silicon ingot during casting process were discussed in this work. According to the dynamics analysis of the decomposition of Si_3N_4 coating, the model of N diffusion was proposed. The iron concentration distribution in the bottom of casting multicrystalline silicon ingot was studied by EDS, and which is compared with the N distribution model.

Experimental

Materials and experimental process

The silicon materials with the purity of 6N were used for casting ingot in this work, the main impurities contents in the raw material were tested by ICP-MS and the results were shows in Table 1.

Table1 The contents of main impurities in raw material（mass fraction，10^{-6}）

Impurity	Fe	Al	Cu	Li	B
Content	343.3	327.2	94.3	17.1	664.5

The silicon materials were loaded into the ingot, and the system was evacuated to 1 Pa to eliminate air, the silicon raw materials was heated to 1500 ℃ under the time of 2.5 h, and then high temperature preservation 3 h for making the silicon materials completely melt. The second process was the crystal growth. In order to obtain the excellent columnar crystal, cooling down the temperature to 1440 ℃ and keep about 400 ℃ temperature difference between TC1 and TC2, while keep the heat cover slow rising at the rate of $10\mu m \cdot s^{-1}$ to the end of crystal growth. The third step is the annealing process. The furnace body temperature reduced to 1200 ℃ and kept 3 h for stress relief. The silicon ingot was cooled to room temperature at the fourth stage. The final ingot had a size of $156\times156\times56$ mm^3.

The ingot was cut from the middle along the longitudinal and a 2 mm thick silicon wafer was selected for sample. The silicon sample was mechanically polished and then etched under hot concentrated alkaline solution to remove the mechanical damage layer with the thickness of about 20μm on the surface, finally, the impurities contamination on the wafer surface was removed by the standard RCA cleaning fluid. The MWPCD (Semilab model WT2000) was used for measurement of minority carrier lifetime, and the result was shown in Fig.1.

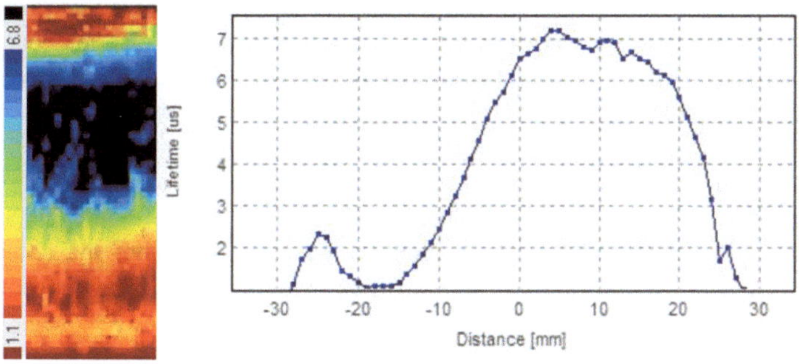

Fig.1 Longitudinal section of minority carrier lifetime scan

Results and discussion

Relations between minority carrier lifetime and Fe concentration

From the Fig. 1, one can find that the minority carrier lifetime in the bottom of ingot is lower than that in the middle. The minority carrier lifetime isn't continued increase in the bottom of the silicon ingot, and which shows a peak value at the distance of 5 mm.

The Fe concentration distribution along the direction of crystal growth line can be calculated according to the formula of $[Fe_i]=C_{MWPCD}\cdot(1/\tau_{before}-1/\tau_{after})$; $C_{MWPCD}=3.4\times10^3 s/cm^3$. YU[5] used this formula to draw Fe distribution along the crystal growth direction of casting multicrystalline silicon ingot. And the Fe distribution results were corrected with the Fe diffusion coefficient measured at 1050 ℃

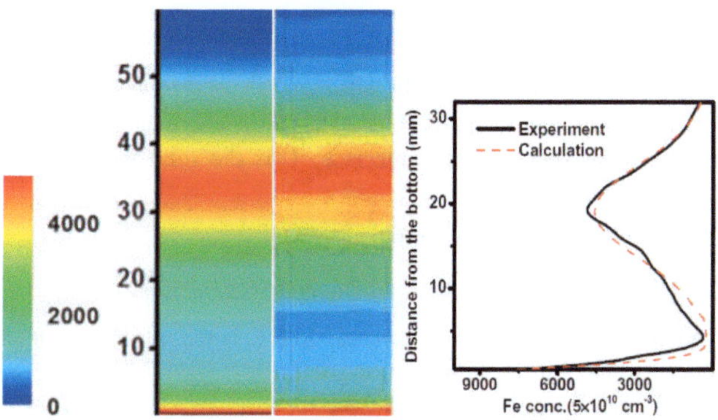

Fig.2 The Fe concentration distribution in the ordinary multicrystalline silicon ingot[5]

The Fe concentration distribution curves were most in conformity with the empirical experimental values. But Fe diffusion distance is slightly smaller than the expected values in the bottom of the crucible. YU [5] proposed that the second peak in the bottom of the crucible was caused by the Fe diffusion from the Si_3N_4 coating into the silicon melt, but the function of the coating itself is to prevent the spread of the impurities from crucible to melt. Buonassisi[6] found that tremendous amount of Fe exist in Si_3N_4 precipitation for the multicrystalline silicon ingot. Therefore, the distribution of Fe in the bottom of the crucible might be associated with the decomposition of Si_3N_4 under high temperature. To confirm the conjecture, the Si_3N_4 decomposition and diffusion length is simulated, and the Fe content and diffusion length were detected and recorded with Si_3N_4 decomposition along the direction of crystal growth in the efficient multicrystalline silicon ingot.

Diffusion of silicon nitride coating

In the phase of crystal growth, argon gas is inlet into casting furnace to promote the remove of volatilization impurity which is driven to the surface of silicon ingot by directional solidification effect. [7] Meanwhile, argon gas can promote the nitrogen converted into the nitrogen atoms, and promote silicon nitride reaction on the surface of the melt [8]. Under the argon gas protection, the argon gas flow lead to more evenly distribution of N elements on the surface of the melt which can react with silicon melt or react with volatile silicon steam to generate Si_3N_4 particles and fall back into the crucible surface.

Suppose the total surface area of silicon sample is *A*: The density of nitrogen is ρ_{Si3N4}, nitriding rate can be expressed as:

$$v = AdNdx / dt \tag{1}$$

dN is nitrogen volume concentration in the silicon nitride, total nitriding rate can be expressed as:

$$v = \frac{AdN}{\rho_{Si3N4}(1 - \frac{3M_{Si}}{M_{Si3N4}})} \square \frac{d(\Delta w / A)}{dt} \tag{2}$$

1) At the beginning of the reaction, the nitrogen released by coating decomposition reacts with silicon melt, and which is controlled by the chemical reaction. Suppose V_c is the nitride reaction speed on the interface, *V* is the total rate of silicon nitride reaction.

Dynamics equations can be expressed as:

$$v = v_c = ACk_c \tag{3}$$

K_C is rate constant of nitrogen and silicon reaction, *C* is the nitrogen concentration on silicon surface, can be represented as:

$$C = \frac{\Delta w}{2(\pi dt)^{1/2}} \exp(-\frac{x^2}{4dt}) \tag{4}$$

2) In the intermediate reaction stage, a layer of nitride has been formed on silicon wafer surface, thus, the only restriction link of the nitride reaction is in existence at this stage. The nitride reaction rate is controlled by the diffusion and chemical reaction rates and which shows different control strength at different reaction times. Suppose interface chemical reaction rate is V_c, diffusion velocity is V_D, D is diffusion coefficient of nitrogen in the silicon nitride, z is the diffusion thickness at time t.
Bring the equation (2) into (1) and simplify the equation (1) as:

$$\frac{d}{c}z-\frac{ad-c}{2c^2}z^2=t \qquad (5)$$

$$a=(1-\lambda)\frac{Ck_c}{d_N};\quad c=\frac{CD}{\lambda d_N};\quad d=\frac{D}{\lambda k_c}$$

λ is the coupling effect of chemical reaction and diffusion, which is close to 1 when the interactions of chemical reaction and diffusion is relatively weak.
The parabolic equation shows that N diffusion rate would slow with time prolonging, which is the general equation of direct nitride process and can well explain the nitride process
3) In the late stage, the nitride layer had been so thick that the diffusion path of N through Si_3N_4 is rather long and the diffusion had become the determining link for the overall reaction rate, that is $v_C >> v_D$ (v_D is the diffusion rate). Ignoring the influence of the chemical reaction, then

$$v=v_D=A\cdot C\cdot D\cdot k_D \qquad (6)$$

K_D is the rate constant of nitrogen in the silicon nitride thin films.
According to the formula (3), (5) and (6), the N concentration distribution model was established in the bottom of the crucible, as shown in Fig. 3

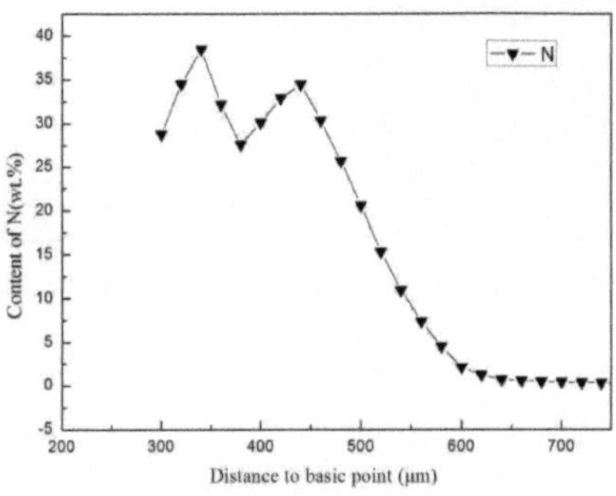

Fig. 3 N concentration distribution model in the bottom of efficient monocrystalline silicon ingot

It can be seen from the Fig.3, the limiting spread thickness of N decomposed from Si_3N_4 is 700 μm in the silicon ingot at high temperatures, and two peaks are present at the distance of 350 and 450 μm. The peak at the distance of 350 μm is caused by a silicon nitride membrane formed by the reaction of silicon melt and N released from Si_3N_4 coating decomposition. Meanwhile, the N and the silicon substrate would be separated with the formation of the silicon nitride membrane. The reaction between the N and silicon can continuously occur when they spread to the interface of Si/Si_3N_4, and which is both controlled by the diffusion and reactivity. The thin film is gradually growth with the duration of the reaction of N and silicon, and the interface of Si/Si_3N_4 continuously extends into the silicon bulk. The diffusion of N atom should become more difficult with the formation of thick and dense silicon nitride membrane. In this case, the reaction is mainly controlled by the diffusion process.

The distribution of Fe in the diffusion layer

Based on the above inference results, scanning electron microscopy (SEM) was used to observe the contact area with the bottom of crucible, and the result was shown in Fig.4. It can find that the numerous rod-like materials are existed in the contact field. The EDS is used to characterize the component of the rod-like matter and it mainly includes Si and Fe. According to the report by Buonassisi et al.[9], the Fe is existed with the form of $FeSi_2$. The Fe content distribution in the bottom of silicon ingot was recorded, as shown in Fig. 5. Within the diffusion distance of 450 μm, the Fe content is increasing with the increase of diffusion distance, but, which has a dramatically decrease when the diffusion distance exceeds 450 μm and the change of Fe content trends to flat when the diffusion distance exceeds 700 μm.

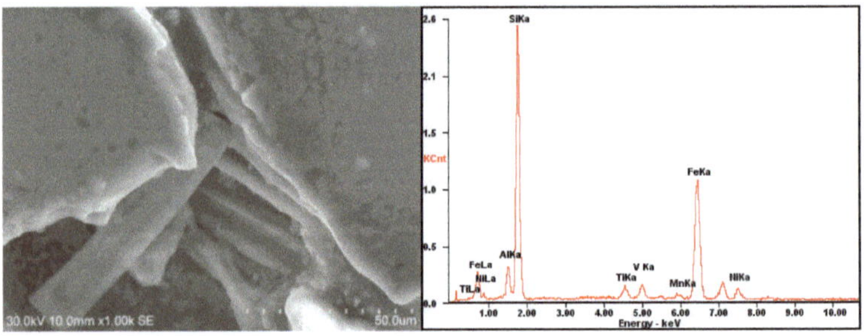

Fig.4 SEM and EDS analysis of silicon in the bottom of ingot

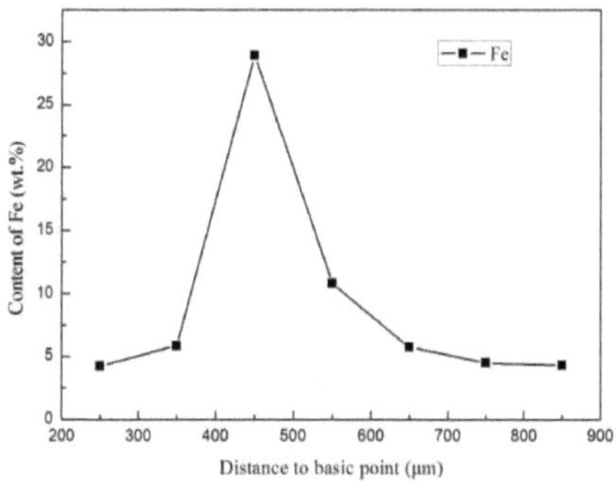

Fig.5 The Fe content distribution in the bottom of silicon ingot

According to the results above, the similar peaks, both for Fe and N content distribution are observed at the diffusion distance of 450 μm. Thus, we deduced that the diffusions of Fe and N might have some connection, which is the change of Fe content is effected by the diffusion of N. The unsaturated bond formed by the N diffusion and Si_3N_4 crystal nucleus would promote the Fe deposition. Some of Fe impurities introduced by the crucible and coating will be deposited on the unsaturated bond of the silicon nitride particles surface and the internal defects of the Si_3N_4 crystal nucleus.

At the same time, the Si crystal nucleuses are gradually formed with the decreasing temperature. As we known, the Fe-Si eutectic point temperature (1480 K) is lower than the melting point (1693 K) of Si, thus, the Fe impurity would be enriched at the interface of silicon and Si_3N_4 crystal nucleus and eventually form inclusion in the grain boundary between Si and Si_3N_4. Therefore, Fe impurity content and N content in the coating show same change tendency.

The free Fe would be enriched in the interface of silicon melt because its segregation coefficient is less than 1. Numerous defects would be caused by the formation of new Si_3N_4 crystal nucleus, which leads to the Fe deposition in the intermediate layers between the silicon melt and bulk. Thus, the diffusion of Fe would be deferred.

Conclusions

Based on the kinetics analysis result of the diffusion of N generated by Si_3N_4 decomposition, we can obtain that the limit spread thickness of N is 700 μm, and two N content peaks appear at the distance of 350 μm and 450 μm, respectively.

The Fe content distribution is analyzed in the bottom of silicon ingot, and the results show that the Fe content distribution is mainly effected by the N diffusion, and they distributions is similar. ,

Numerous defects would be caused by the formation of new Si_3N_4 crystal nucleus, it leads to the Fe deposition in the intermediate layers between the silicon melt and bulk, and which causes the impediment of Fe diffusion into the silicon ingot.

Acknowledgments

This work is supported by the National Natural Science Foundation of China (No. 61404063, 51466005 and U1137601), the National Natural Science Foundation of Yunnan (No. 2012FB125), the Key Project of Chinese Ministry of Education (No. 212159).

References

[1] G. Flamant, V. Kurtcuoglu, J. Murray, A. Steinfeld. Purification of metallurgical grade silicon by a solar process. J. Solar Energy Materials & Solar Cells 90 (2006) 2099-2106.

[2] A. A. Istratov, T. Buonassisi, M.D. Pickett, M. Heuer, E.R. Weber, Control of metal impurities in "dirty" multicrystallineSilicon for solar cells. Material Sciences and Engineering B 2006, 134(2-3) 282-286.

[3] T. U. Nrland, L. Arnberg, A. Holt, Origin of the low carrier lifetime edge zone in multicrystalline PV silicon, Progress in Photovoltaics: Research and Applications 17 (2009) 289.

[4] B. Gao, S. Nakano, K. Kakimoto. Influence of Back-Diffusion of Iron Impurity on Lifetime Distribution near the Seed-Crystal Interface in Seed Cast-Grown Monocrystalline Silicon by Numerical Modeling, Crystal Growth & Design 2011, 12(1) 522-525.

[5] X. G. Yu, X. Gu, S. Yuan, K. X. Guo, D. R. Yang, Two-peak characteristic distribution of iron impurities in the bottom of cast quasi-single-crystalline silicon ingot, Scripta Materialia 68 (2013) 655–657

[6] T. Buonassisi, A. A. Istratov, M. D. Pickett, J.-P. Rakotoniaina, O. Breitenstein, M. A. Marcus, S. M. Heald, E. R. Weber. Transition metals in Photovoltaic-grade ingot-cast multicrystalline Silicon: Assessing the role of impurities in Silicon nitride crucible lining material, Journal of Crystal Growth 287 (2006) 402-407.

[7] K. Fujiwara, W. Pan, K. Sawada, M. Tokairin, N. Usami, Y. Nose, A. Nomura, T. Shishido, K. Nakajima, Directional growth method to obtain high quality polycrystalline silicon from its melt, Journal of Crystal Growth 292 (2006) 282-285.

[8] B. Prasad, S. Bhattacharya, A. K. Saxena, S. R. Reddy, R. K. Bhogra, Performance enhancement of mc-Si solar cells due to synerdetic effect of plasma texturization and SiNx:H AR coating, Solar Energy Materials & Solar Cells, 94 (2010) 1329-1332.

[9] T. Buonassisi, M. Heuer, A. A. Istratov, M. D. Pickett, M. A. Marcus, B. Lai, Z. Cai, S. M. Heald, E. R. Weber, Transition metal co-precipitation mechanisms in silicon, Acta Materialia， 55 (2007) 6119-6126.

EPD Congress 2015
Edited by: James Yurko, Antoine Allanore, Laura Bartlett, Jonghyun Lee, Lifeng Zhang, Gabriella Tranell, Yulia Meteleva-Fischer, Shadia Ikhmayies, Arief Suriadi Budiman, Prabhat Tripathy, and Guy Fredrickson
TMS (The Minerals, Metals & Materials Society), 2015

THERMAL FIELD DESIGN AND OPTIMIZATION OF DIRECTIONAL SOLIDIFICATION FOR MULTICRYSTALLINE SILICON GROWTH

Wenhui Ma [1,2,3*] Xi Yang [1,2,3], Guoqiang Lv [1,3]

1. Faculty of Metallurgical and Energy Engineering, Kunming University of Science and Technology, Kunming 650093, China;
2. State Key Laboratory of Complex Nonferrous Metal Resources Cleaning Utilization in Yunnan Province/The National Engineering Laboratory for Vacuum Metallurgy, Kunming University of Science and Technology, Kunming 650093, China;
3. Key Laboratory of Non-Ferrous Metals Vacuum Metallurgy of Yunnan Province/ Engineering Research Center for Silicon Metallurgy and Silicon Materials of Yunnan Provincial Universities, Kunming 650093, China;

Keywords: Numerical simulation, Directional solidification, Heat transfer, Crystal growth, Thermal stress

Abstract

Cast multicrystalline silicon ingots are widely used in photovoltaic manufacturing. A key issue to achieve high solar cell efficiencies is to attain an optimized temperature field during directional solidification (DS) process. This paper reports numerical investigation of multicrystalline silicon (mc-Si) ingot production using two major types of DS furnace. Specific examination is made on thermal distribution, interface shape and stress field. Evaluation is performed for the applicability of thermal system design to reduce thermal stress, improve crystal quality and enhance energy efficiency. The effects of procedure parameters and geometric configuration on temperature distribution are discussed as well to provide the viable solutions for systems optimization.

Introduction

Silicon in its pure crystalline form is the dominant semiconductor material used in photovoltaics (PV). Currently, more than 90% of commercial PV modules are made from crystalline silicon, with the majority share of silicon grown as multicrystalline ingots. [1,2] DS process is the mainstream technology for production of mc-Si. A key issue in the production of mc-Si ingots is to obtain a sufficient control of the heat transport during DS process in order to obtain satisfactory crystal quality as well as higher energy efficiency. Better control of thermal field will improve the quality and yield of the mc-Si ingots by influencing the temperature gradient in a beneficial way.

Numerical simulation is a powerful tool to help understand and depict the temperature distribution inside the growth furnace and to allow optimizing of the growth conditions to achieve high quality with a minimum of defects. Bellmanna et al. [3] established a transient CFD model of the Bridgman furnace, and investigated the energy changes under different boundary conditions. Shur et al. [4] confirmed that different cooling paths and coolant flow rates could change the directional solidification effect during the crystal growth. Black et al. [5] numerically investigated the Si casting process through DS technology, and proposed an optimized mono-like Si production method after they studied the relationship of thermal field and m/c interface. Liu et al. [6-8] performed numerical simulations and reported several improvement measures of DS furnace from the convection and mass transfer based on the thermal field. He et al. [9] numerically investigated the temperature field and grain growth process.

In this paper, we focus on the heat exchange system for crystal growth. Two major types of casting furnaces, direct solidification system and heat exchange method system, will be compared from the aspects of thermal field, interface shape and stress field according to related numerical simulations. Through the results, we estimated the optimized design in present crystal growth system to get the better solidification technology and crystal quality. It will help to enhance the application of DS furnace as well as the mc-Si materials in industry.

Model descriptions and simulation method

Direct solidification system (DSS) and heat exchange method (HEM) are two of the most widely used methods for multicrystalline silicon growth in the PV industry. The semi-industrial directional solidification furnace for both types as shown in Fig. 1, were used in this work. During the growth in a DSS furnace, the side insulation and heater are moved up at same time to allow heat extraction from the bottom while the crucible remains stationary. In HEM furnace, the crucible and the heat exchange block move downwards during growth so that the crucible has a direct radiation path to the cold wall. Both systems extract heat through the bottom of the crucible in order to initiate solidification. While the basic functions of the two systems are the same, nevertheless, the different moving parts and specific geometries of the systems lead to quite different temperature distributions. Particularly, the different temperature profiles in the crucible result in significant differences in the shape of the solidification interfaces, the distribution of thermal stresses and the formation of defects. Consequently, the characteristics and properties of the ingots grown from DSS and HEM furnaces as well as the control ways of two furnaces are different.

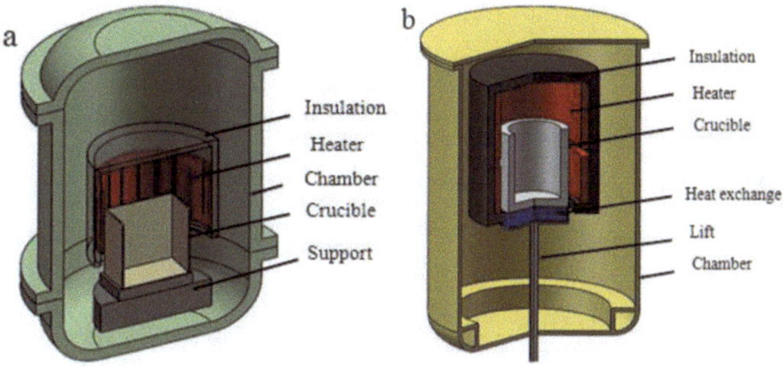

Fig. 1 The DS system: (a) DSS furnace and (b) HEM furnace

As shown in Fig. 1, the DSS furnace uses square crucible and the HEM furnace uses cylindrical crucible, so we established 3D and 2D-axisymmetric simulation models for the above two furnaces respectively. The height of the ingot is about 0.15 m and the mass is 4 kg approximately. Heat transfer depends mainly the conduction and radiation in the vacuum directional solidification process, so the gas flow effect in the furnace and convective heat transfer of the silicon were ignored. The mathematical model was developed by considering the conductive heat transfer in all solid components as well as surface radiation, the effective conductivity coefficients have been used for approximating the heat transfer by convection, and the treatment of the latent heat and analysis of thermal stresses in the silicon ingot has been described in detail in [10], the same method and equations as mentioned are applied to our research systems. The numerical solutions to the governing equation are realized using the commercial software COMSOL Multiphysics 4.2a. The materials' thermophysical properties are provided by manufacturers.

Results and discussion

Temperature distribution and interface shapes

The transient simulation was used to realize DS processes of mc-Si based on actual operation. The thermal distribution and interface shape of silicon during solidification process at different solid fractions are shown in Figs. 2 and 3. It is found that the temperature distribution and interface shape of silicon in two cases are different. For the DSS furnace, due to the role of square crucible, crucible edges place closer to the heater, so the edges of the silicon material solidified slower. In addition, the heat exchange block at the bottom of crucible in this furnace is without cooling water, the moving-up rate of side insulation and heater is slower to ensure the interface is relatively flat. While in the HEM furnace, because of the symmetrical structure of cylindrical crucible and the effect of cooling water at the bottom of crucible, the interface keep slight convex under faster pulling-down rate.

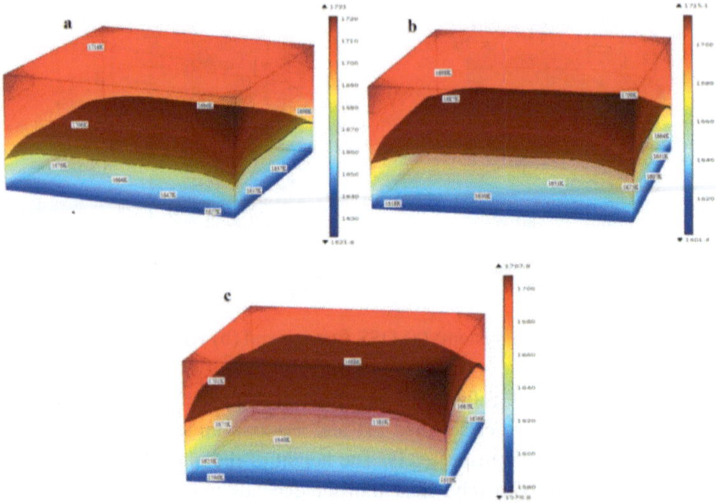

Fig. 2 temperature distribution and interface shapes of silicon in DSS furnace: (a) 0.25 solidification fractions, (b) 0.5 solidification fractions, (c) 0.75 solidification fractions

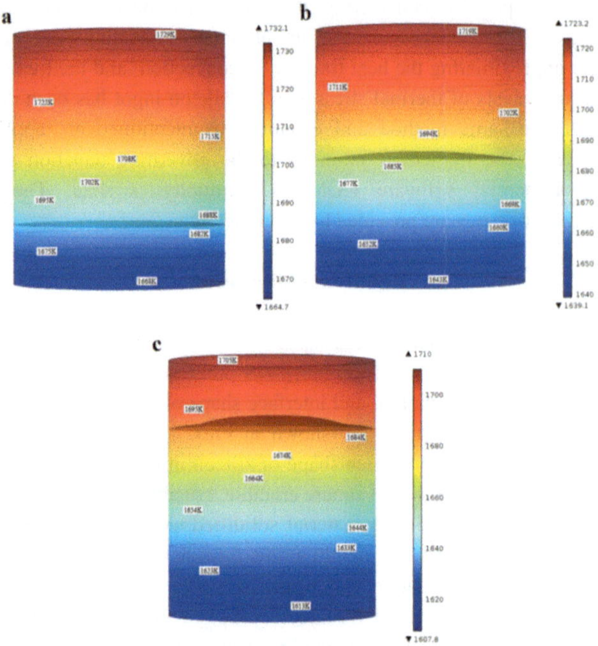

Fig. 3 temperature distribution and interface shapes of silicon in HEM furnace: (a) 0.25 solidification fractions, (b) 0.5 solidification fractions, (c) 0.75 solidification fractions

It is well known that crystal growth proceeds in a direction perpendicular to the interface, and a slightly convex growth front is desired, strongly convex shapes should be avoided. The shape of interface depends on the situation of the heat dissipation of the solidification process. When the heat dissipation from the crucible bottom is greater than that from the wall, the interface will be convex; otherwise it will be concave. Thus, it needs proper moving-up rate of the heater and insulator for the DSS furnace while a combination of pulling-down rate of crucible and flow of bottom cooling water for the HEM system to control the interface shape.

Thermal stress in the silicon ingot

It expands about 8% in volume during the phase change of silicon from melt to crystal. The volume expansion is compensated by an upwards movement of the liquid in the solidification process. So we only comparatively study the effect on thermal stress of silicon after the complete solidification in the DSS furnace and HEM furnace, as shown in Fig. 4. Due to uneven temperature distribution inside the silicon ingot, uneven expansions of adjacent parts of the crystal also induce stresses by mutual restraints. That is, when temperature changes unevenly, different parts of the silicon ingot should expand with different quantities. It is found from Fig. 4 that the thermal stresses value is higher, but it is evenly distributed in square ingot compared with cylindrical ingot, and the maximal stress and the stress concentration occurred on upper edge of ingot. This is advantageous to the material quality and utilization, also indicated that the overall temperature distribution of silicon ingot is more uniform in DSS furnace.

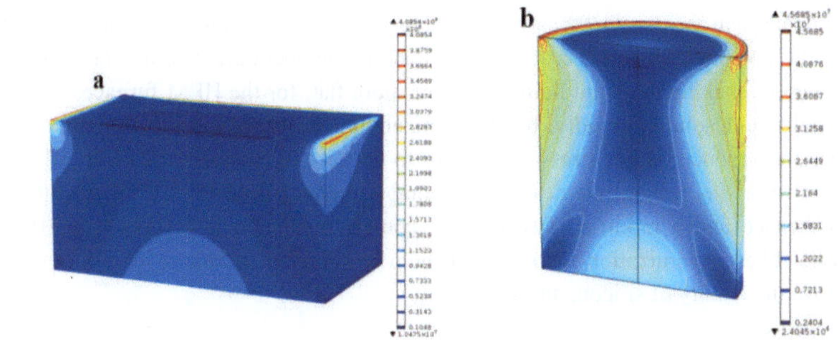

Fig. 4 The thermal stresses of silicon ingot longitudinal section after completion of solidification (unit: Pa): (a) in the DSS furnace, (b) in the HEM furnace

Thermal field design and optimization

According to the results of the above, for DSS furnace, the purpose of thermal field design and optimization is mainly to solve the problem that silicon solidification slower at the edges and corner of crucible, and the means is to make the interface more flat. As shown in Fig. 5, we put forward the optimization of the square crucible suitable for thermal field, which is, the heater and the insulation cover are designed in the shape of octagon, and every side consistent with same distance to crucible. In this case, we can achieve the aim only needing a proper moving-up rate. Similarly, for HEM furnace as shown in Fig. 6, the distinguished design is conical insulation unit, which allows a better control of temperature field in the solidification process of silicon. It provides a growing heat dissipation space, which could effectively control the lateral heat dissipation. It works as a heat preservator and reduces the silicon's internal temperature difference for the improvement of the crystal quality, as we have detailed described in [10].

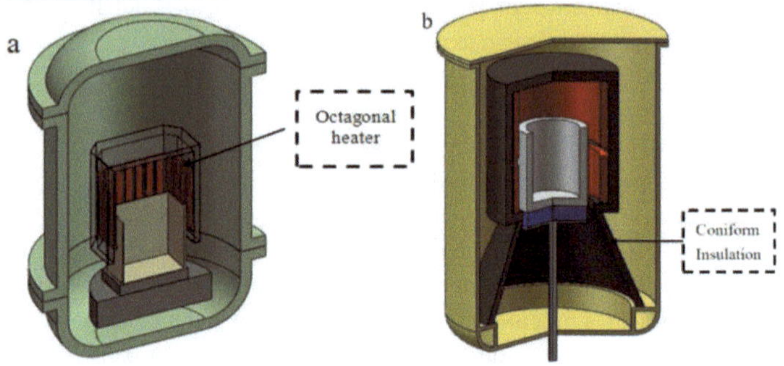

Fig. 5 The optimization design of DS system: (a) the DSS furnace and (b) HEM furnace

Conclusion

The transient simulations were performed to investigate the effects of silicon solidification in two main types of DS furnace, and put forward the improvement methods. For the DSS furnace, it needs to raise the temperature of crucible at edges and corners, making the solid liquid interface more flat; for the HEM furnace, silicon material could solidify rapidly, but the temperature difference inside the silicon is bigger and produces larger stress, so it needs to uniform the temperature of the silicon material. For this reason, we designed a suitable thermal field for crystal growth in DSS furnace and a modified system with conical insulation is proposed in HEM furnace. These optimization methods will help to solve the problems of DS device, improve the quality of silicon, and save energy.

Acknowledgments

This work is supported by NSFC project (Nos. 51466005 and U1137601), the National Key Technology R&D Program (No. 2011BAE03B01), the Program for Innovative Research Team in University of Ministry of Education of China (Grant No. IRT1250).

References

[1] B. Wu, N. Stoddard, R. Ma, R. Clark, Bulk multicrystalline silicon growth for photovoltaic (PV) application, J. Cryst. Growth 310 (2008) 2178–2184.
[2] Jiuan Wei , Hui Zhang , Lili Zheng, Chenlei Wang, Bo Zhao, Modeling and improvement of silicon ingot directional solidification for industrial production systems, J. Solar Energy Materials & Solar Cells 93 (2009) 1531–1539.
[3] M.P. Bellmanna,, E.A. Meese, M. Syvertsen, A. Solheim, H. Sørheim, L. Arnberg, Silica versus siliconnitride crucible: Influence of thermophysical properties on the solidification of multi-crystalline silicon by Bridgman technique, Journal of Crystal Growth 318 (2011) 265–268.
[4] J.W. Shur, B.K.Kang, S.J.Moon, W.W.So, D.H.Yoon, Growth of multi-crystalline silicon ingot by improved directional solidification process based on numerical simulation, J. Solar Energy Materials & Solar Cells 95 (2011) 3159–3164.
[5] A. Black, J. Medina, A. Piñeiro, E. Dieguez. Optimizing seeded casting of mono-like silicon crystals through numerical simulation, Journal of Crystal Growth 353 (2012) 12–16.
[6] Z.Y. Li, L.J. Liu, W.C. Ma, K. Kakimoto, Effects of argon flow on impurities transport in a directional solidification furnace for silicon solar cells, Journal of Crystal Growth 318(2011) 304-312.
[7] Z.Y. Li, L.J. Liu, X. Liu, Y.F. Zhang, J.F. Xiong. Effects of argon flow on melt convection and interface shape in a directional solidification process for an industrial-size solar silicon ingot, Journal of Crystal Growth 360(2012) 87-91.
[8] Z.Y. Li, L.J. Liu, X. Liu,Y.F. Zhang, J.F. Xiong. Heat transfer in an industrial directional solidification furnace with multi-heaters for silicon ingots, Journal of Crystal Growth 385(2014)9-15.
[9] HE Liang, WANG Gang, RONG Yi-ming, Modeling on Directional Solidification of Solar Cell Grade Multicrystalline Silicon Ingot Casting, J. Shanghai Jiaotong Univ. (Sci.), 2011, 16(3): 316-319.
[10] X. Yang, W.H. Ma, G.Q. Lv, K.X. Wei, T. Luo, D.T. Chen, A modified vacuum directional solidification system of multicrystalline silicon based on optimizing for heat transfer, Journal of Crystal Growth 400 (2014) 7-14.

EPD Congress 2015
Edited by: James Yurko, Antoine Allanore, Laura Bartlett, Jonghyun Lee, Lifeng Zhang, Gabriella Tranell, Yulia Meteleva-Fischer, Shadia Ikhmayies, Arief Suriadi Budiman, Prabhat Tripathy, and Guy Fredrickson
TMS (The Minerals, Metals & Materials Society), 2015

MICROSTRUCTURE AND MECHANICAL PROPERTIES OF A SCREEN-PRINTED SILVER FRONT SIDE SOLAR CELL CONTACT

V.A. Popovich[1, 2], M. Janssen[2], I.J. Bennett[3], I.M. Richardson[2]

[1]Materials innovation institute M2i, Delft, The Netherlands, v.popovich@m2i.nl
[2]Delft University of Technology, Department of Materials Science and Engineering, Delft, The Netherlands
[3]Energy Research Centre of the Netherlands, Solar Energy, PV Module Technology, Petten, The Netherlands

Keywords: Silicon solar cells, silver front contact, formation model

Abstract

The most critical processing step during the manufacture of screen-printed crystalline solar cells is firing aluminium and silver contacts, which generates residual stresses and solar cell bowing. In this paper, an alternative Ag contact formation mechanism is proposed and aspects related to electrical contact properties, residual stresses and layer delamination are investigated. It is found that there are two main processing parameters affecting the uniformity and delamination of the Ag/Si interface, namely the peak firing temperature and the silicon surface roughness. Silicon surface polishing gives a better wetting of the silicon surface by the glass layer, resulting in a good contact and lower incidence of large voids, compared to the case of highly textured surfaces. The non-uniformity in the glass layer and large voids at the Ag/Si interface (in the case of a textured surface) are expected to have a negative effect on the mechanical strength of the solar cell.

1. Introduction

Due to pressure from the photovoltaic industry to decrease costs of solar cell production, there is a tendency to reduce the thickness of silicon wafers. Unfortunately, wafers contain defects, created by the various processing steps involved in solar cell production, which reduce the strength of the silicon wafer significantly. A higher breakage rate is to be expected if thinner wafers are produced with identical fracture strength in combination with the same loading during processing. The most critical processing step during the manufacture of screen-printed solar cells is the firing process, during which the screen-printed aluminium and silver layers are simultaneously fired in order to create electrical contacts. Residual stresses are generated within the cell due to mismatch of thermal expansion coefficients and different mechanical behaviour of the materials used in the metallic contacts. The wafer bows and forms a convex or concave body upon cooling, which mechanically loads the cell and may cause fracture [1].
As the thickness of silicon wafers is reduced, cell bowing becomes more pronounced and poses more problems during different processing steps. It is important to find a compromise between electrical properties, strength and costs of the solar cell. To achieve this, it is necessary to have a better understanding of microstructure, stress development and mechanical properties of the cell.
Front-side Ag thick-films metallization is an important step in standard crystalline Si (c-Si) industrial cell production. Silver metallization pastes primarily consist of Ag powder, glass frits, and vehicles for the desired rheological properties of the pastes. Although only a tiny quantity of

glass frits is added to the Ag paste, these frits enable the electrical connection of the metal film to the Si wafer during firing. Thus, it is important to understand the effect of firing parameters on the contact formation.

The main focus of this work is the analysis of silver metallic contact on a macroscopic as well as on a microscopic level. In this section the experiments and results are presented and summarised in an improved model of Ag contact formation. The effect of the processing parameters on the formation and homogeneity of the Ag contact layer is described.

2. Experimental conditions

2.1. Material preparation

Silicon wafers of 156×156 mm^2 and a thickness of 200 μm were sliced off a single multi-crystalline (mc) silicon block. In this study, only wafers from the middle of the block contributed to the results. A standard industrial cell process was used and the screen printing on the rear or front sides was performed in the conventional H-pattern manner with a 165 mesh screen. The alloying was performed (after drying the paste) by firing the wafers in a lamp-heated conveyor-belt furnace. The effect of firing conditions was examined by applying three different peak firing temperatures (750 °C, 850 °C and 950 °C). Silver contact was applied to the full silicon wafer, thus covering the entire front side (for these samples the Al back contact was not applied), in order to increase the surface area tested and to obtain a uniform stress distribution.

To analyse the effect of the silicon wafer's surface roughness on the Si-Ag front contact formation, three types of specimens were prepared from neighbouring mc-Si wafers and fired at 850 °C. All the samples were taken from the as-cut state and divided into three groups depending on the pre-processed surface condition:

- the as-cut state, thus including the saw-damage layer.
- the as-cut wafers, which are subsequently textured by an industrial acid texturing/etching process, that includes a two-step texturing in a $HF/HNO_3/CH_3COOH$ acid bath (to remove the damaged layer and to create a highly textured silicon surface in order to trap the light).
- the as-cut wafers, which are subsequently chemically polished (15 μm removal from both wafer sides in a $HF+HNO_3$ bath for 60 s).

In order to investigate the Ag/Si interface layer and glass phase distribution within the Ag layer, the following etching steps were applied: a) no etching and initial configuration: as-fired condition; b) etching with Aqua Regia (AR): the Ag finger was removed, the glass and the silver beneath remained; c) subsequent etching with AR and HF: the glass was also removed, but the silver underneath the glass still remained.

2.2. Microstructural analysis

A JEOL JSM 6500F scanning electron microscope (SEM) with energy-dispersive spectroscopy (EDS) was used for microstructural analysis.

Phase identification of the as-dried and sintered Ag paste was conducted by X-ray diffraction (XRD) using a D8-discover diffractometer (Cu K_α radiation) equipped with an Euler cradle.

Measurements of the amount of cell bowing resulting from screen printing were made by an optical method, using a Quick Vision Mitutoyo system. Five cells from neighbouring wafers were prepared and measurements were performed over the full length of the solar cell (156 mm).

3. Results and Discussion

3.1. Microstructure and formation model of silver front contact

A layer of conventional Ag paste was screen printed on top of the SiN_x antireflection coating and fired at 850 °C through the SiN_x layer onto the emitter surface. Figure 1 shows EDS mapping of a polished cross section of the fired Ag layer, revealing 3 distinct areas: silver; silicon; ZnO, and PbO rich phases, which are residues of the initial glass phases.
Between the silicon and the porous bulk Ag structure a continuous layer containing oxide phases was detected. A closer look into this Ag/Si interface reveals a non-uniform glass layer separating bulk Ag and Si and large conjugated Ag particles (~150 nm in width) on the silicon wafer, see Figure 2.

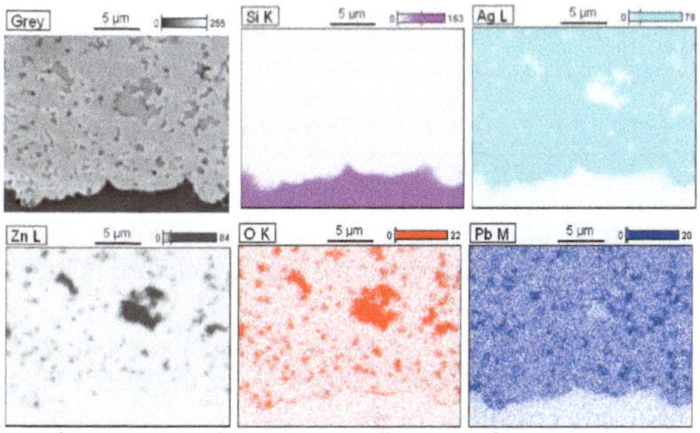

Figure 1. EDS mapping of a cross section of an as-fired Ag front contact.

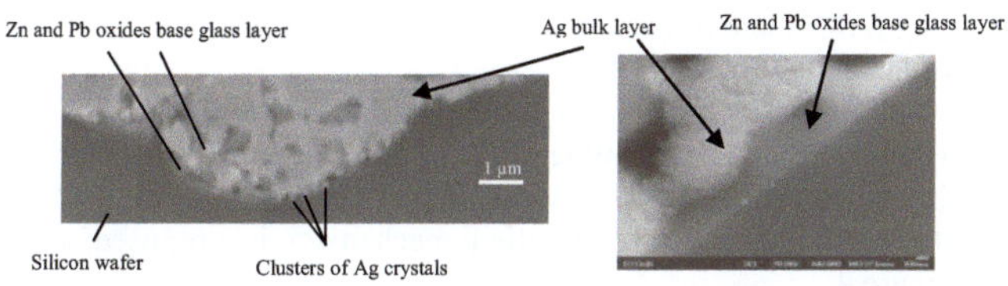

Figure 2. SEM micrograph of Ag paste fired at 850 °C: Ag crystallites and ZnO/PbO-based glass layer at the Ag/Si interface (left image) and an enlarged view of glass layer and sintered Ag particles (right image).

Further EDS analysis shows that the Ag/Si interface has Si and Ag as major constituents, together with Pb and Zn. Therefore, it has the composition of an alloy Ag-Si-M (where M is Zn, Pb). The glass particles in the Ag paste are expected to etch through the passivation layer into the silicon surface, facilitating wetting of the Ag on the Si and hence alloying.
In this work, no clear Ag crystallites etched into the Si emitter surface were found. However, it is believed that, at elevated temperatures (> 550 °C), the glass frit is molten and etches the SiN_x antireflective coating (ARC) [2, 3]. In this way the glass frit brings the Ag particles in direct contact with the Si. The glass, depending on the grain orientation, etches the silicon isotropically or anisotropically.
Metal oxides (*e.g.* PbO, ZnO) present in the glass react with the outer layer of the Ag particles and create a Ag-solvent metal alloy. This alloy fills the pits etched by the glass on the silicon surface. Hence, on isotropically etched silicon, it appears as a round shape and on anisotropically

etched silicon it appears as an inverted pyramid, which can explain pyramidal Ag imprints observed by other authors [2, 3].

An XRD analysis was performed on the front surface of the cell in order to identify phases present in the fired Ag contact layer. Measurements were performed for both as-dried and as-fired Ag paste, Figure 3 a.

a)

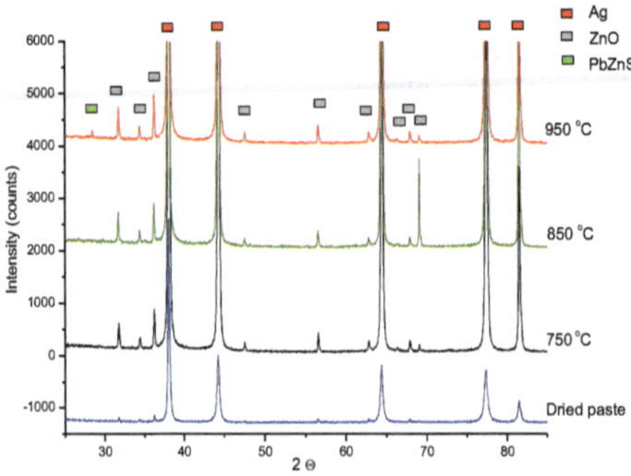

b)

Firing temperature (°C)	Cell bowing (mm)
750	1.8
850	2.5
950	1.8

Figure 3. a) XRD spectrum of Ag paste fired on a silicon wafer under different firing temperatures b) Effect of peak firing temperature on the maximum bowing.

Besides the sharp Ag peaks, two extra phases were detected in the fired Ag layer spectrum, namely ZnO and $PbZnSiO_4$. The latter two are residues of the initial glass frit present in the Ag paste to obtain better sintering properties of the contact layer. However, it is known that initially only two oxide phases were present in the Ag paste, namely ZnO and PbO. It is suggested, that ZnO and PbO form a complex ceramic crystallite phase due to etching through the SiN_x antireflective coating (ARC) and a reaction with the Si wafer during the firing process, resulting in crystallization of larsenite ($PbZnSiO_4$).

It is suggested that in the investigated cells Ag crystallites are formed at the paste/Si interface without the aid of liquid Pb formation. When the Ag particle/agglomerates are in direct contact with the molten glass, the redox reaction between the PbO in the glass and the Si is suppressed.

The following alternative Ag contact formation mechanism is proposed and is also shown schematically in Figure 4.

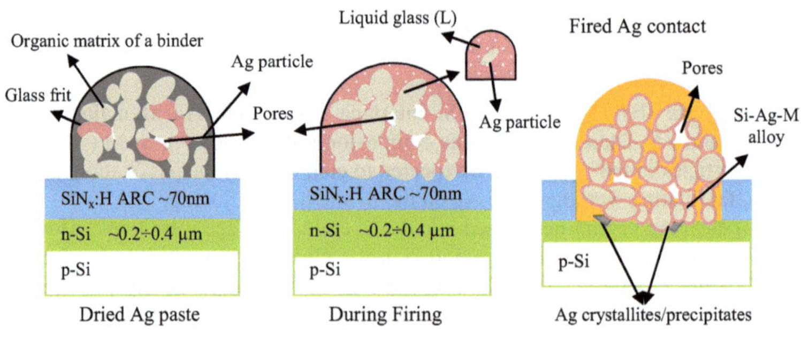

Figure 4. Proposed schematic model for the ionic reduction mechanism during Ag front contact formation under normal industrial firing conditions (850 °C).

After Ag paste deposition the solvents are evaporated in a drying step, leaving behind the metal and glass particles embedded in the organic matrix of the binder. In the subsequent firing step the organics are combusted. As the firing temperature increases above the softening temperature of the glass frit (613 °C for lead silicate glass), the glass frit becomes fluidized, wets the Si surface and a thin layer of liquid glass (L) surrounds the Ag particles. Because Ag and L react to form an alloy of much lower melting point (than Ag), this mechanism leads to the formation of a melt around the Ag particle. With further heating to the peak firing temperature, some of the Ag powder dissolves in the glass and sintering of the Ag powder occurs (an example of sintered Ag particles can be seen in Figure 2).

Ag can dissolve in glass as Ag atoms, however it is a slow process, which can be suppressed by simultaneously oxidation of Ag to Ag+ ions through an interaction with oxygen from the ambient atmosphere [4]:

$$4\,Ag_{(in\,glass)} + O_{2\,(g)} \rightarrow 4\,Ag^{+}_{(in\,glass)} + 2O^{2-}_{(in\,glass)}\,. \qquad (1)$$

Subsequently, silver and oxygen ions dissolved in the liquid glass (L) formed by reaction (1) react with the SiN_x layer and etches it away:

$$SiN_{x\,(s)} + 4Ag^{+}_{(in\,glass)} + 2O^{2-}_{(in\,glass)} \rightarrow SiO_{2\,(in\,glass)} + 4Ag^{o}_{(in\,glass)} + x/2\,N_{2\,(g)}. \qquad (2)$$

As a result of reaction (2), the silicon wafer comes into direct contact with the liquid glass and the silver and oxygen ions diffuse further toward the glass/Si interface and create inverted pyramidal pits on the Si wafer surface via the redox reaction:

$$4\,Ag^{+}_{(in\,glass)} + 2O^{2-}_{(in\,glass)} + Si_{(s)} \rightarrow 4\,Ag^{o}_{(in\,glass)} + SiO_{2\,(in\,glass)}\,. \qquad (3)$$

In this model, metal oxides (PbO and ZnO) in the glass located close to the silicon wafer could also be reduced to liquid Pb or Zn (M), leading to the formation of a Ag-M-Si alloy around the particles. However, some of the liquid Pb or Zn phase (M) may be dissolved back into the glass via:

$$4Ag^{+}_{(in\,glass)} + 2O^{2-}_{(in\,glass)} + Pb_{(l)} \rightarrow 2\,Ag^{o}_{(in\,glass)} + PbO_{(in\,glass)}\,. \qquad (4)$$

When during cooling the molten Si-Ag-M alloy solidifies, it is likely to produce a grading in the composition of the interface between Si and Ag.

As a result of reaction (3), the glass in the etch pits is enriched by Ag atoms and SiO_2. It is known, that Ag^+ ions diffuse faster in fluidized glass than do Ag atoms [4]. Therefore, there is a steady flux of Ag^+ ions toward the Si surface that allows reaction (3) to occur continuously, whereas the Ag atoms reduced via reactions (2) and (4) accumulate in the glass located at the etch pits on the Si surface.

The Ag atoms reduced by the reaction with the Si substrate can precipitate as Ag particles in the molten glass during firing or during the subsequent cooling process, normally by air cooling, precipitate at the apexes of the inverted pyramidal pits. Subsequently, the Ag precipitates grow into inverted pyramidal Ag crystallites as the Ag solubility of the glass decreases with decreasing temperature, resulting in the microstructure shown in Figure 2.

3.2. Effect of Peak Firing Temperature

Figure 5 shows SEM images of cross-sectioned samples fired at different peak temperatures.

As can be seen, the cell conventionally fired at 850 °C shows several features:

1. large agglomerates of Ag particles fused together into a region of continuous Ag,
2. some Ag particles separated by regions of glass, and
3. some isolated particles of Ag that have the appearance of precipitates.

At a low firing temperature of 750 °C the Ag layer is porous and there are some discontinuities in the Ag layer coverage of the silicon wafer. Furthermore, the glass layer is thin and only a small amount of silver is expected to be dissolved at this temperature. Our investigations showed that there is no or very weak Ag crystallite formation at 750 °C, which can be explained by the lack

of a reaction between the glass frit and the SiN_x, which in previous studies was found to start at temperatures of ~680 °C [5]. Hence, the low firing temperature of 750 °C (the actual cell temperature will be ~650 °C) is most likely not sufficient to fire through (*i.e.* penetrate) SiN_x and allow the formation of Ag crystallites. This would also explain the relatively low amount of bowing found after firing (see Figure 3 b), which is probably the result of delamination of the weak Ag/Si contact interface.
A higher firing temperature promotes Ag particle agglomeration as well as the aggressiveness of the etching reaction between the glass frit and the SiN_x layer. Thus, for the 850 °C and 950 °C firing condition the SiN_x is completely etched, allowing the glass frit with dissolved Ag to reach and partly dissolve the underlying Si. With increasing firing temperature the amount of generated glass increases too, since more SiN_x is removed and hence more Si from the SiN_x reacts with the glass frit to SiO_x. Thus, the thickness of the glass is expected to increase with temperature, as observed in this work.

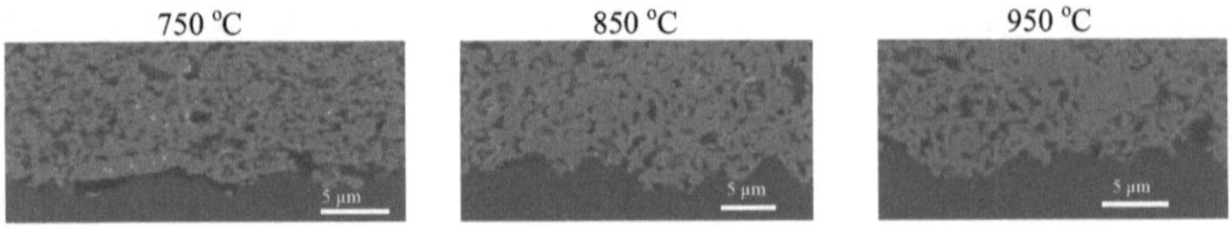

Figure 5. SEM micrographs of cross sections of Ag paste fired on a silicon wafer of different firing temperatures.

Firing the cell at 950 °C results in only local coverage of the silicon surface with glass globules, and hence this non-uniform glass layer could explain low bowing at 950 °C. A substantial delamination was observed at the edges of the fired cell, which is inherent to a large shear stress during bowing. Edge delamination and poor glass coverage explain a reduction of bowing as compared to firing at 850 °C, where a more uniform contact with no delamination is created.
The XRD analysis performed on the samples fired at 750 °C, 850 °C and 950 °C showed crystallite peaks of Ag, ZnO and $PbZnSiO_4$, see Figure 3 a. The cell fired at 850 °C seems to have the largest amount of ZnO phase, according to the peak intensities. Conversely, firing at 950 °C creates more $PbZnSiO_4$, which is consistent with the increase in aggressiveness of the etching reaction between the glass frit and the SiN_x layer. The eutectic reaction of Ag-Si is expected to start at 835 °C [6]. When ZnO-PbO reaches 850 °C, which occurs at a firing temperature of 950 °C, Ag precipitation from the molten glass, as shown in Figure 4, can be suppressed, resulting in a weak mechanical contact (in accordance with the bowing results).
The crystallization of the $PbZnSiO_4$ in the glass probably occurred only due to the temperature rising above 850 °C, as no $PbZnSiO_4$ phase was identified in the 750 °C and 850 °C firing spectra. The crystallization in the glass during firing can affect the formation of Ag on and penetration of Ag into the Si emitter.
It is concluded that firing at 950 °C creates a mechanically weak contact, due to the high crystallization of glass (at temperatures above 800 °C), which prevents the viscous flow of frits and due to the penetration of Ag into the Si emitter in combination with nearly no glass layer formed at the Ag/Si interface. Hence, only local Ag crystallites form in the places where Ag particles make direct contact with the Si wafer. This provides a low degree of bonding between the layers, resulting in a low bowing of the fired cell.

3.3. Effect of Silicon Wafer Surface Treatment

At temperatures typically above 600 °C, glass frit particles melt and exhibit a sufficiently low viscosity to allow a significant rearrangement of the particles and their compaction by capillary action. As a result, wetting occurs and the underlying SiN_x layer is penetrated, allowing interaction with Si to take place. This leads to the formation of Ag crystallites and a good mechanical bonding at the Ag/Si interface. Thus, it can be expected that the silicon surface condition may affect the wettability, the spatial distribution of the glass layer and ultimately the bonding of the layer.

Figure 6 a shows SEM micrographs of polished cross sections of the Ag contact layer created with different Si wafer surface conditions: as-cut, textured and polished. As can be seen, a textured surface creates large voids and inhomogeneities at the Ag/Si interface. However, a polished silicon surface results in a rather uniform distribution of Ag agglomerates. Finally, there are cracks visible running from the interface into the as-cut silicon wafer.

In order to investigate the effect of surface roughness on the quality (bonding) of the Ag/Si interface, the bulk of the Ag layer was etched off using Aqua Regia, leaving behind the glass layer (Figure 6 b).

As can be seen, for the polished surface the glass layer coverage is higher and much more uniform compared with the textured surface. Furthermore, for the textured surface the glass layer is only continuous within the texture valleys. This indicates that melted glass initially fills up the texture valleys. Any excess glass will spill over the valleys and then build up. However in this case, the edges between the valleys are too high to cause any spill over.

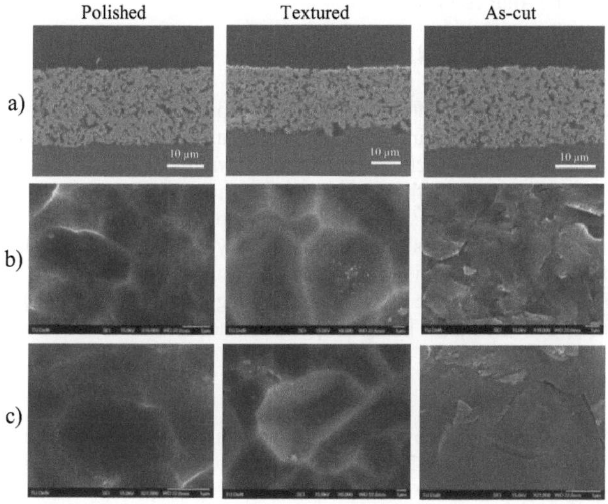

Figure 6. SEM micrographs of: a) cross section of Ag paste fired on different silicon surfaces; b) Ag layer after etching with Aqua Regia, revealing glass phase distribution underneath; c) Ag layer after etching with Aqua Regia+HF, that removes the glass layer to reveal the Ag crystallite.

Etching away the glass layer reveals the distribution of the Ag crystallites underneath the glass, Figure 6 c. For the polished surface, imprints of numerous Ag crystallites can be seen underneath the glass. This type of Ag/Si interface is expected to create a good mechanical bonding between the layers. Conversely, the textured surface shows that Ag crystallites have formed underneath the glass in the texture valleys. The area where Ag particles can be in direct contact with the Si surface also shows formation of Ag crystallites, but now without the aid of a glass phase. This is not found on the polished surface, as on this surface the silicon is almost entirely covered with a glass layer.

Thus, the glass-wetting behaviour on the Si surface plays a crucial role in determining the glass layer distribution. Based on the results, a simplified model for the glass layer and Ag crystallite distribution on textured and polished silicon surfaces is proposed in Figure 7.

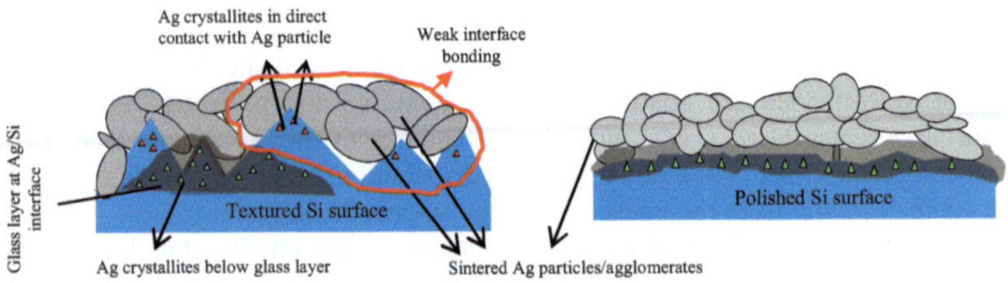

Figure 7. A schematic of the glass layer and Ag crystallite distribution on textured and polished silicon surfaces.

In this model polishing of the silicon surface gives a better wetting by the glass layer, resulting in a lesser incidence of large voids, compared to highly textured surfaces. The non-uniformity of the glass layer and large voids at the Ag/Si interface are expected to have a negative effect on the mechanical strength of the solar cell [7].

4. Conclusions

An alternative mechanism for Ag contact formation was proposed, where under an oxidizing environment Ag dissolves as Ag^+ ions into the molten glass and there is a redox reaction between diffused Ag and silicon substrate, which creates inverted pyramidal pits on the Si surface. The Ag atoms reduced by the reaction with the Si substrate can precipitate as Ag particles in the molten glass during firing or as Ag crystals in the inverted pyramidal pits during the subsequent cooling process. This model competes with the previously proposed one [2, 3], where metal oxides contained in the glass interact with the silicon wafer.

Furthermore, it was found that there are two main processing parameters affecting the uniformity of the Ag/Si interface, namely the peak firing temperature and the silicon surface roughness.

References

1. T. Amstel, V.A. Popovich, P.C. de Jong & I.G. Romijn, "Modelling mechanical aspects of the aspire cell", Proceedings of the 23rd European Photovoltaic Solar Energy Conference, Valencia, Spain, September 1-5, 2008.
2. G. Schubert, Thick film metallization of crystalline silicon solar cells: Mechanisms, models and applications, Dissertation, University of Konstanz, 2006.
3. M. Hilali, A. Rohatgi, & B. To, A review and understanding of screen printed contacts and selective emitter formation. Proceedings of 16th Workshop on Crystalline Silicon Solar Cells & Modules: Materials and Processes. Ed. by B. Sopori et al, pp. 109-116, 2004.
4. K. Yata and T. Yamaguchi, J. Mater. Sci., 27 (101), pp. 1731-1734, 1992.
5. M. Hörteis, et al., High-temperature contact formation on n-type silicon: basic reactions and contact model for seed-layer contacts. Adv. Funct. Mater., p. 476, 2010.
6. R. Olesinski, A. Gokhale, & G. Abbaschian, The Ag-Si (silver-silicon) system. Bulletin of Alloy Phase Diagram, 10 (6), pp. 635-640, 1989.
7. V. Popovich, Microstructure and mechanical aspects of multicrystalline silicon solar cells, PhD dissertation, Delft University of Technology, 2013.

2015 EPD Congress

HIGH-TEMPERATURE ELECTROCHEMISTRY II

2015 EPD Congress

SYMPOSIUM: HIGH-TEMPERATURE
ELECTROCHEMISTRY II

Molten Salt Technology

**Jerome Downey,
Montana Tech of the
University of Montana**

**Steven Herrmann,
Idaho National Laboratory**

EPD Congress 2015
Edited by: James Yurko, Antoine Allanore, Laura Bartlett, Jonghyun Lee, Lifeng Zhang, Gabriella Tranell, Yulia Meteleva-Fischer, Shadia Ikhmayies, Arief Suriadi Budiman, Prabhat Tripathy, and Guy Fredrickson
TMS (The Minerals, Metals & Materials Society), 2015

STUDIES ON THE PURIFICATION OF Na_3AlF_6-K_2SiF_6-AlF_3 MELT

Zhongliang Tian, Shu Yang, Yanqing Lai, Xun Hu, Jie Li

School of Metallurgy and Environment, Central South University;
Changsha, 410083, China

Keywords: impurity P, impurity Fe, purification of the electrolyte, electrodeposited silicon.

Abstract

The purity of the electrolyte has an important influence on the content of the impurities in electrodeposited silicon. In this paper, effects of electrolysis temperature, current density and the time on the concentrations of two important impurities: phosphorus and iron in the electrolyte were investigated during the purification of Na_3AlF_6-10%K_2SiF_6-AlF_3 melt. It was observed that the concentrations of P and Fe in the electrolyte decreased with the electrolysis temperature, current density and the electrolysis time increasing. X-ray fluorescence (XRF) of the electrolyte shows that the concentrations of P and Fe decrease from 4.2ppmw to 1.3ppmw and 204.0ppmw to 42.6ppmw, respectively. The removal rates of P and Fe are 69.08% and 79.12% when the electrolysis was performed for the duration of 3 hours and at a current density of 20mA·cm^{-2} while the operating temperature was 1225K.

Introduction

Crystalline silicon used in photo-voltaic (PV) cells are typically fabricated from high purity silicon (>11N) commonly produced by so-called Siemens process or from scrap electronic silicon (SEG-Si), resulting the purity of silicon usually exceeds actually required (6N) for solar-grade silicon (SOG-Si). Although the Siemens process is advantageous for the purity of silicon and is applicable to the SOG-Si production, it is rather costly and energy consuming, which is the major inhibiting factor to the widespread adoption of solar power. Therefore, new production processes for SOG-Si, including modification of the Siemens and past processes for SEG-Si, have been investigated [1.2]. Among them, the molten salt electrolysis is one of the most extensively investigated one [3-6].

It is known that impurities in polycrystalline silicon materials, significantly affect the stability and efficiency of the photovoltaic cells [7.8]. For impurities P and Fe, the maximum admissible contents in SOG-Si should be less than 1ppmw and 10ppmw respectively [9.10]. Based on the electrode potential difference between the phosphorus (or iron) and silicon, it is advantageous to adopt the method of molten salt electrolysis to prepare polycrystalline silicon with low concentrations of P and Fe [11.12].

As pointed out by previous studies [13-16], the content of the impurities in the electrolyte has much effect on the purity of production silicon at the cathode. The results from reference [13] showed that the purity of electrolyte resulted in the decrease of P content in the obtained silicon. To ensure the concentration of P in the production silicon was less than 1ppmw, the allowable highest concentration of P in electrolyte was 3.1ppmw. The results from Olsen et al [14.15] showed that there was an obvious decreasing trend of impurities in the silicon electrodeposited from MG-Si. The concentrations of Fe and P in electro-deposited silicon were 1.5ppma and 3ppma respectively, if the electrolyte of LiF-KF-K_2SiF_6 was purified to remove dissolved oxides before electrorefining [15]. In addition, with the process of pre-electrolysis for K_2SiF_6-Na_3AlF_6 melt, the concentrations of Fe and P in the refining silicon decreased to 95ppmw and 4.3ppmw, while the concentrations of Fe and P in MG-Si were 1745ppmw and 98.6ppmw, respectively [16]. Therefore, it is significant for electrolytic preparation of SOG-Si to remove some impurities such as P, Fe by the purification treatment of the electrolyte, which their electrode potentials are more positive than that of silicon.

In this work, the effects of electrolysis temperature, current density and the time on concentration of impurities P and Fe in the electrolyte were investigated during the purification of Na_3AlF_6-K_2SiF_6-AlF_3 melt when Cu-Si alloy was adopted as liquid anode.

Experimental

Material chemical and bath preparation

The bath was made up of raw materials Na_3AlF_6, K_2SiF_6 and AlF_3. Na_3AlF_6 and K_2SiF_6 were reagent grade. AlF_3 was purified by sublimation three times and its purity was above 99.5%. All the components were dried at 120℃ for at least 48 hours in the vacuum oven to remove the water before being used. The mole ratio of NaF and AlF_3 was 2.2 and 10 wt% K_2SiF_6 was added as the intermedium of electrorefining. The powdered components were mixed and the initial weight of the electrolyte was 600g.

Cell design and electrolysis procedure

Fig. 1 is the sketch of the experimental cell for the purification of Na_3AlF_6-K_2SiF_6-AlF_3 melt. A hole was drilled at the bottom of alumina crucible and Cu-20 wt% Si alloy was added. Thus, a steady anode surface could be obtained during electrolysis. A graphite rod of high purity was used as cathode. Alumina sleeve was set in the graphite crucible and about 600g electrolyte was contained. Because the cell could not be thermally self-sustaining, it was necessary to provide extra heat by placing the experimental cell in a vertical laboratory furnace and the furnace was heated to the desired temperature.

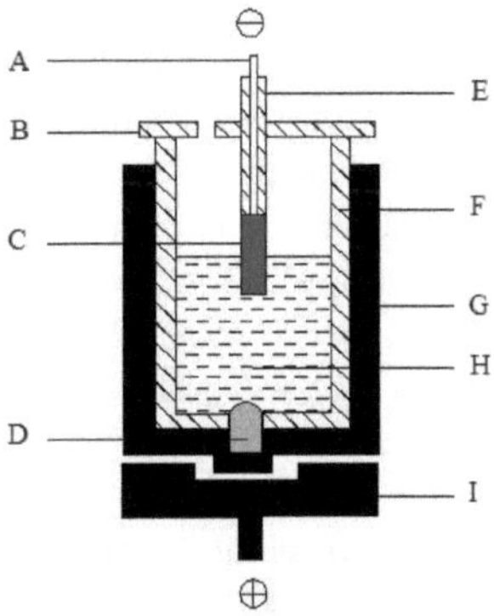

Fig1. The sketch of electrolytic cell for the purification process
A-Stainless steel lead B-Alumina lid C-Graphite cathode
D-Cu-Si alloy anode E-Alumina tube F-Alumina crucible
G-Graphite crucible H-Electrolyte I-Stainless steel plate

The cell was heated to the required temperature and kept for 2 hours before the current was supplied. The temperature of the furnace was controlled to ±1°C by TCE-II programmable temperature control unit. The temperature of the melt was measured with Pt/Pt-10%Rh thermocouple once an hour and maintained over a range of ±3°C during testing. The constant current was supplied throughout the experiment and the cell voltage was monitored by a Multi-Purpose Potentiostat/Galvanostat (model 273A/10, Perkin-Elmer Instruments).

The removal rate of impurity was applied to quantify the actual situation of purification of the electrolyte with respect to the processing parameters as the following Eq. (1).

$$R=\frac{C_0-C_1}{C_0}\times 100\% \qquad (1)$$

Where, C_o and C_1 are the concentration of impurity in the electrolyte before and after the purification of the electrolyte with different processing parameters, respectively.

During purification process, the electrolyte was sampled every hour. The concentrations of impurities P and Fe in the sample were measured by X-Ray Fluorescence spectrum (Philips 8424 TW2424) (analytic error below 5%).

Results and Discussion

Electrolysis temperature

Table 1 shows that the concentration of impurity P in the electrolyte decreases with the temperature increasing at 20 mA·cm^{-2} after being purified by the electrolysis for 3 hours. At the temperature of 1205K, the concentration of impurity P decreases from 4.2ppmw to 1.6ppmw, and R_P is 61.90%. When the temperature is 1225K, its concentration is 1.3ppmw after electrolysis for 3 hours, which is lower than that at 1205K. And at this time R_P is 69.05%, which increases by 7.15%. At 1245K, its concentration is 1.2ppmw after electrolysis for 3 hours and R_P is 71.43% which increases by 2.38% compared with that at 1225K.

From Table 1, for the impurity Fe, its concentration change caused by the electrolysis temperature is as same as that of impurity P. After electrolysis for 3 hours, the concentrations of impurity Fe in the electrolyte are 47.8ppmw, 41.2ppmw and 40.6ppmw when the temperatures are 1205K, 1225K and 1245K respectively. Accordingly, the removal rates R_{Fe} are 76.57%, 79.80% and 80.10%.

Table 1. The concentration of impurities in the electrolyte

Electrolysis Temperature /K	Before electrolysis		After electrolysis		Removal rate of impurities	
	C_P/ppmw	C_{Fe}/ppmw	C_P/ppmw	C_{Fe}/ppmw	R_P/%	R_{Fe}/%
1205	4.2	204.0	1.6	47.8	61.90	76.57
1225			1.3	41.2	69.05	79.80
1245			1.2	40.6	71.43	80.10

Note: Current density 20 mA·cm^{-2}, and the time 3 hours.

The results presented in Table 1 also indicate that the degree of change in the concentration of impurities P and Fe in the electrolyte is relatively small when the electrolysis temperature increases from 1225K to 1245K. After electrolysis for 3 hours, the removal rate of impurity P at 1245K only is increased by 2.38% compared with that at 1225K. For impurity Fe, it is increased by 0.30%. In addition, considering the evaporation of the electrolyte which will result in the loss of the melt, the temperature of electrolysis is determined to be 1225K.

Current density and electrolysis time

From Fig.2, it is concluded that the concentration of impurities P and Fe in the electrolyte after being purified decreases with the current density increasing when the electrolysis time is less than 2 hours at 1225K. But the process of purifying is continued until 3 hours, their concentration is substantially equal.

For impurity P (Fig.2 (a)), after electrolysis for 1 hours, its concentration decreases from 4.2ppmw to 2.6ppmw at the current density of 20 mA·cm^{-2}, and its removal rate (R_P) is 38.10%. When the current density is increased to 50 mA·cm^{-2}, its concentration is 2.4ppmw, which is lower than that at the current density of 20 mA·cm^{-2}. And at this time, R_P is 42.86%. However, when the processes of electrolysis at the current density of 20 mA·cm^{-2} and 50 mA·cm^{-2} lasts for 3 hours, both of the concentrations of impurity P are 1.3ppmw, and R_P is 69.08%.

For impurity Fe (Fig.2 (b)), its concentrations are 99.7ppmw and 88.1ppmw respectively at the current density of 20 mA·cm^{-2} and 50 mA·cm^{-2} after electrolysis for 1 hours. And accordingly, its removal rates (R_P) are 51.13% and 56.81%. If the process is continued until 3 hours, the concentrations of impurity Fe are 42.6ppmw and 43.9ppmw at the current density above-mentioned.

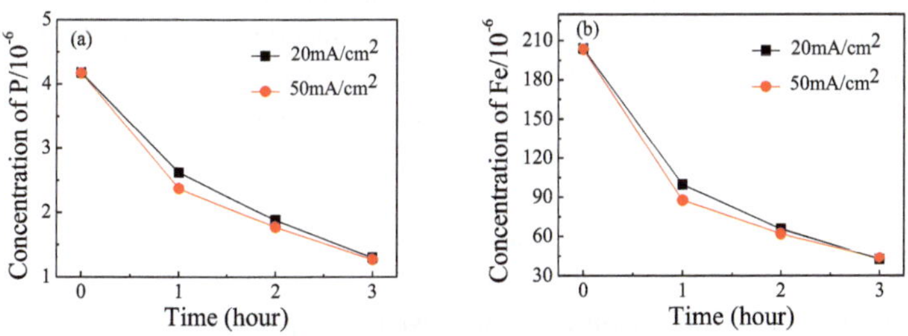

Fig2. The concentration of impurities at the different current densities
(a) Concentration of impurity P (b) Concentration of impurity Fe

A possible explanation for the decrease in the concentration of impurity P in the electrolyte with the current density rising may be that the increase in the rate of reduction reaction. The electrode reaction occurs only on the interface of electrode and electrolyte solution. So, the reaction rate v is related to the area of interface of electrode and electrolyte solution, as shows in Eq. (2) [17].

$$v=\frac{I}{nFA}=\frac{i}{nF} \quad (mol/s{\cdot}cm^2) \qquad (2)$$

Where i is current density (A·cm^{-2}), F is the Faraday's constant, n is the number of exchanged electrons.

From Eq. (2), the reaction rate is proportional to the current density. So when the current density increases from 20 mA·cm^{-2} to 50 mA·cm^{-2}, the reaction rate which P^{3+} near cathode is reduced to P increases.

Besides, according to Eq. (3), the mass of a substance altered at an electrode during electrolysis is directly proportional to the quantity of electricity transferred at that electrode [18].

$$M = k \cdot Q = k \cdot I \cdot t \tag{3}$$

Where M is the mass of a substance (g), k is electrochemical equivalent (g·A^{-1}·h^{-1}), Q is the total electric charge passed through the substance (A·h). I is current intensity, t is the electrolysis time (h).

Because of the same area of electrode, increasing current density leads to the increase of current intensity. According to Eq. (3), the amount of impurities P and Fe deposited at the cathode increases. This means that the concentrations of impurities P and Fe in the electrolyte decreases. However, as long as the electrolysis time is enough, the concentrations of these impurities in the electrolyte are equal though the current densities are different. For example, the concentrations of impurity Fe are 42.6ppmw and 43.9ppmw respectively at the current densities of 20 mA·cm^{-2} and 50 mA·cm^{-2} after electrolysis for 3 hours. However, when the electrolysis time is 2 hours, their corresponding concentrations are 65.8ppmw and 62.4ppmw.

Conclusions

The purification of Na_3AlF_6-K_2SiF_6-AlF_3 melt with the liquid anode of Cu-Si alloy can be realized. The concentrations of P and Fe in the electrolyte decreased with the temperature increasing. An appropriate increase in electrolysis temperature was helpful for the purification of the electrolyte, and the temperature of purifying Na_3AlF_6-K_2SiF_6-AlF_3 melt by the electrolysis should be determined to be 1225K. The concentrations of P and Fe in the electrolyte after being purified decreased with the current density increasing when the electrolysis time was less than 2 hours. However, as long as the electrolysis time was enough, the concentrations of these impurities would not be affected by the current density. X-ray fluorescence (XRF) of the electrolyte shows that the concentrations of P and Fe in the electrolyte decrease from 4.2ppmw to 1.3ppmw and 204.0ppmw to 42.6ppmw, respectively. The removal rates of P and Fe are 69.08% and 79.12% at 1225K, current density of 20 mA·cm^{-2} when the time reaches 3 hours.

Acknowledgments

The authors gratefully acknowledge the financial supports from the State Natural Science Fund (No. 51334002).

References

1. T.Oishi et al., "Process for Solar Grade Silicon Production by Molten Salt Electrolysis Using Aluminum-Silicon Liquid Alloy," *Journal of The Electrochemical Society*, 158(9) (2011), 93-99.
2. M.D.Johnston and M.Barati, "Calcium and titanium as impurity getter metals in purification of silicon," *Separation and Purification Technology*, 107(2013), 129-134.
3. A.L.Bieber, et al., "Silicon electrode position in molten fluorides," *Electrochimica Acta*, 62(2012), 282-289.
4. Meng Tao, "Impurity segregation in electrochemical processes and its application to electrorefining of ultrapure silicon," *Electrochimica Acta*, 89 (2013), 688–691.
5. Hong Youl Ryu et al., "Formation of high purity Si nanofiber from metallurgical grade Si by molten salt electrorefining," *Materials Chemistry and Physics*, 137 (2012), 160-168.
6. Kouji Yasuda et al., "Direct electrolytic reduction of solid SiO_2in molten $CaCl_2$for the production of solar grade silicon," *Electrochimica Acta*, 53 (2007), 106–110.
7. S. De Wolfa et al., "Solar cells from upgraded metallurgical grade (UMG) and plasma-purified UMG multi-crystalline silicon substrates," *Solar Energy Materials & Solar Cells*, 72 (2002), 49–58.
8. S. Zheng et al., "Elimination of phosphorus vaporizing from molten silicon at finite reduced pressure," *Transactions of Nonferrous Metals Society of China*, 21 (2010), 697-702.
9. T. Buonassisi et al., "Transition metal co-precipitation mechanisms in silicon," *Acta Materialia*, 55 (2007), 6119-6126.
10. L. Massot et al., "Silicon recovery from silicon–iron alloys by electrorefining in molten fluorides," *Electrochimica Acta*, 96 (2013), 97-102.
11. Ming Jia et al., "Study on the removal of impurities in silicon by electrorefining," *Acta Phys.Sin*, 59 (2010), 1938-1944.
12. Ji-guang Yi et al., "Pre-electrolysis of electrolyte for silicon electrorefining," *Supplemental proceedings Volume 2: Materials characterization, computation and modeling, TMS* (2009), 281-288.
13. Jiang-sheng Xie et al., "Study on removal of impurities in silicon by direct electrolystic reduction of solid SiO_2," *Journal of Chinese Society of Rare Earths*, 8 (2012), 288-291. (in Chinese)
14. Jerry M. Olson and Karen L. Carleton, "Process for producing silicon," US Patent: 4448651, 1984.

15. Jerry M. Olson and Karen L. Carleton, “A Semipermeable Anode for Silicon Electrorefining,” *Journal of the Electrochemical Society: Electrochemical Science and Technology*, 128(12) (1981), 2698-2699.
16. Yan-qing Lai et al., “Study on the Morphology Evolution and Purification of Electrorefined Silicon,” *Metallurgical and Materials Transactions A*, 41(4) (2010), 929-935.
17. Bard A.J and Faulkner L.R. *Electrochemical Methods Fundamentals And Applications* (JOHN WILEY & SONS, INC.1980), 16.
18. Ehl, Rosemary Gene and Ihde, Aaron, "Faraday's Electrochemical Laws and the Determination of Equivalent Weights," *Journal of Chemical Education*, 31(1954), 226–232.

2015 EPD Congress

SYMPOSIUM: HIGH-TEMPERATURE ELECTROCHEMISTRY II

Nuclear and Rare Earth Technology

Uday Pal,
Boston University

Ramachandran Vasant Kumar,
University of Cambridge

EPD Congress 2015
Edited by: James Yurko, Antoine Allanore, Laura Bartlett, Jonghyun Lee, Lifeng Zhang, Gabriella Tranell, Yulia Meteleva-Fischer, Shadia Ikhmayies, Arief Suriadi Budiman, Prabhat Tripathy, and Guy Fredrickson
TMS (The Minerals, Metals & Materials Society), 2015

ELECTROCHEMICAL SEPARATION OF Dy AND Nd FROM Nd MAGNET SCRAP IN A MOLTEN LiCl-KCl

Hirokazu Konishi[1], Hideki Ono[1], Eiichi Takeuchi[1], Toshiyuki Nohira[2], Tetsuo Oishi[3]

[1]Osaka University; 2-1 Yamadaoka; Suita, Osaka 565-0871, Japan
[2]Kyoto University; Yoshidahonmachi; Sakyo-ku, Kyoto 606-8501, Japan
[3]National Institute of Advanced Industrial Science and Technology; 16-1 Onogawa; Tsukuba, Ibaraki 305-8569, Japan

Keywords: Rare earth, Electrolysis, Molten salt, Separation

Abstract

The Dy-Nd-Pr-Ni alloy sample was prepared by cathodic potentiostatic electrolysis at 0.65 V (vs. Li^+/Li) for 1 h using a Ni plate in a molten LiCl-KCl-$DyCl_3$-$NdCl_3$-$PrCl_3$ system at 723 K. The highest mass ratio (Dy/Nd+Pr) in the alloy sample was observed to be 50 at 0.65 V. Anodic potentiostatic electrolysis at 2.20 V for 12 h was conducted using the Nd-Fe-B magnet electrode in a molten LiCl-KCl system. All elements were almost dissolved from the magnet, and the original form of the magnet disintegrated. After anodic potentiostatic electrolysis at 2.20 V, cathodic potentiostatic electrolysis was conducted at 1.00 V for 5 h using a Mo plate in order to remove the dissolved Fe from the bath. Finally, cathodic potentiostatic electrolysis was conducted at 0.65 V for 12 h using a Ni plate. The mass ratio of Dy/Nd in the alloy sample was determined to be about 18.

Introduction

The use of rare earth (RE)-iron group (IG) alloys has increased significantly in a number of industrial fields over the past few decades. In particular, the demand for Dy-added Nd-Fe-B magnets is rapidly increasing because these magnets are indispensable for high-performance motors in electric vehicles (EVs) and hybrid electric vehicles (HEVs). These magnets need to possess sufficient thermal stability for use in such motors in high-temperature environments. The addition of Dy is necessary to improve the thermal stability of Nd-Fe-B magnets. However, there is the concern about a shortage of RE metals in general, and Dy, in particular, of the RE resources. Thus a worldwide need is being increasingly felt to augment the primary production of Nd and Dy by combining a suitable recycling method in order to reclaim these metals from their recyclable resources. In this context, it is worthwhile to mention that large Nd-Fe-B magnets are the only secondary resource materials as far as Dy is concerned. That is why, it is necessary to develop an inexpensive and environmentally friendly recovery/separation process for the recovery of Nd and Dy from a variety of scrap/waste magnets.

We are proposing a new separation and recovery process for the recovery of Nd and Dy from the RE scraps using molten salt and an alloy diaphragm concept in a molten salt system (Figure 1) [1-3]. This process is based on our previously discovered phenomena, i.e., "electrochemical implantation" and "electrochemical displantation" [4-6]. RE containing scrap is used as the anode. A RE-transition metal (TM) alloy is used as the diaphragm, which functions as a bipolar electrode. During electrolysis, all the RE metals in the anode are dissolved in the molten salt as RE ions. One or several specific RE ions are selectively reduced to form RE-TM alloys on the alloy diaphragm according to their formation potentials and/or alloying rates. Subsequently, the RE atoms chemically diffuse through the alloy diaphragm and are re-dissolved into the molten salt as RE ions in the cathode compartment. The re-dissolved RE ions are finally deposited on either Mo or Fe cathode as RE metals. The RE ions remaining in the anode compartment can be electrodeposited by employing a second cathode leaving behind the impurities as anode mud/residue.

This new process was first applied to chloride melts, and the separation of Dy from Nd was investigated using Ni and Cu cathodes in molten LiCl-KCl-$DyCl_3$-$NdCl_3$ systems [7-9]. The highest mass ratio of Dy/Nd in Dy-Nd-Ni alloy sample was found to be 72 by ICP-AES.
The present study was the electrochemical formation of RE-Ni (RE=Dy, Nd, Pr) alloys using Ni electrodes in a molten LiCl-KCl-$RECl_3$ (0.50 mol% added) at 723 K. Furthermore, the anodic dissolution of RE (Dy, Nd, etc) using Nd-Fe-B magnet electrodes and electrowinning of Dy using Ni electrodes were carried out in a molten LiCl-KCl system at 723 K.

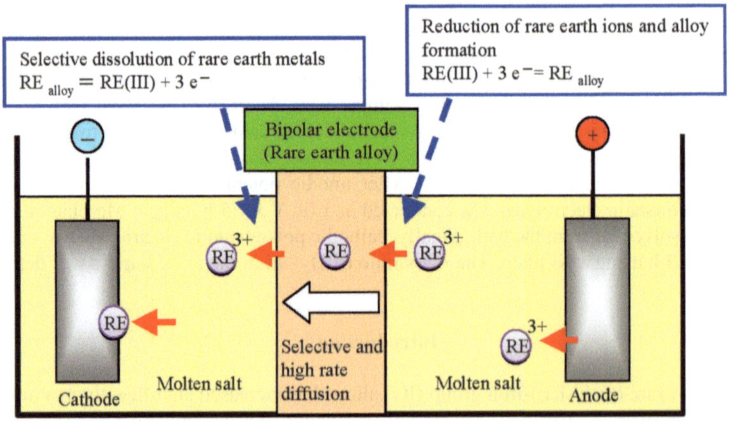

Figure 1. Schematic drawing of the process for separation and recovery of rare earth metals.

Experimental

Anhydrous Reagent (AR) grade eutectic LiCl-KCl salt (58.5 mol% LiCl-41.5 mol% KCl) was placed in a recrystallized alumina crucible, which, in turn, was vacuum-treated for more than 24h at 473K, in order to remove moisture, prior to melting. Both the mixing of the electrolyte constituents and electrochemical measurements were carried out under dry argon atmosphere. 0.5 mol% $DyCl_3$, $NdCl_3$ and $PrCl_3$ (99.9% pure, Kojundo Chemical Laboratory Co. Ltd.) was added to the eutectic salt to prepare the electrolyte. A chromel-alumel thermocouple was used to measure the temperature. A three electrode set up consisting of Nd-Fe-B plate (dimensions: 20mm long, 10 mm wide and 1.5mm thick, composition, mass%: 5Dy+20Nd+5.9Pr+65Fe+1B)/Ni wire (99% pure, 5mm long, 1mm dia.) WE, Mo/Ni plate type (99% pure, 20 mm long, 10 mm wide, 0.2 mm thick) CE and Ag-(1mol%) AgCl RE was used for the electrochemical measurements. The potential of the RE was calibrated against the in situ Li^+/Li electrode, prepared by depositing Li metal on a Mo wire (during the electrochemical measurements). Potentiostatic electrolysis was performed at 723K to prepare the electrodeposited samples. The samples were characterized and evaluated by SEM-EDS, and ICP-AES instruments.

Results and Discussion

Electrochemical Formation of RE-Ni (RE-Dy, Nd and Pr) Alloys

Figure 2 compares the cyclic voltammograms of $DyCl_3$, $NdCl_3$ and $PrCl_3$. Appearance of a relatively larger cathodic current (the peak current being at ~0.40 V) at 0.70 V (with a corresponding anodic peak during the reverse scan) indicated the selective removal of Dy (from Nd and Pr). In contrast, both Nd and Pr exhibited cathode current responses at 0.60 V. Besides, the appearance of smaller cathodic current peaks at ~0.65 V were probably due to the occurrence of Nd(III)/Nd(II) and Pr(III)/Pr(II) reactions. The CV suggests that the application of a voltage in the range 0.60-0.70 V can potentially remove Dy from the mixture of Nd and Pr.

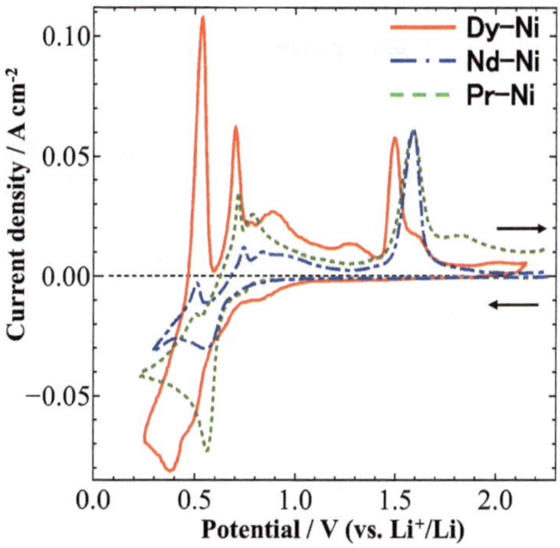

Figure 2. Cyclic voltammograms with Ni electrodes in a molten LiCl-KCl-$DyCl_3$/ $NdCl_3$/$PrCl_3$ (0.50 mol% added) at 723 K. Scan rate: 0.05 V s^{-1}. Surface area: 0.16 cm^2.

Based on the CV response (Figure 2), we have already reported the possibility of separation of Dy from Nd, using a Ni electrode, from the LiCl-KCl-$DyCl_3$-$NdCl_3$ electrolyte system [8]. In the present studies, the specific focus was on the possibility of removal of Dy from Pr and Nd from Pr. Further experiments were carried out, at 0.55-0.70 V for 1 h using Ni plate-type cathode, to examine the extent of removal of Dy from the mixture of Dy, Nd and Pr. Figure 3 shows the Dy/Pr mass ratio in the electrodeposited alloy samples, measured by ICP-AES. The highest value for the ratio was observed to be formed at an operating voltage of 0.70 V. When the studies were extended to examine the separation efficiency between Nd and Pr, it was observed that no separation could be possible as the mass ratio stayed unity at all operating voltages (Figure 4). Experiments aimed at separating Dy from a mixture of Nd and Pr revealed that a highest separation quotient (Dy/Nd+Pr) of 50 could be achieved at an operating voltage of 0.65V (Figure 5).

These results indicated the possibility of separating Dy from Nd and Pr by controlling electrolysis potential in a molten LiCl-KCl-$DyCl_3$-$NdCl_3$-$PrCl_3$ system.

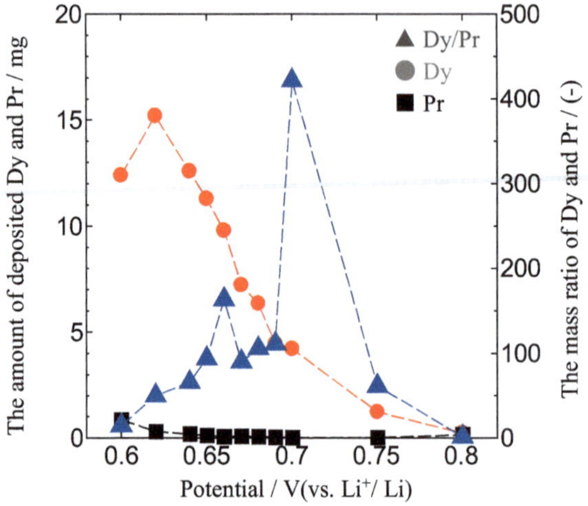

Figure 3. Potential dependences of the amounts of deposited Dy and Pr, and of the mass ratio of Dy/Pr, in a molten LiCl-KCl-$DyCl_3$(0.50 mol% addd)-$PrCl_3$(0.50 mol% added) at 723 K.

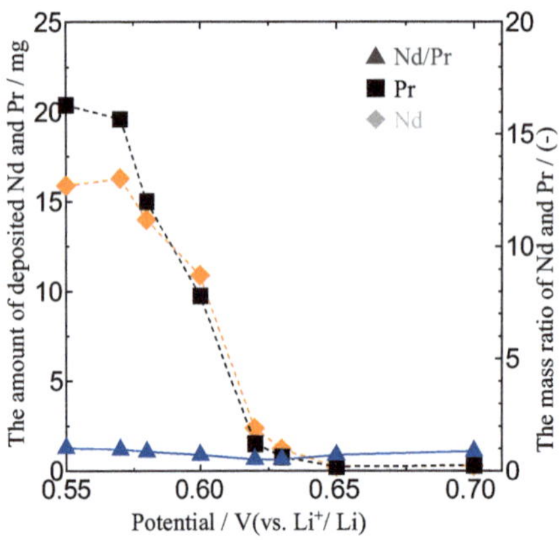

Figure 4. Potential dependences of the amounts of deposited Nd and Pr, and of the mass ratio of Nd/Pr, in a molten LiCl-KCl-$NdCl_3$(0.50 mol% addd)-$PrCl_3$(0.50 mol% added) at 723 K.

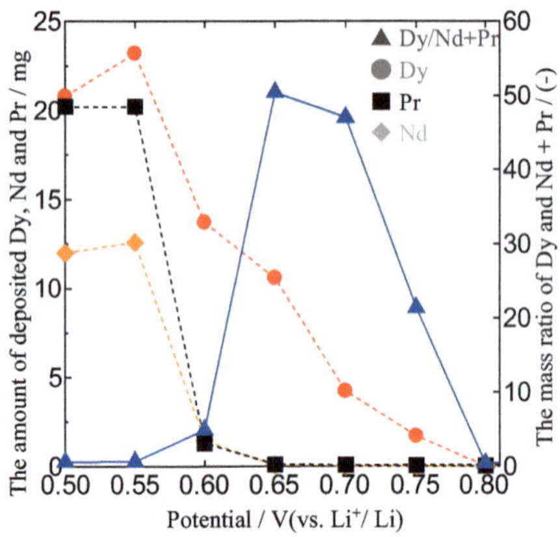

Figure 5. Potential dependences of the amounts of deposited Dy, Nd and Pr, and of the mass ratio of Dy/Nd+Pr, in a molten LiCl-KCl-$DyCl_3$(0.50 mol% addd)-$NdCl_3$(0.50 mol% addd)-$PrCl_3$(0.50 mol% added)at 723 K.

Recovery of Dy under Potentiostatic Electrolysis Condition

A combination of anodic dissolution (at 2.20 V for 12 h) and cathodic deposition (at 1.00 V for 5 h) was carried out to deposit the dissolved Fe (from the electrolyte) onto a Mo plate prior to the separation of Dy from the RE mixtures. Fe, from the Nd-Fe-B magnet, quickly dissolved in the electrolyte upon anodic polarization. The Fe removal was followed up by another potentiostatic experiment, at 0.65 V for 4 and 12 h (Figures 6 and 7 respectively), to remove Dy from the electrolyte and form of Dy-Ni alloy in situ at the Ni plate. The EDS measurements showed the Dy/Nd ratio to be 11 (after 4 h of electrolysis, Figure 6) and 18 (after 12h of electrolysis, Figure 7). As predicted, Pr was observed to be present in both the (Dy-Nd) alloy deposits. These results further confirmed that the separation of Dy, from a mixture of Dy, Nd and Pr, can be achieved under a set of optimum operating conditions.

In order to recover residual Nd and Pr from the electrolyte (after the removal of Dy), further electrolysis was carried out at 0.65 V for 4 h and 12 h respectively. Figure 8 shows the cross-sectional SEM image of the sample obtained at 0.60 V and 1.5 h. The mass ratio of Dy:Nd:Pr in the alloy was determined to be 11:8.4:1. This analysis clearly indicated that quantitative recovery of Dy could not be achieved even after carrying out electrolysis at 0.65 V for the duration of 16 h. A mass ratio of Dy:Nd:Pr as 1.9:4.7:1.0 indicated that it was possible to deposit Nd and Pr onto nickel cathode at an operating voltage of 0.55 V and an electrolysis duration of 3 h (Figure 9).

Figure 6. Cross-sectional SEM image of the sample prepared by cathodic potentiostatic electrolysis using a Ni electrode at 0.65 V for 4 h in a molten LiCl-KCl system at 723 K.

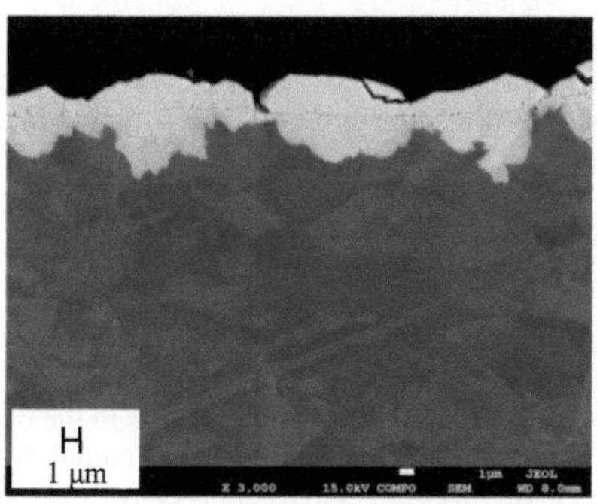

Figure 7. Cross-sectional SEM image of the sample prepared by cathodic potentiostatic electrolysis using a Ni electrode at 0.65 V for 12 h in a molten LiCl-KCl system at 723 K.

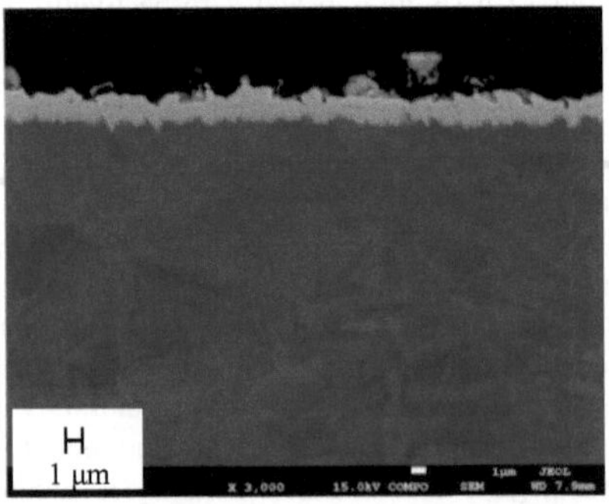

Figure 8. Cross-sectional SEM image of the sample prepared by cathodic potentiostatic electrolysis using a Ni electrode at 0.60 V for 1.5 h in a molten LiCl-KCl system at 723 K.

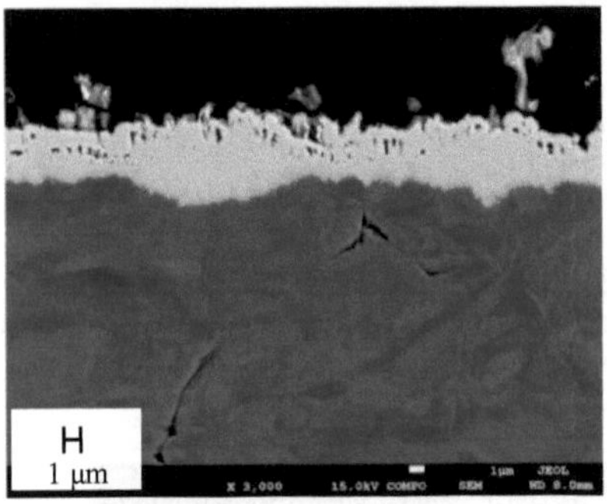

Figure 9. Cross-sectional SEM image of the sample prepared by cathodic potentiostatic electrolysis using a Ni electrode at 0.55 V for 3 h in a molten LiCl-KCl system at 723 K.

Anodic Dissolution Behavior of the Nd-Fe-B Magnet

The Nd-Fe-B magnet was wrapped in a nickel wire and then subjected to anodic dissolution at 1.70-2.20 V for 12 h. Figure 10 shows the cross-sectional SEM image of the sample retrieved after dissolution at 1.70 V. The EDS analysis showed a variation in the dissolution pattern. It was observed that the Nd, present in the outer layer underwent preferential dissolution leaving the Nd present in the inner layer untouched. This might have happened because of the large cracks, found between inner and outer layer, generated during the electrochemical dissolution process. When the anodic dissolution was carried out at a comparatively higher voltage (2.20 V), all the three elements underwent complete dissolution and there was hardly any magnet sample left in the nickel wire.

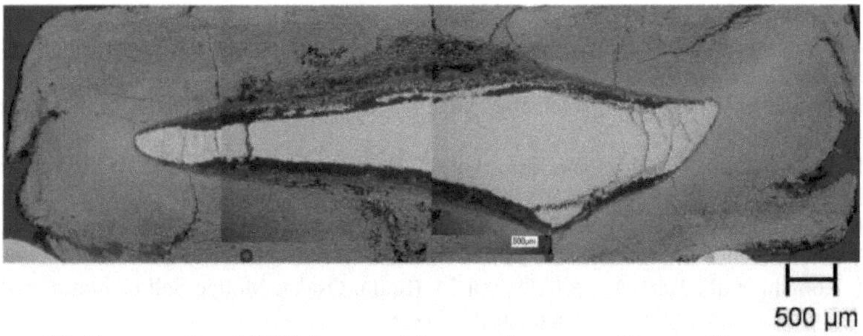

Figure 10. Cross-sectional SEM image of the sample prepared by anodic potentiostatic electrolysis using a Nd-Fe-B magnet electrode at 1.70 V for 12 h in a molten LiCl-KCl system at 723 K.

Conclusions

The electrochemical formation of RE-Ni (RE=Dy, Nd, Pr) alloys using Ni electrodes was investigated in a molten LiCl-KCl-$RECl_3$ (0.50 mol%) at 723 K. Furthermore, the anodic dissolution of RE (Dy, Nd, etc) using Nd-Fe-B magnet electrodes and electrowinning of Dy using Ni electrodes were carried out in a molten LiCl-KCl system at 723 K. The results could be summarized as follows:

1. The alloy sample was prepared by potentiostatic electrolysis at 0.65 V for 1 h using a Ni cathode in a molten LiCl-KCl-$DyCl_3$-$NdCl_3$-$PrCl_3$ system. The highest mass ratio of Dy/Nd+Pr in the alloy sample measured by ICP-AES was determined to be 50 at 0.65 V.
2. Anodic potentiostatic electrolysis at 1.70 V and 2.20 V for 12 h were conducted using Nd-Fe-B magnet electrodes wrapped Ni wires. It was found that RE in the outer layer was selectively dissolved but RE in the inner layer remained undissolved from the EDX analysis of the cross-section of the sample obtained at 1.70 V. The sample obtained at 2.20 V was almost dissolved, and the original form of the magnet disintegrated.
3. After anodic potentiostatic electrolysis at 2.20 V for 12 h using the Nd-Fe-B magnet and cathodic potentiostatic electrolysis at 1.00 V for 5 h using a Mo plate, cathodic potentiostatic electrolysis was conducted at 0.65 V for 12 h using a Ni plate. From the EDX analysis of formed alloy, the mass ratio of Dy/Nd in the alloy sample was about 18.

Acknowledgments

This work was supported by Grant-in-Aid for Scientific Research (B) from Japan Society for the Promotion of Science (JSPS).

References

1. T. Oishi et al., *Kagaku Kogaku Ronbunshu*, vol.36, (Bunkyo-ku, Tokyo: Soc Chemical Eng Japan, 2010), 299-303.

2. S. Kobayashi et al., *J. Electrochem. Soc.*, vol.158, (Pennington, NJ: Electrochemical Soc Inc, 2011), E142-146.
3. S. Kobayashi et al., *J. Electrochem. Soc.*, vol.159, (Pennington, NJ: Electrochemical Soc Inc, 2012), E193-197.
4. H. Konishi et al., *J. Electrochem. Soc.*, vol.148, (Pennington, NJ: Electrochemical Soc Inc, 2001), C506-511.
5. H. Konishi et al., *Electrochim. Acta*, vol.48, (Kidlington, Oxford OX5 1GB: Pergamon-Elsevier Science Ltd, 2003), 1403-1408.
6. H. Konishi et al., *Electrochim. Acta*, vol.48, (Kidlington, Oxford OX5 1GB: Pergamon-Elsevier Science Ltd, 2003), 563-568.
7. H. Konishi et al., *MOLTEN SALTS*, vol.54, (Suita, Osaka: Molten Salt Committee of The Electrochemical Society in Japan, 2011), 21-28.
8. H. Konishi et al., *ECS Transactions*, vol.50, (Pennington, NJ: Electrochemical Soc Inc, 2012), 463-472.
9. H. Konishi et al., *ECS Transactions*, vol.53, (Pennington, NJ: Electrochemical Soc Inc, 2013), 37-46.

EPD Congress 2015

SYMPOSIUM: HIGH-TEMPERATURE ELECTROCHEMISTRY II

Sensors and Advanced Materials

Michael Simpson,
University of Utah

Hirokazu Konishi,
Osaka University

EPD Congress 2015
Edited by: James Yurko, Antoine Allanore, Laura Bartlett, Jonghyun Lee, Lifeng Zhang, Gabriella Tranell, Yulia Meteleva-Fischer, Shadia Ikhmayies, Arief Suriadi Budiman, Prabhat Tripathy, and Guy Fredrickson
TMS (The Minerals, Metals & Materials Society), 2015

Production of Titanium Oxycarbide from Titania-Rich Mineral Sands

Farzin Fatollahi-Fard[1] and Petrus Christiaan Pistorius[1]

[1]Materials Science and Engineering, Carnegie Mellon University
5000 Forbes Avenue, Pittsburgh, PA 15213

Keywords: Titanium, Oxycarbide, Electrowinning, MER Process, Chinuka

Abstract

The production of titanium oxycarbide – a consumable anode material used in molten-salt processes for electrowinning titanium – from heavy mineral concentrate, such as a high-titania leucoxene and natural rutile is described. The oxycarbide is prepared by carbothermic reduction of the oxide at elevated temperature. Particle size distribution (of the titanium oxide raw material) and temperature have large effects on the kinetics of the reaction, which seem to follow a "shrinking-core" model. It is anticipated that low porosity (in the anode pellets) would be advantageous during electrolysis. Particle size distribution, extent of reaction during carbothermic reduction, and sintering temperature are expected to affect pellet porosity after sintering, and are being tested experimentally.

Introduction

Many groups have been working on an electrochemical process to develop an alternative process to Kroll reduction for the production of titanium [1]. The process developed by Materials and Electrochemical Research (MER) Corporation, more commonly referred to as the MER process [2] (or the similar Chinuka process [1,3]), is one such method that is currently being investigated. The MER process begins with a source of relatively pure TiO_2, which also can include grades as low as titania-rich slags [2]. This source of titania is then reduced by carbon to form a TiO-TiC solid solution, referred to as titanium oxycarbide (approximately Ti_2OC), by the carbothermic reaction shown. Titanium oxycarbide is assumed to be a regular solution of TiO and TiC with a heat of mixing of -24.7 kJ/mol [4].

$$2\,TiO_2 + 4\,C \rightarrow Ti_2OC + 3\,CO \quad (T = 1400 - 1600\ °C,\ \Delta G° = -0.518*T(K)+764.16\ kJ) \quad (1)$$

$$Ti_2OC \rightarrow 2\,Ti + CO \quad (T = 800\ °C,\ \text{cell potential} = 1.17\ V) \quad (2)$$

Table I. Half-cell potentials for possible reactions in the Chinuka/MER process

Half-cell reaction	Potential relative to Cl_2 evolution (V)
$2\,TiCl_2 + CO + 4\,e^- \rightarrow Ti_2OC + 4\,Cl^-$ (anode)	-0.61
$FeCl_2 + 2\,e^- \rightarrow Fe + 2\,Cl^-$	-1.11
$SiCl_4 + 4\,e^- \rightarrow Si + 4\,Cl^-$	-1.35
$MnCl_2 + 2\,e^- \rightarrow Mn + 2\,Cl^-$	-1.79
$TiCl_2 + 2\,e^- \rightarrow Ti + 2\,Cl^-$ (cathode)	-1.79

Oxycarbide, so produced, can be used as an anode in a molten salt electrochemical cell (such as eutectic NaCl-KCl) to produce titanium metal. Eqn. 2 describes the overall electrochemical transformation of the Ti_2OC into Ti metal. The presence of some of the impurity elements (such as Fe, Si, Mn), can also initiate some additional electrochemical reactions, listed in Table I.

Table I suggests that as the reduction of impurity elements (Fe, Si, Mn) would occur at less negative potentials (than the reduction of Ti), clearly there is a need to remove them from the anode feed prior to its electrochemical reduction to titanium metal. The current practice has been to use an acid leach to remove the impurities [3].

Although inclusion of an extra processing step prior to electrochemical treatment of iron-rich titania may not be entirely in favor of the process economy, these materials have an inherent advantage of being inexpensive raw materials (Table II) [5].

Table II. Cost of various titanium ores in 2012.

Concentrate	Cost/ton ($/ton)	%TiO_2 [6]	Cost/ kg. TiO_2 ($/kg.)
Ilmenite	300	54	0.56
Titaniferous slag	637.50	88	0.73
Natural rutile	2650	95	2.79

Although some of the iron-rich ores are being used by the Kroll process in order to prepare titanium ($TiCl_4$ being the feed stock in the Kroll process, separated from iron during the prior step of carbochlorination), the Becher process can be adopted to effectively reduce the Fe content from low grade ilmenite ($FeTiO_3$) and lecoxene (geologically weathered ilmenite with up to ~90% TiO_2) and upgrade the TiO_2 content to a value as high as ~95%. In the Becher process, carbon is used as the reductant to chemically convert ilmenite to metallic iron and TiO_2 (Eqn. 3).

$$FeTiO_3(s) + C(s) \rightarrow Fe(s) + TiO_2(s) + CO(g) \quad (3)$$

The iron, so produced, can be removed from the mixture by treating the reduced mass with an aqueous solution of ammonium chloride. Such a treatment removes bulk of the iron from the mixture. Under optimized conditions, a sufficiently pure TiO_2 can be produced by this process which then can be directly chlorinated to prepare $TiCl_4$, without taking recourse to any other purification process.

A slight modification of the Becher process can be adopted to prepare relatively pure titanium oxycarbide (Ti_2OC) from ilmenite for its subsequent electrochemical conversion to titanium metal (Eqn. 4).

$$2\ FeTiO_3(s) + 6\ C(s) \rightarrow Ti_2OC(s) + 5\ CO(g) + 2\ Fe(l) \quad (T = 1600\ °C) \quad (4)$$

Equation 4 shows that the resulting iron will be molten at the reaction temperature. Given molten iron's propensity to dissolve other elements such as silicon, manganese, carbon, and chromium, molten iron can be used as a sink for other impurities that are reduced in the course of the reaction. Upon cooling, this molten iron will solidify, following which the solid iron may be removed by magnetic separation from the rest of the oxycarbide anode precursor, along with any other impurities that are dissolved in the iron. Other than magnetic separation, this process does not require any additional processing step(s) or costly equipment to prepare relatively pure Ti_2OC. Besides, unlike acid dissolution process, this preparative method does not generate large quantities of liquid waste.

Because of the relatively coarser size ranges, associated with the naturally occurring leucoxene and ilmenite, the unprocessed raw materials can pose challenges in terms of the overall reaction kinetics. However, given the higher density of the intermediate phases and reaction product (4.17, 4.49, and 4.94 g/cm^3 for rutile, Ti_2O_3, and Ti_2OC, respectively), the reaction product is expected to be porous. This porosity allows gas to reach the reaction front deeper within the particle. The main byproduct of the oxycarbide reaction shown in Equation 1 is carbon monoxide (CO), which is itself a reducing agent that can participate in the reaction. Equation 1 is rewritten as the set of reactions Equation 5(a-c), reflecting the role of CO as the reducing gas. A reaction mitigated by rate-limiting gaseous diffusion of a reducing gas through a porous product layer is called the "shrinking-core model" [7].

$$C + CO_2 \rightarrow 2\ CO \quad (5a)$$
$$2\ TiO_2 + CO \rightarrow Ti_2O_3 + CO_2 \quad (5b)$$
$$Ti_2O_3 + 4\ CO \rightarrow Ti_2OC + 3\ CO_2 \quad (5c)$$

Given the high reaction temperature required by the oxycarbide production, the CO_2 produced in Reaction 5b and 5c will react with carbon to form CO, although its propensity to do so depends on temperature and the extent of reaction, as illustrated in Figure 1. The newly reduced CO can either participate in the reaction again or leave the system; there is a net production of gas (2 moles of CO produced per 1 mole of CO_2), so some CO must leave the system.

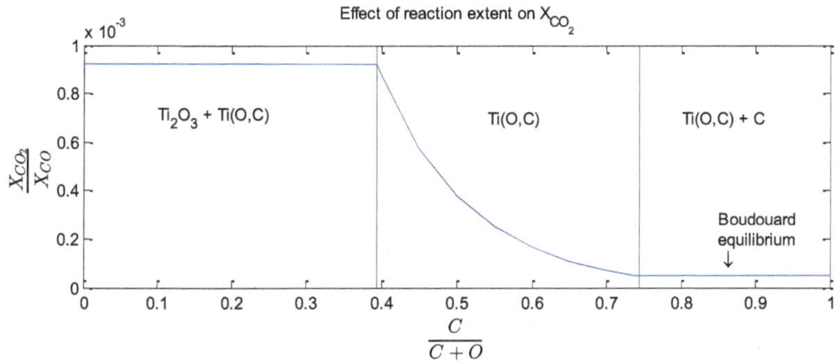

Figure 1. Effect of reaction extent on equilibrium CO_2 concentrations at T = 1600 °C, P = 1 atm.

Methods

The raw materials used in these investigations were leucoxene mineral sand (Iluka, Eucla HyTi 90 Grade, 100 μm particle size), natural rutile (Iluka, MB Rutile 95, 100 μm particle size), and carbon black (Fisher Chemical), mixed in the molar ratio 1:2.2 molar ratio (TiO_2:C, 1-10 grams total). An alumina crucible was used for the reduction experiments.

The crucible, containing the powder mixture, was lowered into a vertical furnace and the furnace, in turn, was heated to 1600 °C at a rate of 5 °C min^{-1} under continuously flowing Ar (99.995% pure) gas at a rate of 0.435 dm^3 min^{-1}. The furnace off gas, generated during the reduction reaction, was analyzed by an IR gas analyzer. The reduced mass was evaluated and characterized by powder XRD (PANalytical X'Pert Pro x-ray diffractometer) and SEM-EDS

(Phillips XL-30/FEI Quanta 600) instruments. The x-ray data was used to compute the lattice parameters of the TiO:TiC mixture as per the procedure given in literature [8]. SEM-EDS measurements were primarily focused on identifying the regions richer in heavier impurities, such as iron. Leucoxene sand (Table III) was extensively used to prepare the oxycarbide. To examine the effect of particle size distribution, the leucoxene was ground into smaller size fractions in a "ring and puck type" tungsten carbide grinder (SPEX SamplePrep ShatterBox 8530). The ground powder was divided into three size ranges (>63μm, 50 μm and <38 μm). Some of the impurity elements, picked up during grinding operation, were removed from the ground powder by a magnetic separator.

Table III. Composition of the leucoxene (Eucla HyTi 90 Grade)
Constituents with wt.% < 0.1 not shown

Constituent	Composition (wt.%)
TiO_2	91.0
Fe_2O_3	2.17
Al_2O_3	0.56
Cr_2O_3	0.144
SiO_2	2.22
ZrO_2	2.42

Results

In our early experiments, titanium oxycarbide was successfully produced by using pigment-grade TiO_2 at 1400 °C. However, similar results with either leucoxene or natural rutile could not be obtained even at temperatures as high as 1600 °C. Powder XRD patterns of these samples showed Ti_2O_3 as the major phase. Small oxycarbide peaks were found to be formed but were mostly hard to distinguish them from the background noise. Also the appearance of very small peaks indicated the formation of insignificant quantities of oxycarbide. Analysis of the mass loss data suggested that the conversion of Ti_2O_3 to Ti_2OC, under these conditions, was < 25%.

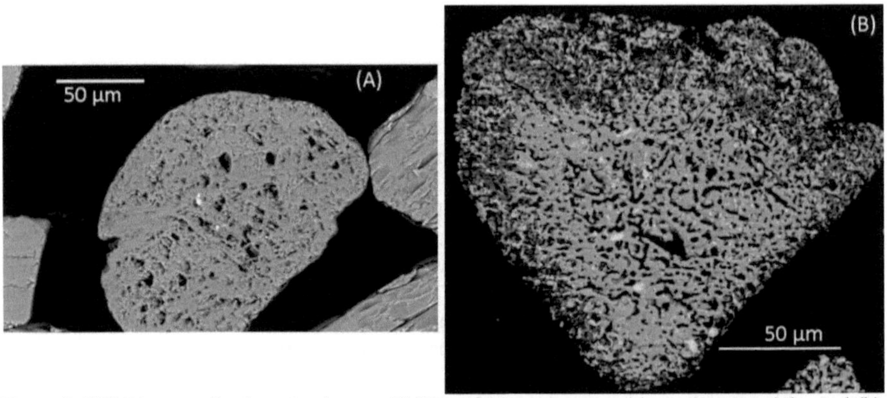

Figure 2. SEM images (backscatter image, 20 kV) of (a) an unreacted leucoxene particle, and (b) a leucoxene particle subjected to normal oxycarbide reduction conditions

SEM images (Figs. 2a and b) suggested that although the reduced powder appeared to have some porosity it nevertheless maintained its original overall shape. The particle core was observed to be more dense (with an approximate composition resembling Ti_2O_3) while the less dense outer layer was observed to have a relatively lower oxygen concentration and hence is most likely the oxycarbide phase. SEM image (2b) further suggested that the reaction pattern seemed to have followed the shrinking-core model with a moving reaction front.

For shrinking-core reduction, particle size strongly affects the reaction rate. To test this, ground leucoxene particles with an approximate size of 50 μm (38 μm < d < 63 μm) were reacted with carbon at a peak temperature of 1600 °C. X-ray diffraction on this sample showed almost complete reduction to titanium oxycarbide, as shown in Figure 3.

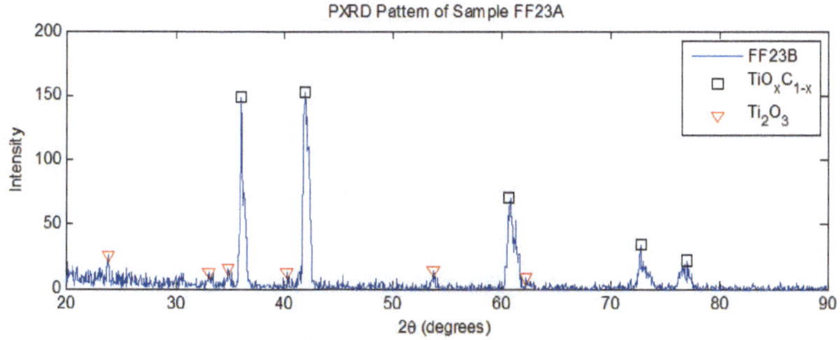

Figure 3. X-ray diffraction pattern of an oxycarbide made from ~50 μm leucoxene particles (after removal of magnetic particles)

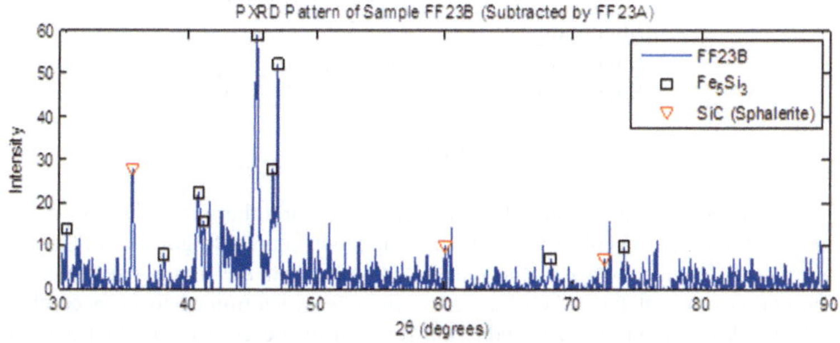

Figure 4. X-ray diffraction patterns of the magnetically separated particles. Pattern is subtracted by the titanium oxycarbide pattern from Figure 3.

One of the characteristics, associated with the shrinking core reaction model, is the strong dependence of the reaction rate with the particle size distribution. The reduction reaction of the 50 μm size fraction (of leucoxene) at 1600 °C indicated the formation of oxycarbide as the major phase (Fig. 3), which confirms the assumption of a shrinking core model to appropriately describe the mechanism of the reduction reaction. Peak broadening at higher angles ($2\theta > \sim 60^0$) indicated the presence of relatively larger volume of the oxycarbide phase with a smaller fraction tending towards larger lattice parameters (an indication of relatively higher oxygen content). Further, SEM images suggested the formation of uniformly porous powder with Ti_2O_3 in the core in relatively coarser size fractions (~60 μm). The reduced oxycarbide powder was subjected to magnetic separation and the XRD profile of the removed powder indicated the presence of two major phases (Fe_5Si_3 and SiC respectively) (Fig. 4). It is interesting to note here that none of the peaks in Figure 4 matched that of α-Fe.

Backscattered SEM images of the magnetically separated powder were seen to have oxycarbide phase present (Fig. 5). Also, the presence of non-porous powder in the mixture indicated phases other than oxycarbide (bright phases, Fig. 5). SEM image also revealed the segregation between brighter and darker phases. EDS measurements showed the compositional variation between these phases (Table 4). While Si was found in all the phases, the grey/dark regions had very little iron (Table 4).

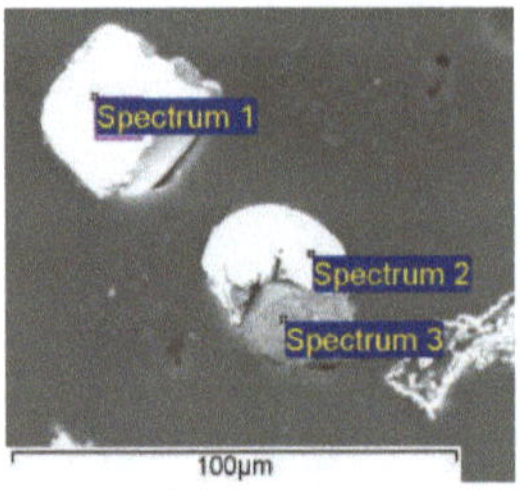

Figure 5. SEM image of magnetically separated particles.
Locations of analyzed EDS spectra also shown (see Table IV).

Table IV. EDS measurements of regions shown in Figure 9. Concentrations in wt.%

	C	Si	Cr	Mn	Fe
Spectrum 1	-	24.4	3.9	0	71.7
Spectrum 2	-	24.4	2.2	0	73.4
Spectrum 3	34.1	65.2	0	0	0.7

Discussion

Particle size distribution was observed to influence the carbothermic reduction of leucoxene sand. While the reduction was observed to be incomplete in coarser fragments, in finer fragments a concentration difference, between the unreduced oxide and the moving reaction front, was found. The composition at the moving reaction front was determined to be a combination of oxides and carbides of titanium. The reduction reaction in presence of product gases (CO and CO_2) can be thought of to have proceeded in the following manner.

CO reacted with TiO_2 to form Ti_2O_3 and to Ti_2OC (Equations 5b and c). The CO_2 released during this process diffused through the product layer to react with the excess carbon to form CO (via Boudouard reaction, Eqn. 5a). As the oxycarbide phase forms at a comparatively high temperature, the Boudouard reaction is not expected to be the rate determining step. The rate determining step, in the present study, is a combination of diffusion of CO into the reactant particles and out-diffusion of the CO_2 from the product layer. Relatively higher CO_2 partial pressure in the vicinity of Ti_2O_3-Ti_2OC phase region (calculated to be 0.9×10^{-3} atm., Fig. 1) as compared to that generated by the Boudouard reaction (estimated to be ~0.05×10^{-3} atm., Fig. 1) suggests that diffusion of CO_2 out of the product layer might have provided the driving force for the occurrence of the reduction reaction.

Looking back at Figure 3, neither Fe_5Si_3 nor SiC are evident in the x-ray pattern of the full sample. It is likely that many of the impurity oxides present in the leucoxene were reduced by the carbon and dissolved in the molten metal, where they precipitated out along with the iron. Fe_5Si_3 is a high temperature phase which is only stable down to 825°C (under equilibrium conditions), but can also exist as a metastable phase at room temperature. This phase has a Curie temperature around 373 to 385 K [9].

The source of the silicon carbide is likely to be the molten iron itself, which was expected to be high in dissolved silicon and carbon. As the molten iron cooled, silicon carbide precipitated out first, lowering the silicon content of the iron, forming Fe_5Si_3, rather than another Fe-Si phase, such as FeSi. This is fortuitous, due to the fact that the SiC is physically attached to the Fe_5Si_3, which is magnetic, allowing both impurities to be removed. There is no SiC evident in Figure 3, so if there is any SiC present, it would only be present in trace amounts. There are also no visible peaks corresponding to Fe_5Si_3.

Conclusions

It is possible to produce titanium oxycarbide from leucoxene and remove the bulk of the iron and silicon impurities. The silicon was reduced along with the iron, as was chromium which formed magnetic particles that were removed via magnetic separation. The remaining impurities Al_2O_3 and $ZrSiO_4$ are expected to be inert in the electrochemical reaction and would not significantly affect the quality of the electrodeposited titanium. The successful reduction of titania-rich mineral sands is strongly dependent on the particle size. This is due to the shrinking core behavior exhibited by the reaction.

References

1. D.J. Fray, "Novel methods for the production of titanium," *International Materials Reviews*, 53 (6) (2008), 317-325.

2. J.C. Withers, R.O. Loufty, J.P. Laughlin, "Electrolytic process to produce titanium from TiO_2 feed," *Materials Technology*, 22 (2) (2007), 66-70.

3. A.M. Martinez et al. "New Method for Low-Cost Titanium Production," *Key Engineering Materials*, 436 (2010), 41-53.

4. B. Jiang et al. "Thermodynamic Study of Titanium Oxycarbide," *Metallurgical and Materials Transactions A*, 43 (10) (2012), 3510-3514.

5. G.M. Bedinger, "2012 Minerals Yearbook - Titanium, 2014," United States Geological Survery. http://minerals.usgs.gov/minerals/pubs/country/2012/myb3-2012-nz.pdf (accessed September 10, 2014).

5. G. Auer et al., "Pigments, Inorganic, 2. White Pigments," *Ullmann's Encyclopedia of Industrial Chemistry*, (2012), 257-291.

6. H.Y. Sohn, "Process Modeling in Non-Ferrous Metallurgy," *Treatise on Process Metallurgy: Industrial Processes, Part A*, ed. S. Seetharaman (Elsevier, 2013), 758-771.

7. G. Neumann, R. Kieffer, P. Ettmayer, "Über das System TiC–TiN–TiO," *Monatshefte für Chemie*, 103 (4) (1972), 1130-1137.

8. V. Johnson et al., "Magnetic and Mössbauer effect studies of Mn_5Si_3:Fe_5Si_3 solid solutions," *Journal of Solid State Chemistry*, 4 (2) (1972), 311-323.

EPD Congress 2015
Edited by: James Yurko, Antoine Allanore, Laura Bartlett, Jonghyun Lee, Lifeng Zhang, Gabriella Tranell, Yulia Meteleva-Fischer, Shadia Ikhmayies, Arief Suriadi Budiman, Prabhat Tripathy, and Guy Fredrickson
TMS (The Minerals, Metals & Materials Society), 2015

EXPERIMENTAL THERMODYNAMIC STUDY ON THE Ag–Sb SYSTEM AT ELEVATED TEMPERATURES

Markus Aspiala, Fiseha Tesfaye and Pekka Taskinen

Aalto University School of Chemical Technology, Department of Materials Science and Engineering, Metallurgical Thermodynamics and Modeling (TDM), Vuorimiehentie 2 K, Espoo, PO Box 16200, FI-00076, Finland

Keywords: Ag_3Sb, Ag_6Sb, EMF technique, Electrochemistry, Gibbs energy of formation

Abstract

In this paper, the standard thermodynamic properties of Ag_3Sb and Ag_6Sb intermetallic compounds have been studied by the solid state EMF-method, using Ag^+ ion conducting β-alumina and AgI. The intermetallic compounds were synthesized from pure substances in evacuated silica ampoules and their formation were confirmed by the SEM-EDS analyses. The EMF measurements were made on electrochemical cells of the type [Ag | β-alumina | Ag-Sb] and [Ag | AgI | Ag-Sb]. The usage of β-alumina electrolyte, in this study, enabled thermodynamic measurements of the Ag-Sb-intermetallic compounds above 710 K, hitherto unreported. Based on the new experimental data, thermodynamic properties of the Ag-Sb-intermetallic compounds have been determined. The obtained experimental values have been compared with the available literature data.

Introduction

Pb-Ag-Sb-Te, in general, and Ag-Sb-Te based intermetallic alloys, in particular, offer several advantageous features in pyrometallurgy, geochemistry and materials science related applications and as a result these materials are being investigated extensively for a variety of applications, mostly as thermoelectric materials [1-4]. Some of the other applications of silver antimonides include as anode in lithium ion batteries [5-7] and fabrication of lead-free solder alloys [8-11]. Both from the standpoints of their recovery, from silver and copper ores and anode slimes, respectively [12], as well as evaluation of their properties, it is important to obtain precise key thermodynamic parameters.

Both calorimetric and electrochemical techniques have been extensively used to estimate the thermodynamic parameters of the intermetallic alloys in the Ag-Sb system [13-17]. Wallbrecht et al have reported the temperature dependence of heat capacities of $Ag_{0.77}Sb_{0.23}$ and $Ag_{0.9}Sb_{0.1}$ in the temperature range 250-800 K [18]. Kleppa used calorimetric technique to estimate the heat of formation of Ag-Sb intermetallic compounds in liquid tin at 723 K [19]. Although Oh et al, on the basis of available literatures data, have assessed high temperature thermodynamic data, their studies lack the estimation of low temperature (<500 K) thermodynamic data [11]. In order to construct the complete binary equilibrium phase diagram, it is imperative to determine the experimental thermodynamic data over a relatively wide temperature regime. The objective of the present research was to experimentally determine the thermodynamic parameters of two different alloy compositions in the temperature range 425.86 – 752.64 K, hitherto unreported.

Phase Relations in the Ag-Sb System

The Ag-Sb system consists of two intermetallic phases known as the minerals allargentum (Ag_6Sb) and dyscrasite (Ag_3Sb) [13,20,21]. According to the Ag-Sb phase diagram proposed by Feschotte et al. [20], there exist solid solutions of antimony in silver, Ag_6Sb and Ag_3Sb. The solubility of silver in antimony is extremely low [22]. The binary compounds form peritectically at temperatures of 975.65 ($Ag + L = Ag_6Sb$) and 835.15 K ($Ag_6Sb) + L = Ag_3Sb$). They have considerable homogeneity ranges of 7 and 5 at.%, respectively, at low temperatures. The solid solubility of antimony into silver is 7.2 at.% at T = 975.7 K and decreases to 5.5 at.% at T = 573 K. At $T = 757$ K and $x_{Sb} = 38$ at.% there exist one eutectic point, i.e. $L = Ag_3Sb + Sb$ [22].

Experimental Section

Several electrochemical cells were prepared with different materials to determine the most efficient cell arrangement and improve the accuracy of the experiments. The methods used to prepare the electrodes are described in the following sections. The provenance and mass fraction purity of the materials used are listed in Table I.

Table I. Provenance and purity of the materials used in this study.

Substance	Manufacturer	Country of manufacture	Mass fraction purity
Ag (plate)	Alfa Aesar	Germany	0.999
Ag (powder)	Alfa Aesar	Germany	0.9995
Sb	Alfa Aesar	Germany	0.99999
AgI	Alfa Aesar	Germany	0.99999
β-alumina	Ionotec	United Kingdom	0.9978
Ar	AGA	Finland	0.99999
Re	Alfa Aesar	Germany	0.9997
Pt	Johnson-Matthey Noble Metals	England	0.9999
α-Al_2O_3	Friatec	Germany	0.995

Preparation of the pelletized electrodes

The binary phases were synthetized separately in evacuated quartz glass ampoules. Stoichiometric amount of fine powdered elements were mixed and pressed into pellets (460 MPa pressure) of different composition. Each pellet was inserted into different ampoules, which were flushed ten times with argon before the final evacuation and sealing. The sealed ampoules were heated above the melting point of antimony inside a Lenton chamber furnace and annealed at 723 K for two days. The homogeneity of the synthetized compounds was confirmed by SEM-EDS analysis.

The synthetized compounds were ground in an agate –mortar and mixed with other fine powders to form experimental sample electrodes (in test electrodes of cells (A) and (B), the Sb content was 30 at.% and 17.5 at.%, respectively). The test electrodes were finalized by pressing each sample into different pellets. In case of Pt–lead wire connection to the test electrode, the test electrode was pressed together with the contact Pt–lead wire. After the press, the end of the Pt wire was gouged out so that the second end of the wire remained inside the pellet with an unrestrained electric connection to the electrode. The Ag powder reference electrodes were pressed into a pellet with Pt lead wire as well. When Re contact was used, the electric contact to the electrode was formed by pressing a spiral like shape of Re-wire towards the electrode during the experiments. There was no noticeable difference in the EMF values when different contact lead wire materials were used.

Preparation of solid electrodes

The solid electrodes were prepared by measuring substances to a quartz ampoule with such ratio that two phases coexisted in the test electrode in equilibrium (in test electrodes of cells (A) and (B), the Sb content was 30 at.% and 17.5 at.%, respectively). The ampoules were flushed with argon, evacuated and heat treated as described in the previous topic. The solid electrodes were finalized by grinding the contact surfaces of the electrode to improve the electric contact to the electrolyte and to the contact wire. The composition of a small piece of the sample was determined by the SEM-EDS. In case of solid pieces as the test electrode, the electric contacts to the electrodes were formed by pressing a spiraled shape of Re or Pt -wire on the electrode during the experiments. The pressing force was produced by steel springs outside the furnace.

The reference electrode was made of powdered or plate type Ag. Pt contact wire was attached to the Ag plate by bending the Ag plate twice around the Pt wire and squeezing the end of the Pt wire between Ag sheets.

Experimental apparatus and measurements

In most of the EMF cells β-alumina plates were used as a Ag^+ ion conducting electrolyte. β-alumina electrolytes were in a form of 0.5 mm thick disc with 10 mm diameter. To prove the functionality of the β-alumina for Ag-Sb systems as well as to extend the themodynamic measurements to low temperature region, solid AgI electrolyte was used. Fine powder of AgI was pressed into pellets. The pellets were either pressed with the test electrode or with both electrodes to form unrestrained connections between electrodes and the electrolyte.

In the present study, measurements were carried out on the cells (A) and (B):

(+)Pt or (+)Re, Sb, Ag_3Sb | (Ag)β-alumina or AgI | Ag, Pt(-) (A)

(+)Pt or (+)Re, Ag_3Sb, Ag_6Sb | (Ag)β-alumina or AgI | Ag, Pt (-) (B)

Each cell arrangement was compressed between two spring loaded alumina tubes during the experiments. Both closed end alumina tubes held resistance thermometers for temperature measurements, at both ends of the experimental cell. With the help of real time temperature readings from the two resistance thermometers, temperature gradient over the EMF cell was reduced to less than ±0.2 K. A symmetrical cell with two Ag electrodes was measured to determine parasitic voltages contribution to the overall EMF as a function of cell temperature and the temperature gradient over the experimental cell. The measured data of the symmetrical cells did not exceed 0.133 mV or go under -0.296 mV in the experimental conditions.

The electrochemical cell was placed in a quartz tube, which, in turn, was lowered into the Lenton resistance furnace. All of the experiments were carried out in protective Ar atmosphere. The EMF produced by the experimental cell was measured using Keithley 6517B electrometer. The temperature measurements were accomplished with two Keithley 2000 DMMs connected to the calibrated resistance thermometers. All of the measurement devices were connected to a PC, which collected data of all the devices every 5 seconds.

The only collected data point by experimental cell (III) was measured to examine the lower temperature limits of the pure ionic conductivity of β-alumina. Despite remarkable scatter in the measured data, the average of the collected data corresponded very well with the data observed by the other cells with identical sample electrode set up.

Results and Discussion

The obtained EMF data with the experimental cells (A) and (B) and experimental uncertainties were compiled in Table II. The equilibrium was considered reproducible when the heating and cooling curves coincided and the measured data was consistent with that of the congruent experimental cells.

Table II. The measured experimental data of this study presented with the estimated uncertainty. Experimental cells containing Ag_3Sb-Sb –electrode (cell (A)) are in the left side of the table and cells with Ag_6Sb-Ag_3Sb electrodes (cell (B)) are in the right side of the table.

T/K	U(*T*)/K[a]	*E*/mV	U(*E*)/mV[a]
I.	(+)Pt, Sb, Ag_3Sb | β-alumina | Ag, Pt(-)		
752.64	0.134	58.10	0.065
742.62	0.056	57.37	0.049
732.54	0.035	56.49	0.044
II.	(+)Pt, Sb, Ag_3Sb | β-alumina | Ag, Pt(-)		
737.12	0.023	56.74	0.111
727.02	0.024	55.88	0.086
716.93	0.021	55.02	0.082
706.83	0.047	54.19	0.099
676.52	0.030	51.61	0.074
III.	(+)Pt, Sb, Ag_3Sb | β-alumina | Ag, Pt(-)		
483.65	0.406	37.24	1.933
IV.	(+)Pt, Sb, Ag_3Sb | β-alumina | Ag, Pt(-)		
674.79	0.024	50.82	0.091
684.86	0.017	51.94	0.071
694.92	0.025	52.83	0.071
704.98	0.115	53.70	0.069
720.08	0.028	54.89	0.075
710.02	0.011	54.05	0.066
674.79	0.023	50.82	0.092
732.15	0.026	55.80	0.064
699.95	0.019	53.25	0.054
727.10	0.022	55.29	0.060
722.06	0.014	54.92	0.066
709.98	0.069	53.89	0.117
V.	(+)Pt, Sb, Ag_3Sb | AgI | Ag, Pt(-)		
665,15	0,019	51,35	0,012
644,93	0,007	49,68	0,007
425,86	0,005	32,52	0,005
VI.	(+)Re, Sb, Ag_3Sb | AgI | Ag, Pt(-)		
573.59	0.041	44.22	0.009
624.40	0.055	48.26	0.010

T/K	U(*T*)/K[a]	*E*/mV	U(*E*)/mV[a]
1. (+)Pt, Ag_6Sb, Ag_3Sb | β-alumina | Ag, Pt(-)			
727.58	0.171	21.68	0.012
707.53	0.044	20.98	0.013
677.17	0.028	19.45	0.031
697.44	0.043	20.42	0.018
717.66	0.033	21.39	0.013
737.99	0.045	22.07	0.010
2. (+)Re, Ag_6Sb, Ag_3Sb | AgI | Ag, Pt(-)			
524.79	0.052	12.93	0.009
545.07	0.049	13.91	0.010
565.46	0.060	14.84	0.011
585.79	0.046	15.69	0.009
606.09	0.041	16.54	0.009
626.34	0.039	17.39	0.009
646.52	0.024	18.12	0.008
666.74	0.028	18.90	0.008
3. (+)Re, Ag_6Sb, Ag_3Sb | AgI | Ag, Pt(-)			
535.29	0.044	13.32	0.005
575.67	0.069	15.20	0.007
616.29	0.056	16.79	0.008

[a] $U=3u_c$, where u_c is standard deviation. Level of confidence is greater than 0.99.

The presented values are averages of 1000 measured data points after reaching equilibrium. Temperatures are averages of measured values of temperatures at both ends of the EMF cells. Isothermal equilibrium EMF values were achieved after continuous measurements between 2 hours and 14 days.

As the electronic conductivity of both AgI and β-alumina, under the given experimental conditions are negligible, their transference numbers being ≥ 0.99 [23,24], the experimental EMF values of the virtual electrochemical cell reactions can be expressed as follows:

$$3Ag + Sb = Ag_3Sb, \quad \text{(i)}$$

$$3Ag + Ag_3Sb = Ag_6Sb. \quad \text{(ii)}$$

Thermodynamic Functions

Thermodynamic functions of electrochemical cell reactions (i) and (ii) were calculated by the following thermodynamic equation:

$$\Delta_r G° = -z \cdot F \cdot E \quad (1)$$

where E is the EMF produced by a galvanic cell, z is the number of electrons involved in the virtual cell reactions (i) and (ii), F is the Faraday constant 96487 $C \cdot mol^{-1}$.

In the experimental cells, the virtual cell reactions can be written as:

$$aA + bB + cC\ldots = mM, \quad \text{(iii)}$$

where a, b, c and m indicates the stoichiometric coefficients of the substances A, B, C and M in the reaction where m moles of M-compound has been formed. Therefore, the standard Gibbs energies of formation of the virtual cell reactions take form:

$$\Delta_f G° = G°_{products} - G°_{reactants} = mG°_M - aG°_A - bG°_B - cG°_C\ldots. \quad (2)$$

The standard Gibbs energies of formation of Ag_3Sb saturated with antimony were calculated directly from the measured EMF vs. T relations according to equation (1). The standard Gibbs energies of formation for antimony saturated Ag_6Sb have been calculated by combining equations (1) and (2):

$$\Delta_f G°_{Ag6Sb} = \Delta_{r(ii)} G° + \Delta_f G°_{Ag3Sb}, \quad (3)$$

where $\Delta_{r(ii)}G°$ is the standard Gibbs energy of reaction (ii). The obtained results are shown in Figure 1 and compared with the literature data. The coefficients of determinations (R^2) for the least square expressions are 0.9999 and 0.9953 for cell (A) and cell (B), respectively.

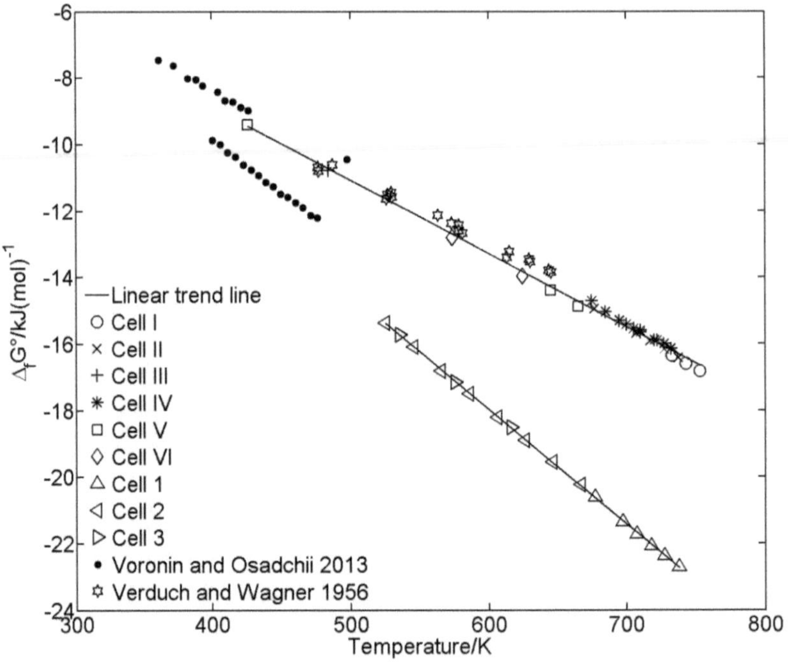

Figure 1. A summary of the obtained results together with the results of Voronin and Osadchii [25] and Verduch and Wagner [14], the least squares fitting of the experimental data in this work are shown with solid lines.

Conclusions

In this study, standard Gibbs energies of formations of Ag_3Sb and Ag_6Sb have been determined by the solid electrolyte EMF technique using two different Ag^+ conducting solid electrolytes β-alumina and AgI. The experimental cell temperatures were measured at both ends of the electrochemical cells, accurately, and the temperature gradient over the cell was controlled to be less than 0.2 K. The reliability of the study has also been proven by the symmetric cells, which confirmed that in the experimental conditions parasitic thermo-EMF was between -0.296 and 0.133 mV.

The standard Gibbs energies of formations of Ag_3Sb and Ag_6Sb which are derived from the cells above are described as linear functions of temperature as:

$$\Delta_f G^\circ_{Ag3Sb}(\text{kJ/mol}) = (-0.0680 \pm 0.2104) - (0.0220 \pm 0.0003) \cdot T(\text{K}), \quad (484 \leq T/K \leq 753)$$

$$\Delta_f G^\circ_{Ag6Sb}(kJ/mol) = (2.7056 \pm 0.0649) - (0.0345 \pm 0.0001) \cdot T(K). \qquad (525 \leq T/K \leq 738)$$

Acknowledgement

Financial support of the Research Foundation of Helsinki University of Technology and Technology Industries of Finland Centennial Foundation ares gratefully acknowledged by the authors. Authors want to thank B. Sc. Petteri Piskunen for SEM–EDS analyses.

References

[1] H. Wu, S. Chen, "Phase Equilibria of Ag-Sb-Te thermoelectric materials," *Acta Mater.,* 59 (2011), 6463-6472.

[2] G.J. Snyder, E.S. Toberer, "Complex thermoelectric materials," *Nature Materials.*, 7 (2008), 105-114.

[3] E.A. Skrabek, D.S. Trimmer, *CRC handbook of thermoelectrics*, ed. D.M. Rowe (Boca Raton, Florida: CRC Press LLC, 1995), 267-275.

[4] F. Hsu Kuei, S. Loo, F. Guo, W. Chen, S. Dyck Jeffrey, C. Uher, T. Hogan, E.K. Polychroniadis, G. Kanatzidis Mercouri, "Cubic $AgPb_mSbTe_{2+m}$: Bulk Thermoelectric Materials with High Figure of Merit," *Science.*, 303 (2004), 818-821.

[5] M. Wachtler, M. Winter, J.O. Besenhard, "Anodic materials for rechargeable Li-batteries," *J. Power Sources.*, 105 (2002), 151-160.

[6] J.T. Vaughey, L. Fransson, H.A. Swinger, K. Edström, M.M. Thackerey, "Alternative anode materials for lithium-ion batteries: a study of Ag_3Sb," *J. Power Sources.*, 119-121 (2003), 64-68.

[7] F. Wang, G. Yao, M. Xu, M. Zhao, P. Zhang, X. Song, "Ag–Sb composite prepared by chemical reduction method as new anode materials for lithium-ion batteries," *Mater. Sci. Eng. B.*, 176 (2011), 442-445.

[8] D.B. Masson, B.K. Kirkpatrick, "Equilibrium solidification of Sn-Ag-Sb thermal fatigue-resistant solder alloys," *J. Elektron. Mater.*, 15 (1986), 349-353.

[9] H. Lee, M. Chen, H. Jao, C. Hsu, "Effect of Adding Sb on Microstructure and Adhesive Strength of Sn-Ag Solder Joints," *J. Electron Mater.*, 33 (2004), 1048-1054.

[10] J. Liu, C. Guo, C. Li, Z. Du, "Thermodynamic description of the Ag-Bi-Sb system," *Thermochim. Acta.*, 539 (2012), 44-50.

[11] C. Oh, J. Shim, B. Lee, D.N. Lee, "A thermodynamic study on the Ag-Sb-Sn system," *J. Alloys Comp.*, 238 (1996), 155-166.

[12] G.G. Pavlova, A.S. Borisenko, "The age of Ag-Sb deposits of Central Asia and their correlation with other types of ore systems and magmatism," *Ore Geol. Rev.*, 35 (2009), 164-185.

[13] F. Weibke, I. Efinger, "Elektrochemische Untersuchungen am System Silber-Antimon," *Z. Electrochem.*, 46 (1940), 61-69.

[14] A.G. Verduch, C. Wagner, "Contributions to the thermodynamics of the system PbS-Sb_2S_3, Cu_2S-Sb_2S_3, Ag_2S-Sb_2S_3 and Ag-Sb," *J. Phys. Chem.*, 61(1957), 558-562.

[15] N. Birks, H. Rickert, C. Wagner, "Berechnung von Metall–Schlacken–Gleichgewichten sowie Gleich gewichten zwischen binären flüssigen oder festen Legierungen und festen Oxyden oder Sulfiden mit einer Anwendung auf das System Ag–Sb–S," *Z. Electrochem.*, 66 (1962), 266-269.

[16] G.M. Lukashenko, V.N. Eremenko, V.R. Sidorko, "A thermodynamic study of the system silver-antimony," *Zh. Neorg. Khim*, 9 (1964), 220-221.

[17] E. Osadchii, D. Boström, S. Lunin, E. Rosen, "Thermodynamic properties of dyscrasite (Ag_3Sb) at isostatic pressures 4000-8000 bar," *Phys. Chem. Miner.*, 25 (1998), 288-291.

[18] P.C. Wallbrecht, R. Blachnik, K.C. Mills, "The heat capacity and enthalpy of some Hume—Rothery phases formed by copper, silver and gold. Part I. Cu + Sb, Ag + Sb, Au + Sb, Au + Bi systems," *Thermochim. Acta.*, 45 (1981), 189-198.

[19] O.J. Kleppa, "Heat of Formation of Solid and Liquid Alloys in the Systems Silver–Cadmium, Silver–Indium and Silver–Antimony at 450°," *J. Phys. Chem.*, 60 (1956), 846-852.

[20] P. Feschotte, F. Monachon, P. Durussel, "The binary system Sb–Ag: A revision of the Ag_3Sb phase boundaries," *J. Alloys Compd.*, 186(1992), L17-L18.

[21] C. Cipriani, M. Corazza, G. Mazzetti, "Reinvestigation of natural silver antimonides," *Eur. J. Mineral.*, 8(1996), 1347-1350.

[22] S. Hassam, Z. Bahari, B. Legendre, "Phase diagrams of the Ag–Bi–Sb ternary system," *J. Alloys Compd.*, 315(2001), 211-217.

[23] J.W. Patterson, "Conduction Domains for Solid Electrolytes," *J. Electrochem. Soc.*, 118 (1971), 1033-1039.

[24] M.S. Whittingham, R.A. Huggings, "Transport Properties of Silver Beta Alumina," *J. Electrochem. Soc.*, 118 (1971), 1-6.

[25] M.V. Voronin, E.G. Osadchii, "Standard Thermodynamic Properties of Ag_3Sb and Ag_6Sb Evaluated by EMF Measurements," *Inorg. Mater.*, 49 (2013), 550-554.

2015 EPD Congress

SYMPOSIUM: HIGH-TEMPERATURE ELECTROCHEMISTRY II

Energy Storage Devices, Corrosion and Molten Salt Science

P. Chris Pistorius,
Carnegie Mellon University

Dihua Wang,
Wuhan University

EPD Congress 2015
Edited by: James Yurko, Antoine Allanore, Laura Bartlett, Jonghyun Lee, Lifeng Zhang, Gabriella Tranell, Yulia Meteleva-Fischer, Shadia Ikhmayies, Arief Suriadi Budiman, Prabhat Tripathy, and Guy Fredrickson
TMS (The Minerals, Metals & Materials Society), 2015

SYNTHESIS OF HIGH PERFORMANCE $LiMn_{0.8}Fe_{0.2}PO_4/C$ CATHODE MATERIAL FOR LITHIUM ION BATTERIES: EFFECT OF CALCINATION TEMPERATURE

Enrui Dai [1,2], Weibing Chen [1,2], Haisheng Fang [1,2,*], Hui Wang [3], Bin Yang [1,2,*], Wenhui Ma [1,2]

[1]State Key Laboratory of Complex Nonferrous Metal Resources Clean Utilization, Kunming University of Science and Technology, Kunming 650093, China
[2]Key Laboratory of Nonferrous Metals Vacuum Metallurgy of Yunnan Province, Kunming University of Science and Technology, Kunming 650093, China
[3]Technical Institute of Physics and Chemistry, Chinese Academy of Sciences, Beijing 100190, China

Keywords: Lithium ion batteries; Cathode material; Lithium manganese phosphate

Abstract

$LiMn_{0.8}Fe_{0.2}PO_4/C$ composite was prepared by an improved solid-state method and the effect of calcination temperature on properties of the obtained materials was investigated. The results showed that increasing calcination temperature from 600 to 700 °C improved the performance of the $LiMn_{0.8}Fe_{0.2}PO_4/C$ due to enhanced crystallinity and increased conductivity, but further increase in calcination temperature to 800 °C led to degraded performance due to particle growth and decrease in porosity. Therefore, the $LiMn_{0.8}Fe_{0.2}PO_4/C$ composite prepared at 700 °C exhibited the best electrochemical performance, and could deliver a high capacity of 152 mAh g^{-1} at 0.1 C, 147 mAh g^{-1} at 1 C and 114 mAh g^{-1} at 10 C. In addition, the performance of the $LiMn_{0.8}Fe_{0.2}PO_4/C$ and $LiMn_{0.8}Fe_{0.19}Mg_{0.01}PO_4/C$ was compared when they were obtained at the optimum calcination temperature.

Introduction

As a cathode material for lithium ion batteries, olivine structured $LiMnPO_4$ has received increasing attention because of its good cyclic performance, excellent chemical and thermal stabilities, low toxicity and low cost. It offers a redox potential of 4.1 V vs. Li^+/Li, which means a higher energy density as compared to that of $LiFePO_4$ (3.4 V vs. Li^+/Li) [1]. However, previous studies showed that the kinetic behavior of $LiMnPO_4$ is too poor to show any reversible capacity [2]. Several approaches have been adopted to improve the performance of $LiMnPO_4$: preparing smaller particles [3-9], carbon coating [10-12] and cation substitution [13-16]. However, no matter whichever approach is used, the improvement of the performance is critically dependent on synthesis method. Recently, we have developed an improved solid-state method to synthesize both pure $LiMnPO_4/C$ and substituted $LiMnPO_4/C$ where Mn was substituted by Zn, Mg and Mg+Fe. The resultant materials showed good electrochemical performance. In the year of 2010 we reported that the Fe&Mg co-substituted $LiMnPO_4/C$ showed better electrochemical performance than the Fe substituted $LiMnPO_4/C$ when they were prepared by the same method at the same calcination temperature (800 °C) [17], but more recently we found out that the optimum calcination temperature may be different for substituted $LiMnPO_4$ even if they were prepared by the same method. For example, the Zn or Mg substituted $LiMnPO_4/C$ showed the best

* Corresponding author. Tel. / fax: +86 871 65107208.
E-mail address: hsfang1981@hotmail.com (H.S. Fang), kgyb2005@126.com (B. Yang).

performance at 700 °C [18], while the Fe&Mg co-substituted $LiMnPO_4/C$ had the optimum performance at 800 °C [19]. Now the question arises as to whether there is a difference in the optimum calcination temperature between Fe and Fe-Mg substituted $LiMnPO_4/C$ and if so does the latter show better performance over the former when both are synthesized at the same (optimum) calcination temperature? If this is the case, then the question is whether the performance of the Fe&Mg co-substituted $LiMnPO_4/C$ is still better than that of the Fe substituted $LiMnPO_4/C$ when they are obtained at the optimal calicination temperature. The paper describes a new solid state synthesis process for $LiMn_{0.8}Fe_{0.2}PO_4/C$ and how the calcination temperature helped improve its performance characteristics. Moreover, the performance of the $LiMn_{0.8}Fe_{0.2}PO_4/C$ and $LiMn_{0.8}Fe_{0.19}Mg_{0.01}PO_4/C$ is compared when they are obtained at the optimal calicination temperature. The objective of this investigation was two-fold: to verify if the (i) new preparative approach can yield high performance $LiMn_{0.8}Fe_{0.2}PO_4/C$ and (ii) Fe&Mg co-substitution is better than just the Fe substitution!

Experimental

High purity LiH_2PO_4, $MnC_4H_6O_4\cdot4H_2O$, $H_2C_2O_4\cdot2H_2O$ and $FeC_2O_4\cdot2H_2O$ in the mole ratio of 1:0.8:0.8:0.2 were mixed and homogenized with ~7 wt.% sucrose in a ball miller for 6 h. During ball-milling process, the oxalic acid induced a room temperature solid-state reaction with $MnC_4H_6O_4\cdot4H_2O$ to form nanosize $MnC_2O_4\cdot2H_2O$. The milled mixture was dried and then heated (heating rate: 2 °C min^{-1}) to 600-800 °C for 10 h under Ar atmosphere.

Powder XRD (D/MaX-3, Rigaku) technique was used to identify the product phases. Lattice parameters of different phases were determined by the refinement of the XRD patterns using silicon (99.9% pure) as the internal standard. Powder morphology and particle size distribution were measured by SEM (XL30, Phillips) and TEM (Tecnai G2 F20, FEI) respectively. The total carbon content was determined by a carbon analyzer (VarioEL III elementar). The surface area and pore size distribution were estimated by a surface adsorption analyzer (NOVA2200e, Quantachrome instrument) using nitrogen as the carrier gas. The electronic conductivity was determined by a four-point resistivity measurement probe (RTS-8, Four Probes Tech.). The samples for the conductivity measurement were prepared in the form of discs by manually pressing the powder under an applied load of 20MPa. Electrochemical measurements were carried out with CR2025 coin cells and lithium metal as the anode. The cathode was prepared in the form of a slurry by mixing the synthesized powder, super P (that is conductive carbon black), and polyvinylidene fluoride (PVDF) (in the ratio 8:1:1) in N-methyl pyrrolidinone (NMP). The slurry was coated with an aluminum foil using a doctor blade coater. The diameter and active mass of the cathode were 1.3 cm and ~2.3mg respectively. 1M $LiPF_6$ in EC/DMC (1:1 ratio) was used as the electrolyte. The electrochemical cell was assembled in an argon-atmosphere glove box and the measurements were carried out under both constant current as well as voltage modes during charging and constant current mode during discharging, using a battery test system (Land CT2001A) at 30 °C.

Results and discussion

Fig. 1 shows the XRD pattern of the synthesized $LiMn_{0.8}Fe_{0.2}PO_4/C$ at different temperatures. All samples were observed to have similar patterns that can be indexed into an orthorhombic structure with a space group of *Pmnb*. The diffraction peaks become more intense with increasing temperature, indicating enhanced crystallinity of the samples at higher temperatures. The refinement of XRD patterns revealed that the lattice parameters of these samples varied negligibly with the temperature, and all were around a = 6.0822 Å, b = 10.4253 Å and c =

4.7318 Å. Besides, the absence of carbon peak in the XRD patterns indicated the formation of amorphous residual carbon by way of pyrolysis of sucrose. Relatively higher temperatures led to an apparent decrease in the carbon content in the composite. However, electronic conductivity increased with an increase in the temperature, which can be attributed to the enhanced electronic conductivity of the residual carbon since raising pyrolysis temperature above 700 °C can dramatically increase the electronic conductivity of the carbon film [20]. These observations suggest that the quality of carbon is more important than its quantity. The carbon content and electronic conductivity of the $LiMn_{0.8}Fe_{0.2}PO_4/C$ prepared at different temperatures are listed in Table 1.

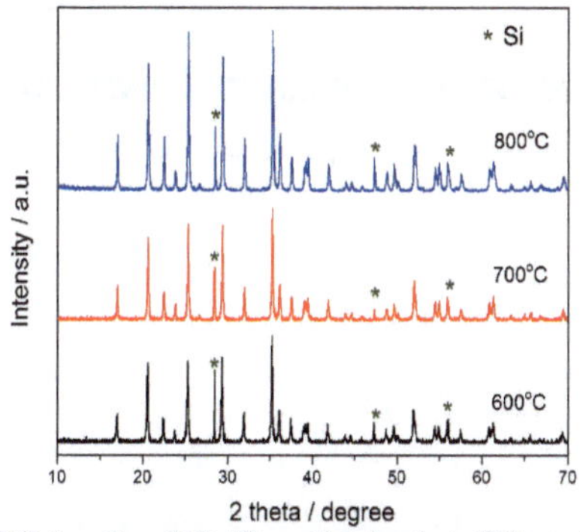

Figure 1. XRD patterns of $LiMn_{0.8}Fe_{0.2}PO_4/C$ synthesized at different temperatures. The asterisk (*) denotes the internal silicon standard.

Table 1. Carbon content and electronic conductivity of $LiMn_{0.8}Fe_{0.2}PO_4/C$ synthesized at different calcination temperatures.

Calcination temperature (°C)	Carbon content (wt.%)	Electronic conductivity (S cm^{-1})
600	9.05	1.3×10^{-4}
700	7.32	1.1×10^{-2}
800	6.64	2.5×10^{-2}

Fig. 2 presents the SEM images of $LiMn_{0.8}Fe_{0.2}PO_4/C$ synthesized at different temperatures. All samples showed a similar morphology with primary particles being agglomerated together. Increased (from 600-700 °C) did not cause obvious coarsening of primary particles, but temperatures > 800 °C led to an apparent increase in the primary particles size. The TEM images (Fig. 3) of $LiMn_{0.8}Fe_{0.2}PO_4/C$ samples showed a porous structure of aggregate particles which have also been observed by others for $LiMnPO_4$ composites, prepared by the same route [21-22]. Formation of such a porous structure has demonstrated to be crucial for the electrochemical performance [21]. Moreover, the coarsening of primary particles was observed for the samples prepared at 800 °C, which is similar to the SEM observation.

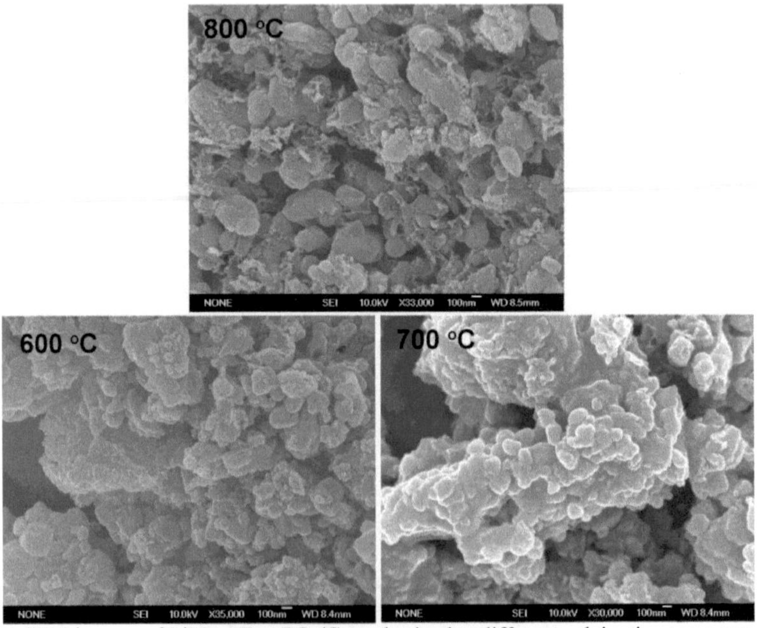

Figure 2. SEM images of $LiMn_{0.8}Fe_{0.2}PO_4/C$ synthesized at different calcination temperatures.

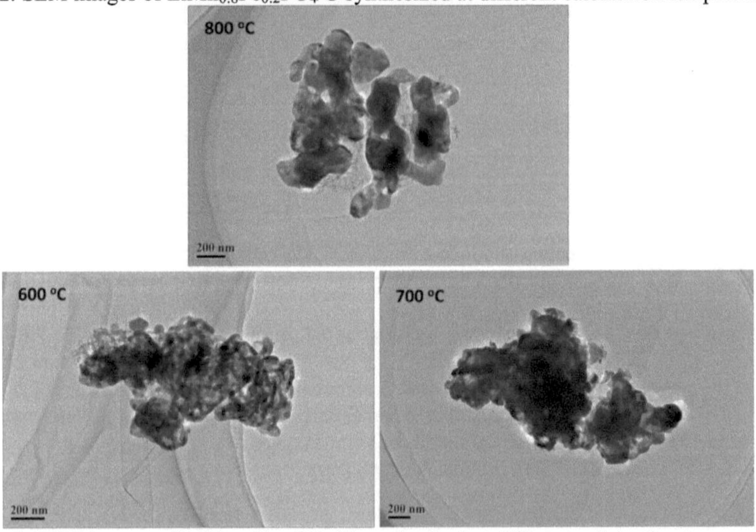

Figure 3. TEM images of $LiMn_{0.8}Fe_{0.2}PO_4/C$ synthesized at different calcination temperatures.

Fig. 4 represents N_2 adsorption and desorption isotherms of $LiMn_{0.8}Fe_{0.2}PO_4/C$ synthesized at different temperatures. All isotherms are of type IV with type H3 hysteresis loops at the higher P/P_o ratio [23], indicating the existence of non-uniform and slit-shaped mesopores. The surface area of the samples, synthesized at 600, 700 and 800 °C were estimated to be 51.5, 45.1 and 25.8

m^2 g^{-1}, respectively. The dramatic decrease of the surface area of the sample prepared at 800 °C meant a rapid growth of primary particles. Interestingly, our previous study showed that there was no evidence of the growth of the particles for Fe&Mg co-substituted $LiMnPO_4/C$ prepared by the same method in the temperature range 600-800 °C [24]. Considering the particle-size dependent performance of $LiMnPO_4$ based material, such a difference may result in a different optimal calcination temperature between the Fe substituted $LiMnPO_4/C$ and the Fe&Mg co-substituted $LiMnPO_4/C$. Besides the particle size change, the N_2 adsorption measurement also revealed a variation in the porosity of the three samples. Based on the Barrett - Joyner - Halenda (BJH) method, the derived surface area of the samples synthesized at 600, 700 and 800 °C was determined to be 52.8, 45 and 35.1 m^2 g^{-1}, respectively. The measurement of the BJH surface area was determined based on the capillary condensation in pores and thus this result indicated that the porosity of samples decreased with the increase in calcination temperature. Indeed, the pore volume of the samples synthesized at 600, 700 and 800 °C decreased to 0.185, 0.160 to 0.133 m^3 g^{-1}, respectively. These results are consistent with the SEM and TEM observations. Our previous work has proved that the presence of sufficient porosity is essential to achieve high performance of $LiMnPO_4$ based material [21].

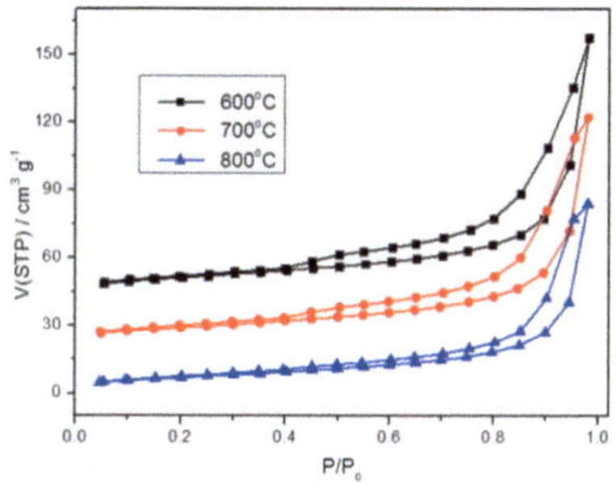

Figure 4. N_2 adsorption and desorption isotherms of $LiMn_{0.8}Fe_{0.2}PO_4/C$ synthesized at different calcination temperatures.

Fig. 5a shows the typical charging and discharging curves obtained for $LiMn_{0.8}Fe_{0.2}PO_4/C$ synthesized at different calcination temperatures. Cells were charged at 0.2 C (equal to 30 mA g^{-1}) to 4.5 V, held at 4.5 V until the current decreased to 0.02 C, and then discharged at 0.2 C to 2.5 V. As can be seen in Fig. 5a, the charging-discharging plateaus around 4.1 V are related to the Mn^{3+}/Mn^{2+} redox couple, and the plateaus around 3.4 V are related to the Fe^{3+}/Fe^{2+} redox couple. Although all samples exhibited similar charging/discharging profiles, it is clear that the $LiMn_{0.8}Fe_{0.2}PO_4/C$ synthesized at 700 °C showed a much higher reversible capacity as compared to other two samples. The specific discharge capacities at 0.2 C were 139 mAh g^{-1}, 150 mAh g^{-1} and 128 mAh g^{-1} for the $LiMn_{0.8}Fe_{0.2}PO_4/C$ synthesized at 600, 700 and 800 °C, respectively. Moreover, all samples exhibited good cycling performance as shown in Fig.5b, and only tiny capacity loss was observed after 50 cycles, especially for the $LiMn_{0.8}Fe_{0.2}PO_4/C$ prepared at 700 °C.

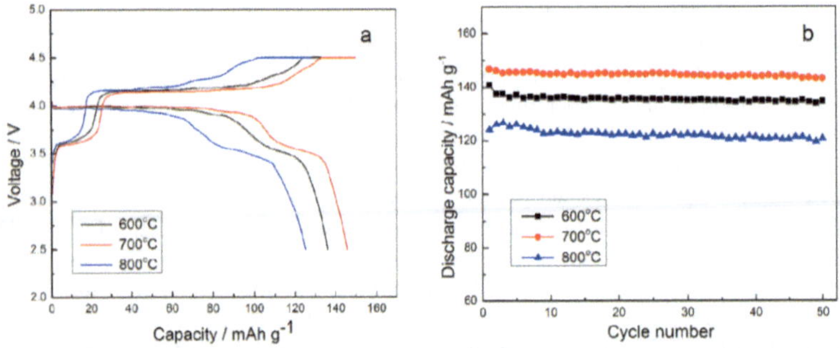

Figure 5. (a) Charging/discharging curves of $LiMn_{0.8}Fe_{0.2}PO_4/C$ synthesized at different calcination temperatures cycled at 0.2 C (equal to 30 mA g^{-1}). (b) Cycling performance of $LiMn_{0.8}Fe_{0.2}PO_4/C$ synthesized at different calcination temperatures cycled at 0.2 C. Cells were charged at 0.2 C to 4.5 V, held at 4.5 V until the current decreased to 0.02 C, and then discharged at 0.2 C to 2.5 V.

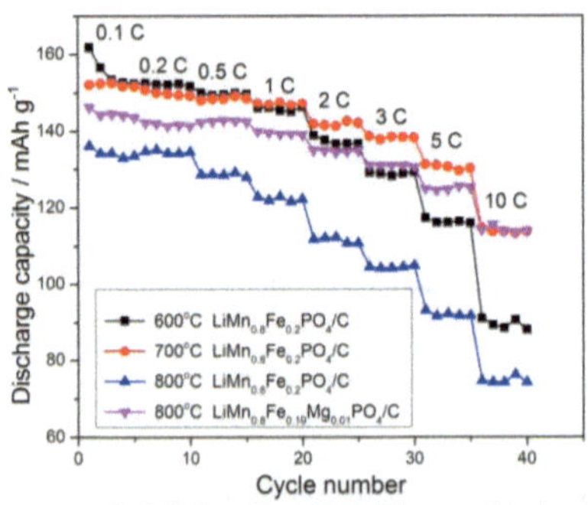

Figure 6. Rate performance of $LiMn_{0.8}Fe_{0.2}PO_4/C$ synthesized at different calcination temperatures and $LiMn_{0.8}Fe_{0.19}Mg_{0.01}PO_4/C$ synthesized at 800 °C. Cells were charged at 0.1 C to 4.5 V, held at 4.5V until the current decreased to 0.01 C, and then discharged at various rates to 2.0 V.

Fig. 6 depicts the rate performance of $LiMn_{0.8}Fe_{0.2}PO_4/C$ synthesized at different calcination temperatures. Cells were charged at 0.1 C to 4.5 V, held at 4.5 V until the current decreased to 0.01 C, and then discharged at various rates to 2 V. It is evident that the $LiMn_{0.8}Fe_{0.2}PO_4/C$ synthesized at 700 °C showed the best rate capability. The reversible capacity could reach 152 mAh g^{-1} at 0.1 C, 147 mAh g^{-1} at 1 C, 142 mAh g^{-1} at 2 C, 138 mAh g^{-1} at 3 C, 131 mAh g^{-1} at 5 C and 114 mAh g^{-1} at 10 C. There have been many reports on the synthesis of $LiMn_{0.8}Fe_{0.2}PO_4/C$. For example, Martha et al. communicated that carbon-coated $LiMn_{0.8}Fe_{0.2}PO_4$ nanoparticles synthesized by a solid-state reaction could deliver a reversible capacity of 165 mAh g^{-1} at 1/20 C and 100 mAh g^{-1} at 10 C [24]. Huang et al. reported a carbon coated mesoporous $LiMn_{0.8}Fe_{0.2}PO_4$ solid solution which could discharge a capacity of 140 mAh g^{-1} at 1/20 C and 80 mAh g^{-1} at 2 C [25]. Du et al. reported that a reversible capacity of 151.1 mAh g^{-1} at 0.1 C, 98.4 mAh g^{-1} at 5 C and 82 mAh g^{-1} at 10 C could be achieved for $LiMn_{0.8}Fe_{0.2}PO_4/C$ prepared by a co-precipitation method [26]. Evidently, the $LiMn_{0.8}Fe_{0.2}PO_4/C$ prepared at 700 °C by us exhibited better performance than all the values reported in literatures thus far.

Thus all the above experimental studies indicated that the better performance characteristics of $LiMn_{0.8}Fe_{0.2}PO_4/C$ could be achieved with increasing calcination temperature from 600 to 700 °C because of the enhanced crystallinity and increased conductivity at 700 °C. The degradation in the performance of the sample, prepared at 800 °C, could be correlated to the coarsening of the particle resulting in a decrease in the porosity. Studies also revealed that the particle size distribution and higher conductivity (because of amorphous carbon coating) are two important parameters responsible for the improvement in performance. Besides, careful experimentation revealed that while the optimum temperatures for (i) $LiMn_{0.8}Fe_{0.2}PO_4/C$, and (ii) Zn/Mg substituted $LiMnPO_4/C$) to be 700 °C, the value for Fe&Mg co-substitution was determined to be 800 °C [19]. It was also observed that the Fe&Mg co-substitution ($LiMn_{0.8}Fe_{0.19}Mg_{0.01}PO_4/C$ with ~7 wt.% C) showed somewhat lower capacity than the compound without Mg ($LiMn_{0.8}Fe_{0.2}PO_4/C$) when the rate was ≤ 5 C but almost has the same capacity at 10 C. This suggests that the Fe&Mg co-substitution can demonstrate superior performance characteristics, over just Fe substitution, at higher rates.

Scale up studies may be performed to establish the superiority of the co-substitution (Fe with Mg) with a view to designing the $LiMnPO_4$ based cathode materials for high power lithium ion batteries.

Conclusions

$LiMn_{0.8}Fe_{0.2}PO_4/C$ composite was prepared by a solid-state method at different calcination temperatures. The lattice parameters of $LiMn_{0.8}Fe_{0.2}PO_4$ changed slightly with calcination temperature, but the conductivity, particle size and porosity of the $LiMn_{0.8}Fe_{0.2}PO_4/C$ were strongly dependent on the calcination temperature. The $LiMn_{0.8}Fe_{0.2}PO_4/C$ obtained at 700 °C had the best electrochemical performance with a discharge capacity of 152 mAh g^{-1} at 0.1 C and 114 mAh g^{-1} at 10 C. However, the $LiMn_{0.8}Fe_{0.2}PO_4/C$ still showed a relatively faster capacity fade at high rates as compared to the $LiMn_{0.8}Fe_{0.19}Mg_{0.01}PO_4/C$.

Acknowledgements

This work was supported by the National Natural Science Foundation of China (No. 51304098), the Specialized Research Fund for the Doctoral Program of Higher Education (No. 20125314120004) and the Program for Innovative Research Team in University of Ministry of Education of China (No. IRT1250).

References

[1] A.K. Padhi, K.S. Nanjundaswamy, et al., "Phospho-olivines as Positive Electrode Materials for Rechargeable lithium batteries," *J. Electrochem. Soc.* 144 (1997), 1188-1194.
[2] M. Yonemura, A. Yamada, et al., "Comparative Kinetic Study of Olivine Li_xMPO_4 (M = Fe, Mn)," *J. Electrochem. Soc.* 151 (2004), A1352-A1356.
[3] Z. Bakenov, I. Taniguchi, "Electrochemical performance of nanocomposite $LiMnPO_4/C$ cathode materials for lithium batteries," *Electrochem. Commun.* 12 (2010), 75-78.
[4] J. Xiao, W. Xu, et al., "Synthesis and Characterization of Lithium Manganese Phosphate by a Precipitation Method," *J. Electrochem. Soc.* 157 (2010), A142-A147.
[5] D. Wang, H. Buqa, et al., "High-performance, nano-structured $LiMnPO_4$ synthesized via a polyol method," *J. Power Sources* 189 (2009), 624-628.
[6] D. Choi, D. Wang, et al., "$LiMnPO_4$ nanoplate grown via solid-state reaction in molten hydrocarbon for Li-ion battery cathode," *Nano Lett.* 10 (2010), 2799-2805.

[7] H. Ji, G. Yang, et al., "General synthesis and morphology control of $LiMnPO_4$ nanocrystals via microwave-hydrothermal route," *Electrochim. Acta* 56 (2011), 3093-3100.
[8] K.T. Lee, J. Cho, "Roles of nanosize in lithium reactive nanomaterials for lithium ion batteries," *Nano Today* 6 (2011), 28-41.
[9] Y. Dong, L. Wang, et al., "Two-phase interface in $LiMnPO_4$ nanoplates," *J. Power Sources* 215 (2012), 116-121.
[10] C. Delacourt, P. Poizot, et al., "One-Step Low-Temperature Route for the Preparation of Electrochemically Active $LiMnPO_4$ Powders," *Chem. Mater.* 16 (2004), 93-99.
[11] S.M. Oh, S.W. Oh, et al., "High-Performance Carbon-$LiMnPO_4$ Nanocomposite Cathode for Lithium Batteries," *Adv. Funct. Mater.* 20 (2010), 3260-3265.
[12] Z. Bakenov, I. Taniguchi, "Physical and electrochemical properties of $LiMnPO_4$/C composite cathode prepared with different conductive carbons," *J. Power Sources* 195 (2010), 7445-7451.
[13] T. Shiratsuchi, S. Okada, et al., "Cathodic performance of $LiMn_{1-x}M_xPO_4$ (M = Ti, Mg and Zr) annealed in an inert atmosphere," *Electrochim. Acta* 54 (2009), 3145-3151.
[14] J.W. Lee, M.S. Park, et al., "Electrochemical lithiation and delithiation of $LiMnPO_4$: Effect of cation substitution," *Electrochim. Acta* 55 (2010), 4162-4169.
[15] J. Kim, D.H. Seo, et al., "Mn based olivine electrode material with high power and energy," *Chem. Commun.* 46 (2010), 1305-1307.
[16] G. Yang, H. Ni, et al., "The doping effect on the crystal structure and electrochemical properties of $LiMn_xM_{1-x}PO_4$ (M = Mg, V, Fe, Co, Gd)," *J. Power Sources* 196 (2011), 4747-4755.
[17] C.L. Hu, H.H. Yi, et al., "Improving the electrochemical activity of $LiMnPO_4$ via Mn-site co-substitution with Fe and Mg," *Electrochem. Commun.* 12 (2010), 1784-1787.
[18] H.S. Fang, H.H. Yi, et al., Unpublished Results.
[19] H.H. Yi, C.L. Hu, et al., "Optimized electrochemical performance of $LiMn_{0.9}Fe_{0.1-x}Mg_xPO_4$/C for lithium ion batteries," *Electrochim. Acta* 56 (2011), 4052-4057.
[20] R. Kostecki, B. Schnyder, et al., "Surface studies of carbon films from pyrolyzed photoresist," *Thin Solid Films* 396 (2001), 36-43.
[21] H.S. Fang, Y.J. Hu, et al., "The essential role of aggregate porosity in improving the performance of $LiMnPO_4$/C," *Electrochim. Acta* 106 (2013), 215-218.
[22] S. Liu, H.S. Fang, et al., "Effect of carbon content on properties of $LiMn_{0.8}Fe_{0.19}Mg_{0.01}PO_4$/C composite cathode for lithium ion batteries," *Electrochim. Acta* 116 (2014), 97-102.
[23] K.S.W. Sing, D.H. Everett, et al., "Reporting physisorption data for gas/solid systems with special reference to the determination of surface area and porosity," *Pure Appl. Chem.* 57 (1985), 603-619.
[24] S.K. Martha, J. Grinblat, et al., "$LiMn_{0.8}Fe_{0.2}PO_4$: an advanced cathode material for rechargeable lithium batteries," *Angew. Chem. Int. Ed.* 48 (2009), 8559-8563.
[25] B. Zhang, X.J. Wang, et al., "Enhanced Electrochemical Performances of Carbon Coated Mesoporous $LiFe_{0.2}Mn_{0.8}PO_4$," *J. Electrochem. Soc.* 157 (2010), A285- A288.
[26] K. Du, L.H. Zhang, et al., "Synthesis of $LiMn_{0.8}Fe_{0.2}PO_4$/C by co-precipitation method and its electrochemical performances as a cathode material for lithium-ion batteries," *Mater. Chem. Phys.* 136 (2012), 925-929.
[27] J. Wolfenstine, U. Lee, et al., "Electrical conductivity and rate-capability of $Li_4Ti_5O_{12}$ as a function of heat-treatment atmosphere," *J. Power Sources* 154 (2006), 287-289.

AUTHOR INDEX
EPD Congress 2015

SUBJECT INDEX
EPD Congress 2015

MIX
Papier aus verantwortungsvollen Quellen
Paper from responsible sources
FSC® C105338

If you have any concerns about our products,
you can contact us on
ProductSafety@springernature.com

In case Publisher is established outside the EU,
the EU authorized representative is:
Springer Nature Customer Service Center GmbH
Europaplatz 3, 69115 Heidelberg, Germany

Printed by Libri Plureos GmbH
in Hamburg, Germany